Designing Digital Work

Stefan Oppl • Christian Stary

Designing Digital Work

Concepts and Methods for Human-centered Digitization

Stefan Oppl
Institute of Business Informatics
Johannes Kepler University Linz
Linz, Austria

Christian Stary
Institute of Business Informatics
Johannes Kepler University Linz
Linz, Austria

ISBN 978-3-030-12258-4   ISBN 978-3-030-12259-1   (eBook)
https://doi.org/10.1007/978-3-030-12259-1

Library of Congress Control Number: 2019933402

© The Editor(s) (if applicable) and The Author(s) 2019 This book is an open access publication
Open Access This book is licensed under the terms of the Creative Commons Attribution 4.0 International License (http://creativecommons.org/licenses/by/4.0/), which permits use, sharing, adaptation, distribution and reproduction in any medium or format, as long as you give appropriate credit to the original author(s) and the source, provide a link to the Creative Commons licence and indicate if changes were made.

The images or other third party material in this book are included in the book's Creative Commons licence, unless indicated otherwise in a credit line to the material. If material is not included in the book's Creative Commons licence and your intended use is not permitted by statutory regulation or exceeds the permitted use, you will need to obtain permission directly from the copyright holder.

The use of general descriptive names, registered names, trademarks, service marks, etc. in this publication does not imply, even in the absence of a specific statement, that such names are exempt from the relevant protective laws and regulations and therefore free for general use.

The publisher, the authors and the editors are safe to assume that the advice and information in this book are believed to be true and accurate at the date of publication. Neither the publisher nor the authors or the editors give a warranty, express or implied, with respect to the material contained herein or for any errors or omissions that may have been made. The publisher remains neutral with regard to jurisdictional claims in published maps and institutional affiliations.

This Palgrave Macmillan imprint is published by the registered company Springer Nature Switzerland AG
The registered company address is: Gewerbestrasse 11, 6330 Cham, Switzerland

Preface and Acknowledgments

Seeking advice and support when digitalizing business operation can easily lead to humans being taken 'off the loop', despite their knowledge on organizing work and accomplishing business processes. Acting in dedicated roles and being technically skilled, we need them to describe the work process when addressing digital challenges. Their knowledge is crucial when using digital technologies to change work processes while moving towards a business model that aims to provide value-producing opportunities in an increasingly digitally driven organizational setting. Transforming transaction knowledge. Workforce needs to become skilled to assess novel developments in an informed way so as to generate beneficial insights for business operation.

Digitized work processes including the human in the loop is becoming mainstream, and not only for the bigger players. As more Small and Medium Enterprises (SMEs) seek to save time and staffing costs, digital work design is becoming a cost-effective necessity for many businesses. Thereby, adjusted digital and organizational stakeholder innovation is what helps companies gain edge for future development. Ensuring consistent articulation, alignment, and enactment of work where tools and instruments interactively reframe workers' behavior is likely to maximize validity and relevance.

Understanding digital work design as continual process of stakeholder articulation, alignment, and enactment as well as the results achieved by this process, we capture its dual character in this book:

- Digital work design is about digital support of eliciting, sharing, and implementing work knowledge—digital systems support the design process, addressing the Gestalt aspect.
- Digital work design is about digital support of running business operation, for example, workflow engines—digital systems support execution of work processes, addressing the implementation aspect.

Presenting a blend of theory, methods, and tools, this book addresses the elicitation of work in organizations, with the purpose to improve or redesign their internal business. We reframe the modeling process as a means to identify and resolve perspectives on collaborative work processes, and integrate methods from Knowledge Management, Business Process Management, and Computer-Supported Co-operative Work. Latest technologies are put into the context of design support while providing the conceptual underpinnings of the articulation and alignment processes occurring during work process elicitation. The methodological inputs refer to transitioning from as-they-are to they-could-be work processes via direct stakeholder involvement.

Providing a unifying framework that guides the design of organizational interventions promotes constructive and structured emergence of novel digital workplace designs and work practices. We want this approach to be understood as an invitation to unfold individual and collective organizational intelligence of concerned stakeholders. Our inputs aim to empower them so that their explication, reflection, and prototyping of work designs in increasingly digital system settings can receive the required appreciation, from both collaborators and management—the latter also held responsible for innovative development and transformation projects.

We are aware of the ambitious undertaking of writing about an interdisciplinary topic, taking into account ecological, technical, cognitive, social, psychological, organizational, and economic aspects of increasingly complex work processes. However, looking for constructively intertwining these different aspects—recognizing relationships as the core carrier of knowledge—we are convinced our findings are an essential trigger to start re-designing socio-technical systems through aligning digital and human capabilities in a resilient way.

While working on the book, we have enjoyed a team spirit, allowing everyone to bring in their different background and experience, both in terms of theory and practice. Our intense collaboration allowed us to come up with a comprehensive picture of subject orientation. We experienced the struggle of streamlining structure and content as a constructive and inspiring moment of our cooperation. We hope the readers are still able to grasp it, in particular when reflecting the systemic nature of Subject-oriented Business Process Management.

For the support we experienced in performing research and development relevant to this book, we want to thank:

- Our families supporting our endeavor
- All project partners allowing us to evaluate research in organizational development projects and various operational settings
- Our students from Johannes Kepler University Linz, Austria (JKU) helping us to gain in-depth insights into our methodological and technical research
- Palgrave Macmillan publishing house, particularly Liz Barlow and Lucy Kidwell, for their constructive support and cooperation

Special thanks go to Christoph Bawart for his effectiveness and efficiency throughout editing and for finishing all figures in time. We are happy that this book is published under an Open Access License and thus is available to everybody to read for free. The book is funded by the Johannes Kepler Open Access Publishing Fund.

In case the readers are interested in background information and application details, we invite them to join us on ResearchGate (see also researchgate.net). There, interested readers will find recent work and original material. When looking for instruments available, readers may look at jku.at/ce and i2pm.net (in particular with respect to subject orientation) for free downloads and case studies in various application areas.

Linz, Steyr, and Vienna, Austria Stefan Oppl
2019 Christian Stary

Contents

1	**Introduction**	1
	1.1 Conceptual Foundations—An Overview	4
	1.2 Knowledge Lifecycle	6
	1.3 Articulation Work	9
	1.4 Model-Centered Learning	10
	1.5 Collaborative Multi-perspective Modeling	13
	1.6 Natural Versus Techno-Centric Modeling	16
	1.7 Taking an Integrated Socio-technical System Perspective	17
	References	19
2	**Elicitation Requirements**	27
	2.1 Setting the Stage—Awareness on Roles and Their Management	28
	2.2 Situation Awareness	31
	2.3 Conceptual Understanding of Complex Systems	38
	2.4 Creating a Reflective Practice for Situations-to-Be	46
	2.5 Focusing While Utilizing Multiple Perspectives	50
	2.6 Articulating Intangible Assets	56
	2.7 Engage in Alignment for Collective Intelligence	61
	2.8 Synthesis	69
	References	71

Contents

3 Value-Oriented Articulation — 83
- 3.1 Shaping Role Identities Through Contextual Behavior Articulation — 85
  - 3.1.1 Start Simple, Using Natural Language — 85
  - 3.1.2 Roles As Semantic and Pragmatic Entities — 88
  - 3.1.3 Acting in a Specific Role—Pragmatic Modeling — 89
  - 3.1.4 Conclusive Summary — 96
- 3.2 Sorting Out: Cards As Carrier of Functions and Interaction — 97
  - 3.2.1 Articulation Concepts — 98
  - 3.2.2 Articulation Process — 101
  - 3.2.3 Mapping to Subject-Oriented Models — 103
- 3.3 On the Go: Capturing Functions and Interactions While Working — 107
- 3.4 Capturing Tangibles and Intangible Exchange Relationships — 110
  - 3.4.1 Organizations As Transactional Networks of Roles — 110
  - 3.4.2 Tangible and Intangible Transactions — 112
- 3.5 Cross-Cutting Issues — 118
- References — 127

4 Alignment of Multiple Perspectives: Establishing Common Ground for Triggering Organizational Change — 133
- 4.1 Alignment Concept and Principles — 134
- 4.2 Towards Direct Stakeholder Support—Minimizing Semantic Distance — 142
- 4.3 Alignment Scheme — 145
- 4.4 Alignment Approaches — 147
  - 4.4.1 Example: Ex-ante Communication Alignment — 152
  - 4.4.2 Example: Ongoing Communication Alignment — 155
- 4.5 Alignment Practice: Ex-post Communication Alignment with CoMPArE/WP — 156
  - 4.5.1 Component 1—Setting the Stage — 159
  - 4.5.2 Component 2—Articulation and Alignment — 161

		4.5.3	Component 3—Refinement via Virtual Enactment	168
		4.5.4	Transition from Modeling to Enactment	170
	4.6	Conclusion		171
	References			174

5 Acting on Work Designs: Providing Support for Validation and Implementation of Envisioned Changes — 179

- 5.1 Creating Executable Models Through Scaffolding Articulation and Alignment — 179
  - 5.1.1 Scaffolding — 181
  - 5.1.2 Scaffolds for Stakeholder-Centric Work Modeling — 183
  - 5.1.3 A Framework for Scaffolding Model Articulation and Alignment — 186
  - 5.1.4 Scaffolding Articulation and Alignment in CoMPArE/WP — 189
  - 5.1.5 Example — 191
- 5.2 Participatory Enactment Support Instrument — 193
  - 5.2.1 Background: Process Walk-Throughs and Enacted Prototypes — 194
  - 5.2.2 Implications of Enacting Dynamically Changeable Prototypes — 196
  - 5.2.3 Tool Support — 199
  - 5.2.4 Conclusive Summary — 214
- 5.3 S-BPM-Driven Execution of Actor-Centric Work Processes — 215
  - 5.3.1 S-BPM Activity Bundles in the Business Processing Environment — 220
  - 5.3.2 S-BPM Activity Bundles in the Knowledge Processing Environment — 222
  - 5.3.3 Tool Support — 227
- 5.4 Synthesis — 234

References — 240

6 Enabling Emergent Workplace Design — 249
- 6.1 Articulation Work and Mental Models — 251
- 6.2 Mental Models Theory and Articulation Work for Organizational Learning — 253
- 6.3 Towards an Integrated Framework — 256
  - 6.3.1 Relevant Concepts — 257
  - 6.3.2 Implementation of Work Processes — 259
  - 6.3.3 Responsibilities and Skills — 259
  - 6.3.4 Towards Instantiation — 264
  - 6.3.5 Behavioral Interfaces for Interaction Coordination — 264
  - 6.3.6 Behavioral Constraints for Individual Actions — 264
  - 6.3.7 Varying Degrees of Freedom in Individual Activity — 269
- 6.4 Articulation Engineered for Organizational Learning — 269
  - 6.4.1 Featuring OL Processes — 273
  - 6.4.2 Support for Repository Access — 274
  - 6.4.3 Process Knowledge Elicitation and Knowledge Claim Development — 276
  - 6.4.4 Process Visualization for Elicitation and Reflection — 279
  - 6.4.5 Process Validation and Simulation for Reflection and Alignment — 279
- 6.5 Conclusion — 281
- References — 281

7 Putting the Framework to Operation: Enabling Organizational Development Through Learning — 287
- 7.1 Sample Actor-Centric Tool Support for Articulation and Elicitation — 291
  - 7.1.1 Comprehand Cards — 292
  - 7.1.2 Comprehand Table — 293
  - 7.1.3 Collaborative Model Articulation and Exploration — 301
- 7.2 Sample Actor-Centric Tool Support for Representation — 303
  - 7.2.1 Representing Role Knowledge and Descriptive Meta-knowledge — 303
  - 7.2.2 Representing Conceptual Meta-knowledge — 304

		7.2.3	Enabling the Assessment of Cognitive Meta-knowledge	305

    7.3 Sample Actor-Centric Tool Support for Intelligent Content Manipulation   307
    7.4 Sample Actor-Centric Tool Support for Processing Work Models   309
    7.5 Towards Seamless Tool Support—A Showcase   310
            7.5.1 Articulation and Elicitation   311
            7.5.2 Representation   312
            7.5.3 Manipulation   313
            7.5.4 Processing   315
    7.6 Conclusions   317
    References   318

8 Case Studies   325

    8.1 Categorical Knowledge Building Support—A Planning Case   326
            8.1.1 Sample Case   330
            8.1.2 Insights   334
    8.2 CoMPArE/WP Facilitating Project-Based Business Operation   335
            8.2.1 Sample Case   337
            8.2.2 Observed Effects   348
            8.2.3 Insights   351
    8.3 Articulating and Aligning Digital Learning Support Features   352
            8.3.1 Articulation Support of Intentional Education   357
            8.3.2 Developing Digital Learning Support Baselines (Course and Content Models)   365
            8.3.3 Semantic Navigation   376
            8.3.4 Alignment in User-/Usage-Oriented Design Spaces   381
            8.3.5 Insights from the Case   388
    8.4 Subject-Oriented Organizational Management   389
            8.4.1 Organizational Management   390
            8.4.2 Subjects As Carrier of Work Behavior   392

8.4.3	Essential Principles	394
8.4.4	Structuring Articulation	399
8.4.5	Sample Applications	404
8.4.6	Insights from the Case	409
References		410

9 Epilogue 419
References 423

Ontological Glossary 425

Index 429

List of Figures

Fig. 1.1	Kernel theories situated in the MTO-framework	5
Fig. 1.2	The Knowledge Lifecycle of Firestone and McElroy (adapted from Firestone and McElroy 2003)	7
Fig. 1.3	Schemes and mental models (translated and adapted from Ifenthaler 2006)	11
Fig. 1.4	Foci of research addressed in this book	18
Fig. 2.1	Awareness on roles	31
Fig. 2.2	The articulation scheme containing trigger, role-specific activity, and effect	34
Fig. 2.3	Customer service actor behavior handling customer product claims	35
Fig. 2.4	Scoping another actor behavior—Idea Provider	36
Fig. 2.5	Situation awareness	38
Fig. 2.6	Conceptual understanding of complex systems	45
Fig. 2.7	Work-agogy (according to Arbeitsagogik.ch)	49
Fig. 2.8	Creating a reflective practice for situations-to-be	50
Fig. 2.9	Focusing while utilizing multiple perspectives	56
Fig. 2.10	Articulating intangible assets	61
Fig. 2.11	Engage in alignment for collective intelligence	69
Fig. 3.1	Natural language description of an application procedure for vacation (released under a Creative Commons Attribution 4.0 International License (CC BY 4.0))	87

Fig. 3.2	Subject identification for the holiday application process, providing subjects and their interaction	91
Fig. 3.3	Employee behavior in holiday application process	93
Fig. 3.4	Manager's behavior in holiday application process	94
Fig. 3.5	HR department behavior in holiday application process	95
Fig. 3.6	A subject with predicates and objects	96
Fig. 3.7	Elements of the card-based modeling language	100
Fig. 3.8	Sample result of individual articulation	102
Fig. 3.9	Result of collaborative consolidation	103
Fig. 3.10	Transformation from card-based to S-BPM model	105
Fig. 3.11	Process capturing	108
Fig. 3.12	Sample holomap for developing Sales and Presales relations	116
Fig. 4.1	Architecture of ontology-based BPM systems (adapted from Jung 2009)	137
Fig. 4.2	Ontology-based alignment (adapted from Jung 2009)	138
Fig. 4.3	Alignment through merging ontology fragments (adapted from Jung 2009)	139
Fig. 4.4	Facilitating resolving semantic ambiguities in process modeling based on ontologies according to Fan et al. (2016)	140
Fig. 4.5	Developing a domain process ontology instance (according to Fan et al. 2016)	141
Fig. 4.6	Alignment of business processes as part of co-developing organizations	141
Fig. 4.7	CoMPArE articulation scheme	146
Fig. 4.8	Example setting of role-distributed models in an intermediate stage during modeling	149
Fig. 4.9	Co-located creation of interaction models on a shared surface	153
Fig. 4.10	Modeling of internal behavior on an interactive surface	154
Fig. 4.11	Multi-surface setup for distributed modeling of subject-oriented models (bold arrows indicate linked messaging ports)	156
Fig. 4.12	The CoMPArE approach represented as a BPMN process	157
Fig. 4.13	Result of individual articulation	163
Fig. 4.14	Result of component 2.2: Collaborative Consolidation	165
Fig. 5.1	Dimensions of scaffolding during work modeling	187
Fig. 5.2	Examples of different forms of scaffolds for work modeling	189
Fig. 5.3	Scaffolds deployed in CoMPArE/WP (references indicate the foundation for design)	190
Fig. 5.4	Top left: model layout template; top right and bottom: modeling results of workshops	191

Fig. 5.5	Platform architecture	202
Fig. 5.6	Enactment UI (released under a Creative Commons Attribution 4.0 International License (CC BY 4.0))	204
Fig. 5.7	Expected messages in subject UI (released under a Creative Commons Attribution 4.0 International License (CC BY 4.0))	205
Fig. 5.8	Process visualizations (released under a Creative Commons Attribution 4.0 International License (CC BY 4.0))	206
Fig 5.9	Prompting sequence for elaboration	208
Fig. 5.10	Example for interactive elaboration prompt (released under a Creative Commons Attribution 4.0 International License (CC BY 4.0))	209
Fig. 5.11	Specification of messages during elaboration (released under a Creative Commons Attribution 4.0 International License (CC BY 4.0))	209
Fig. 5.12	Scaffolding prompts (released under a Creative Commons Attribution 4.0 International License (CC BY 4.0))	211
Fig. 5.13	Example for exploration scaffold (released under a Creative Commons Attribution 4.0 International License (CC BY 4.0))	212
Fig. 5.14	Example for unhandled communication scaffold (released under a Creative Commons Attribution 4.0 International License (CC BY 4.0))	213
Fig. 5.15	The S-BPM activity bundle (adapted from Fleischmann et al. 2012)	217
Fig. 5.16	Integration of the KLC with S-BPM activity bundles	219
Fig. 5.17	Subject-oriented representation schema for three-party process	228
Fig. 5.18	Generic behavior of the start subject "Subject 1"	229
Fig. 5.19	Generic behavior of "Subject 2"	230
Fig. 5.20	Generic structure of the business object "Mail"	231
Fig. 5.21	Instantiating a process scheme	231
Fig. 6.1	Mental model theory and Articulation Work in the KLC	254
Fig. 6.2	Conceptual framework	257
Fig. 6.3	Work processes and areas of responsibility	260
Fig. 6.4	Persons and areas of responsibility	261
Fig. 6.5	Organizational roles clustering areas of responsibility in different work processes	262
Fig. 6.6	Interfaces and behaviors of team members	263
Fig. 6.7	Instantiation of behavior fragment	265
Fig. 6.8	Linking behavioral interfaces	266

Fig. 6.9	Different behavioral requirements for a single behavioral interface	267
Fig. 6.10	Meeting behavioral requirements through different behavioral implementations	268
Fig. 6.11	Conceptual framework for situation-specific interdisciplinary teams	270
Fig. 6.12	Articulation engineered for organizational learning (Chris Stary 2014)	271
Fig. 6.13	Transactive memory concept used for the codified part of the repository (according to Neubauer et al. 2013)	276
Fig. 7.1	Sample model created with modeling cards	292
Fig. 7.2	Comprehand Table overview (top-left: interaction on table surface; top-right: modeling tokens with projected connections; bottom: schematic bird's eye view of tabletop)	294
Fig. 7.3	Labeling and associating	298
Fig. 7.4	Users can open a token and put additional information into it. Additional information is bound to smaller tokens	298
Fig. 7.5	Elements and tools for tabletop concept mapping	300
Fig. 7.6	Exemplifying CMap navigation and content links	305
Fig. 7.7	Architecture of process enactment environment	309
Fig. 7.8	Card-based model (left), interactive surface modeling (right)	311
Fig. 7.9	Card-model recognition for conceptual representation: web-interface (left), recognition results (top right), XML-based model representation (bottom right) (released under a Creative Commons Attribution 4.0 International License (CC BY 4.0))	313
Fig. 7.10	Work process content in the learning environment (released under a Creative Commons Attribution 4.0 International License (CC BY 4.0))	314
Fig. 7.11	Processing and simultaneous manipulation on an interactive modeling tabletop	316
Fig. 8.1	Embodying the planning case into the digital work design framework	327
Fig. 8.2	Leveraging stakeholder knowledge for organizational change	328
Fig. 8.3	Interactive concept mapping (see also Oppl and Stary 2009, 2011)	330
Fig. 8.4	Start map	331
Fig. 8.5	Completing the relevant part of the organization	332
Fig. 8.6	Patient-oriented treatment planning (out-patient department)	333
Fig. 8.7	Finalization of treatment planning (LINAC)	334

Fig. 8.8	Embodying the CoMPArE approach to the digital work design framework	336
Fig. 8.9	Result of component 1—"Setting the Stage"	339
Fig. 8.10	Result of component 2.1—"Individual Articulation" for participants representing "Client" (left) and "Contact Person" (right)	340
Fig. 8.11	Result of component 2.1—"Individual Articulation" for participants representing "Mentor" (left) and "Team Leader" (right)	341
Fig. 8.12	Result of component 2.2—"Collaborative Consolidation"	343
Fig. 8.13	Result of transformation to BPMN	346
Fig. 8.14	Example of refinement (left: original process; right: refined process)	347
Fig. 8.15	Embodying the educator case to the digital work design framework	353
Fig. 8.16	Tabletop concept mapping	359
Fig. 8.17	Tabletop concept mapping for articulating educational design—sample patterns	360
Fig. 8.18	Approaches to progressive education, according to Weichhart and Stary (2014)	362
Fig. 8.19	John Dewey's approach, according to Weichhart and Stary (2014)	363
Fig. 8.20	Helen Parkhurst's approach, according to Weichhart and Stary (2014)	363
Fig. 8.21	Learning principles, according to Weichhart and Stary (2014)	364
Fig. 8.22	Progressive learning environment requirements, according to Weichhart and Stary (2014)	366
Fig. 8.23	Process map for digital learning support content engineering according to Auinger et al. (2007)	367
Fig. 8.24	Content outline map for business process management	368
Fig. 8.25	Annotated structure map	369
Fig. 8.26	Structure map for interviewing and result presentation	370
Fig. 8.27	Educational metadata structure	373
Fig. 8.28	Tagged BPM content—'background information' and 'practical guideline' on the development of process-based organizations (released under a Creative Commons Attribution 4.0 International License (CC BY 4.0))	374
Fig. 8.29	Didactically enriched concept map navigation	377

List of Figures

Fig. 8.30	Relationships between main views according to Neubauer et al. (2011)	378
Fig. 8.31	Linking hierarchical and associative navigation design	380
Fig. 8.32	Categories of design elements	382
Fig. 8.33	A layered approach to a user-/usage-centered learning design space	383
Fig. 8.34	Schematic instance of design map according to Weichhart and Stary (2014)	385
Fig. 8.35	Dalton Plan editor according to Weichhart and Stary (2014) (released under a Creative Commons Attribution 4.0 International License (CC BY 4.0))	386
Fig. 8.36	Feedback graphs according to Weichhart and Stary (2014) (released under a Creative Commons Attribution 4.0 International License (CC BY 4.0))	387
Fig. 8.37	Embodying the organizational management case to the digital work design framework	390
Fig. 8.38	Sample universe of discourse for 'The clock has fallen off the wall'	392
Fig. 8.39	Sample interaction pattern for 'The clock has fallen off the wall'	393
Fig. 8.40	Sample Behavior Synchronization of 2 SBDs	394
Fig. 8.41	Cascading perspectives	400
Fig. 8.42	Sample diagrammatic representation	403
Fig. 8.43	Sample of elicited knowledge and sample of subject-oriented representation	407
Fig. 8.44	Person B's 'management-by-delegation'	408
Fig. 8.45	Person C—getting responsible actors involved	408
Fig. 9.1	System development involving the ground model supported by ASM (Börger and Stärk 2012)	421
Fig. A.1	Ontology of essential terms used in this work	427

List of Tables

Table 2.1	Managing elicited knowledge (according to and translated from F. Fuchs-Kittowski and Fuchs-Kittowski 2007)	54
Table 2.2	Summary of elicitation requirements	70
Table 3.1	Value-oriented articulation approaches	119
Table 3.2	Elicitation requirements and subject-oriented articulation	120
Table 3.3	Elicitation requirements and card-based elaboration	123
Table 3.4	Elicitation requirements and value network-based articulation	125
Table 4.1	Elicitation requirements and CoMPArE/WP	171
Table 5.1	Processing work models for validation and enactment	235
Table 5.2	Elicitation requirements and scaffolding-based validation and virtual enactment	236
Table 5.3	Elicitation requirements and S-BPM-based validation and execution	238
Table 7.1	Learning/design dimensions, activities, and tools	289
Table 8.1	Example of tagging a BPM content structure	373

1
Introduction

Human work in organizations has been influenced and shaped by digital technologies ever since their advent in the mid-twentieth century. In the earlier stages of development, digital systems were mainly used for calculation tasks that were cumbersome or time-intense for humans to perform. Such tasks are found in all domains of industry and have led to a wide-spread penetration of IT systems for planning and control tasks. In a later wave of development, linked to the advent of more powerful and interlinked digital devices, systems were devised to support the coordination and collaboration of actors—independently of whether they were humans, machines, or whole organizations. Such systems, however, mainly adopt a Tayloristic view on organizational work, aiming at top-down division, coordination, and control of work tasks in an organization. Today's digital technologies, however, also allow for a more agile, bottom-up approach to work design and execution support. In this book, we argue for such an actor-centric view on organizational work and propose a set of instruments that supports the design of collaborative work systems in an environment with ubiquitous access to digital communication technologies.

The deployment and use of digital work support systems has increasingly gained importance since the 1980s for implementing organizational work processes (Curtis et al. 1992; Thome 1982). These systems do not solely aim at improving productive, value-adding work. They are also deployed as an instrument for governing and coordinating work to optimize the use of available resources (Orlikowski and Iacono 2001).

The focus on optimizing organizational resources for effective and efficient use is facilitated by conceptualizing organizational reality in enterprise architectures that describe the orchestration of resources to reach organizational goals (Jonkers et al. 2006). This abstraction is usually implemented by encoding and interlinking the social and technical elements of these architectures in conceptual models. These models can be processed by means of Information and Communication Technology (ICT) to provide support in process optimization as well as implementation (Curtis et al. 1992; Herrmann et al. 2002).

When enterprise architecture models are used as organizational artifacts to direct and control organizational work practices, the social and cognitive skills of the involved human actors are usually not explicitly considered (Davidson 2006). This can lead to suboptimal use of resources, as individual improvement of relevant skills might be ignored (Herrmann et al. 2002), and can hamper adequate reactions on changing conditions in the organizational environment (Davidson 2006). Organizational behavior and functions of ICT-based support measures gradually diverge, leading to a misfit between actors' expectations and actually provided support. This ultimately results in actors' ignorance of and resistance against IT-based support and guidance measures (Feldman and Pentland 2003).

Despite these challenges, socio-technical work support instruments such as ERP-systems (Enterprise Resource Planning), SOPs (Standard Operating Procedures), or MES (Manufacturing Execution Systems) are widely deployed in industry (Ragowsky and Somers 2002). Adoption has also risen in Small and Medium Enterprise (SMEs) in the last decade (Haddara and Zach 2012), confronting virtually every organization directly or indirectly with guidance and support measures originating in these systems.

Operative actors in an organization thus have to cope with the potential discrepancy between the support measures provided based on idealized or out-dated models of a work task and the perceived reality of their work situation (Davidson 2006). These perceived mismatches can range from inappropriately designed on-screen forms for data entry, over lacking information required for a specific work step, to work procedures that cannot be implemented in the way prescribed by a support system. They lead to workarounds, which increase the cognitive load and effort required by an organizational actor to complete the respective task, or to an accommodation of one's behavior to the routines and constraints encoded in the support systems (Davidson 2006; Soh et al. 2003).

Still, today's organizational work is shaped and influenced by requirements on standardization and documentation that can hardly be met without deploying socio-technical support systems (Botta-Genoulaz and Millet 2006; Davies et al. 2006). Active involvement of organizational actors in articulating and aligning their collaborative work processes thus has to be embedded in the context of the organizational reality shaped by these systems. Feldman and Pentland (2003) recognize this constraint and conceptualize it by distinguishing ostensive aspects from performative aspects of work in an organization. They argue that, in order to influence the ostensive aspects of organizational work, the performative aspects have to be made visible in a form that is acceptable on all layers of an organization. While Feldman and Pentland (2003) do not detail this requirement any further, it shows that operative organizational actors—being the sources of performative aspects of work—have to be enabled to recognize and understand the ostensive mechanisms influencing their work (Weick et al. 2005), relate them to their performative behaviors (Davidson 2006), and articulate them in a form that allows them to directly influence the way their work is (ostensively) understood within the organization.

The skills necessary to create these commonly acceptable representations of work cannot be taken for granted (Frederiks and van der Weide 2006; Recker and Rosemann 2009). Existing research addressing this issue considers organizational actors as mere sources of information, whose utterances about their work need to be transformed into a form that can be processed by expert analysts (Herrmann and Nolte 2014; Hjalmarsson et al. 2015; Simões et al. 2016). This indirect approach, however, does not facilitate the alignment of different perspectives on and understandings about a work task (Türetken and Demirörs 2011) and might cause modelers' bias that manifests in incomplete or inappropriate representation of the work process (Goncalves et al. 2009). We here consider a work process as a sequence of specific activities to complete a work task. The alignment between the performative and ostensive aspects of organizational work thus is hampered and might lead to the introduction of further discrepancies between expected and actually provided work support measures.

This book introduces support measures and instruments for articulating, aligning, and enacting performative aspects of organizational work. These measures and instruments should allow organizational actors to actively design their collaborative work processes based on their individual views using their own conceptualizations of their work, while ensuring and still leading to a syntactically correct and semantically valid sound conceptual model for further processing in digital work systems.

Since the book addresses and involves knowledge from various disciplines, an ontological glossary has been developed (see appended Ontological Glossary). It provides conceptual and terminological orientation. The remainder of this chapter describes the conceptual foundations informing the methods and framework proposed in this book.

1.1 Conceptual Foundations—An Overview

This book focuses on examining how human actors perceive, understand, articulate, and align their collaborative work in an organizational context. It ultimately aims at supporting this articulation and alignment processes by socio-technical means (Baxter and Sommerville 2011) to ultimately improve operative organizational work processes and work support systems in an increasingly digitized work environment. The theories informing the design of the artifacts to be developed consequently can be found in areas researching human interaction and collaboration in an organizational context. Figure 1.1 situates these theories in the MTO-framework (Mensch-Technik-Organisation—German for human-technology-organization) (Strohm and Ulich 1997) to show their respective foci.

Organizations are viewed as entities in which actors use their knowledge to perform business processes. If they are not able to satisfactorily complete their work, they deploy compensation activities and ultimately

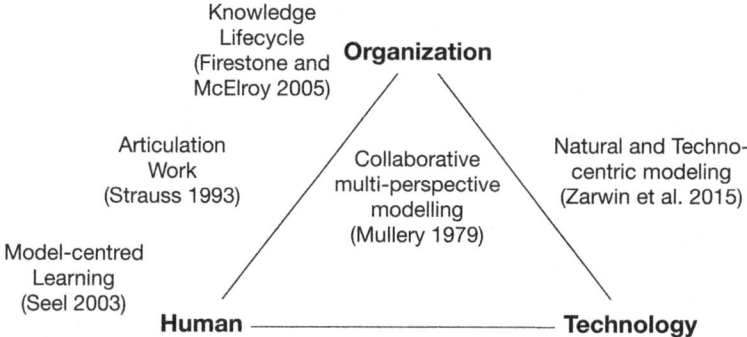

Fig. 1.1 Kernel theories situated in the MTO-framework

question the knowledge foundations they build their decisions on. In such a case, new knowledge is created in the organization that should allow the avoidance of observed problems. The theory explaining and conceptualizing this process for the present work is the *Knowledge Lifecycle* of Firestone and McElroy (2003).

The Knowledge Lifecycle does not explicitly explain the activities of actors that lead to the alignment of operative work in case contingencies arise. This issue is addressed by Strauss (1993) in his theory of *Articulation Work* that offers a descriptive framework of how workers overcome perceived obstacles in their collaborative work processes by implicit or explicit coordination activities (Strauss 1988). In the course of Articulation Work, the involved actors develop new knowledge that shapes their expectations of the behavior of their organizational environment in general and their collaborators in particular.

Neither the Knowledge Lifecycle nor the concept of Articulation Work provides input on the mental processes of actors when developing new knowledge and how to support it. The theory of *model-centered learning* (Seel 2003), however, conceptually describes these mental processes and offers insights into how to facilitate them. Enabling actors to explicitly articulate their mental models leads to their refinement (Ifenthaler et al. 2007), and creates results that can serve as boundary objects for making the mental models understandable for others (Dann 1992), ultimately making them accessible for alignment to create common ground on how to collaborate (Convertino et al. 2008).

The process of articulation and alignment of mental models can be supported by conceptual modeling practices (Recker and Dreiling 2011; Herrmann et al. 2002). In collaborative modeling, one challenge is to make sure that the views of all involved actors are considered in the final result. *Multi-perspective modeling* (Mullery 1979) addresses this issue by splitting the modeling process in a first phase, where the involved actors individually create models of their own perspective on the subject of modeling, and a second phase, where these models are consolidated in a structured way to form a single, agreed upon model.

In order to support operative work processes, the results of articulation and alignment need to be made accessible for processing on an organizational and/or technical level. This poses requirements on the syntactical correctness of conceptual models that might not have been relevant during actor-centric modeling (Zarwin et al. 2014). The theory of the continuum between *natural and techno-centric modeling* (ibid.) enables us to derive requirements on the artifacts to be developed in order to provide a link between articulation and alignment practices and the integration of the results in existing enterprise architectures (Jonkers et al. 2004).

The following subsections summarize the mentioned kernel theories. At the end of each section, the respective theory is linked to its use in the present research.

1.2 Knowledge Lifecycle

The Knowledge Lifecycle (KLC) proposed by Firestone and McElroy (2003) is a process-oriented approach to knowledge management that builds upon different earlier approaches on organizational learning processes (mainly and foremost Argyris and Schön's (1978) concept of single- and double-loop learning). The KLC introduces a fundamental distinction among activities performed in the 'business processing environment' and activities performed in the 'knowledge processing environment'. Figure 1.2 provides an overview of the Knowledge Lifecycle as originally described by Firestone and McElroy (2003). Operative activities directly contributing to achieving a business goal are executed in the scope of the business processing environment. As long as the outcome of

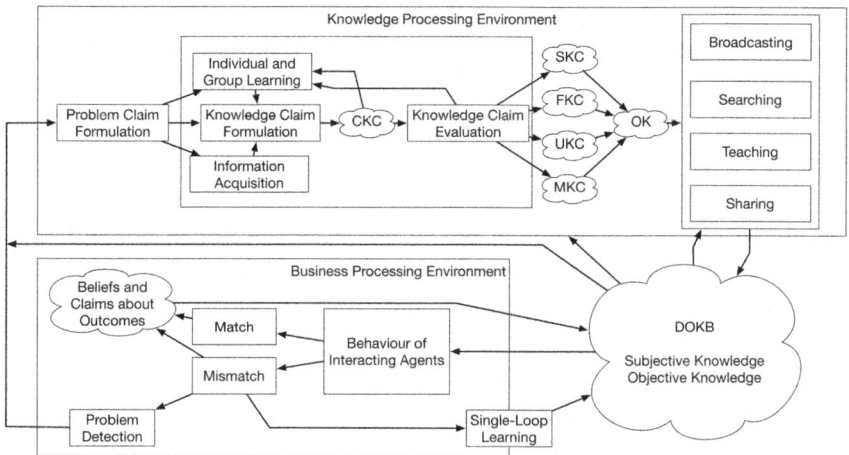

Fig. 1.2 The Knowledge Lifecycle of Firestone and McElroy (adapted from Firestone and McElroy 2003)

all activities and interactions is as expected, organizational actors (referred to as 'interacting agents' in Fig. 1.2) continue their activities in this mode. If problems occur, that is, if some outcome does not comply with the expectations of any actor, learning occurs. Learning here always refers to a change in an organizational phenomenon referred to as the distributed organizational knowledge base (DOKB). The DOKB contains all knowledge an organization builds upon to pursue its aims, in both uncodified and codified form, that is, being anchored in the memory of actors or being explicitly implemented in specified business processes or IT systems.

The content of the DOKB is not altered without reason. If outcomes of particular activities match what has been expected based on knowledge from the DOKB, the beliefs about the correctness of the particular knowledge artifact are strengthened. If mismatches occur (i.e., if the outcome of an activity does not fit the expectations derived from the DOKB), learning occurs and affects the content of the DOKB. Learning conceptually is distinguished in single-loop- and double-loop-learning, following the approach of Argyris and Schön (1978). Single-loop learning does not question the fundamental beliefs the activities that led to the mismatching outcome are based on. Rather, the way such activities are performed is adapted and populated back to the DOKB.

If a more fundamental problem occurs and cannot be incorporated into the DOKB by assimilating a problem solution, the mismatch requires a more fundamental consideration. Detection of such problems triggers a double-loop learning process, which is executed in the knowledge processing environment (cf. Fig. 1.2). Neither Firestone and McElroy (2003) nor Argyris and Schön (1978) specify the decision process that leads to either single-loop or double-loop learning in detail. The theory of model-centered learning provides an approach to describe this decision process from an individual perspective. The concept of Articulation Work allows bridging the conceptual gap between the KLC and model-centered learning and provides a starting point for developing support for this decision. Both theories are described below.

The knowledge processing environment is triggered with the formulation of a problem claim, that is, a description of the problem that needs to be resolved. This problem claim is not necessarily yet agreed upon by all involved or affected actors—involvement of other actors mostly happens during knowledge production activities following later on. Based upon the problem claim, a knowledge claim is formulated. The knowledge claim contains the 'new' knowledge (e.g., a fundamentally new version of a business process) and evolves over time in the iterative process of knowledge production. This process includes knowledge evaluation that takes an already codified (i.e., externalized) knowledge claim and verifies its correctness and applicability in the business processing environment based upon the current contents of the DOKB. As soon as no further revisions of the knowledge claim are considered necessary, Firestone and McElroy (2003) provide no statements on how to decide upon this—again, Articulation Work can be used as a starting point here), knowledge distribution is triggered. Knowledge distribution takes the outcome of the knowledge production activities (which can also be falsified or undecided knowledge claims, that is, knowledge claims that did not solve the problem that occurred in the business processing environment) and makes it accessible to the organization as a whole. The means of distribution are manifold, with the common objective of integrating the new knowledge in the DOKB. Activities here can range from distributing the codified knowledge claim to the relevant actors (as they carry the actual work knowledge and need to apply it when acting in a work process) and stakeholders in the organization to implement it in an IT-system that prescribes new behavior in the business processing environment. The Knowledge Lifecycle is closed via the re-integration of the outcomes of the knowledge-processing activities into the DOKB. New knowledge persisting in the DOKB can be used eventually for future activities in the business processing environment.

1.3 Articulation Work

The Knowledge Lifecycle does not explicitly address how work is organized by interacting actors in the business processing environment and how they react upon observed contingencies. Work is an inherently cooperative phenomenon (Helmberger and Hoos 1962). Whenever people work, they have interfaces with others, either cooperating directly or mediated via shared artifacts of work (Strauss 1985).

Cooperative work requires that participating parties have a common understanding of the nature of their cooperation. This includes dimensions such as when, how, and with whom to cooperate using certain means. The mutual understanding of cooperation has to be developed when cooperative work starts and has to be maintained over time, as changing environment factors may influence cooperation (Fujimura 1987). All activities concerned with setting up and maintaining cooperative work are summarized using the term, "Articulation Work" (Strauss 1985). Articulation Work mostly happens implicitly and is triggered during the actual productive work activities whenever contingencies arise (Gerson and Star 1986). Cooperative practices are established without a conscious act of negotiation in "implicit" Articulation Work, relying on social norms and observation to form a mutually accepted form of working together (Strauss 1988).

Implicit Articulation Work, however, is not sufficient when cooperative work situations are perceived to be 'problematic' or 'complex' by at least one of the involved parties (Strauss 1993). The terms 'problematic' and 'complex' here explicitly refer to individual perceptions, and are intrinsically subjective. As such, they cannot be detailed from an outsider's perspective. Consequently, relying on implicit Articulation Work can influence cooperation substantially. Different understandings of the same work situation impact the way of accomplishing tasks and the quality of work results, as long as Articulation Work remains on an implicit level.

Negotiation and development of a common understanding has to be carried out deliberately and consciously in such cases. This has been termed "explicit" Articulation Work by Strauss (1988). The expected outcome is to enable involved stakeholders starting or continuing their cooperative work towards a shared goal. The roles and activities of stakeholders involved in explicit Articulation Work need to be clarified, as it goes beyond implicit Articulation Work and the prevention of "problematic" (as termed by Strauss) situations.

Conducting Articulation Work facilitates the alignment of individual views about collaborative work. Strauss (1993) argues that these individual views (termed as 'thought processes' and 'mental activities') affect human work and direct individual action. In particular, for problematic or complex work situations, where social means of alignment (Wenger 2000) might not be sufficient, a closer look at the individuals' understandings of their and others' work is of interest. It should enable the design of effective support measures for explicit Articulation Work. From how 'thought processes' are described by Strauss (1993), they correspond to instances of 'schemes' and 'mental models' in cognitive sciences (Johnson-Laird 1981). The modification of mental models in the course of Articulation Work can thus be described using the theory of model-centered learning (Seel 2003).

1.4 Model-Centered Learning

People's activities in a work process, their decisions, and reactions to contingencies are driven by their perception of organizational reality (Weick et al. 2005). How people perceive their work context in an organization and how they derive their reactions on these perceptions is examined in cognitive sciences in the field of mental model theory (Johnson-Laird 1981). Mental model theory has also been used in knowledge management to explain operative triggers of organizational change processes (Firestone and McElroy 2003). Mental model theory here is used to describe individual and collective learning processes, that is, the adaptation of mental models to accommodate perceived changes in the organizational environment (Seel 2003).

Mental models are cognitive constructs that are used by persons to make plausible and assess their perceptions of phenomena in the real world (Seel 1991). Consequently, the alignment of individuals' views on work manifests in changes of the individuals' mental models—these changes are considered a form of learning (Seel 1991). The concept of 'model-centered learning' (Seel 2003) thus provides the foundation to design support instruments for explicit Articulation Work.

Model-centered learning is based on the constructs 'scheme' and 'mental model' (cf. Fig. 1.3). They serve to explain different strategies of humans to cope with external stimuli. Schemes are generalized abstract

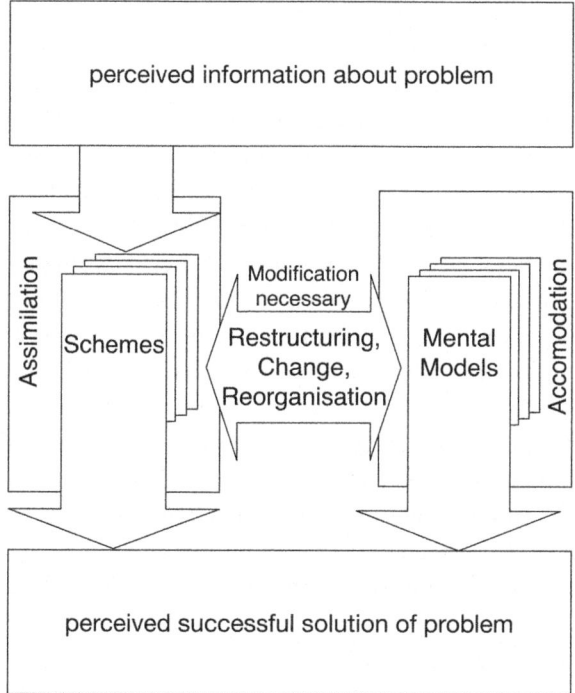

Fig. 1.3 Schemes and mental models (translated and adapted from Ifenthaler 2006)

knowledge patterns that are derived from prior experiences. They are used to immediately react on phenomena in the perceived reality without further planning activities. In situations that differ from prior experiences or are completely new to an individual, schemes are not applicable. Individuals create mental models in these cases to explain their perceptions and derive adequate reactions. Mental models might be incomplete or even be inherently contradictory. Individuals develop mental models for one particular situation only to a point enabling them to react to the stimulus in a way they consider adequate.

Mental models become more elaborate as more and more external stimuli and perceived information about the environment are incorporated. This process of 'accommodation' of mental models is considered a form of learning (Seel 1991). In the course of learning, mental models evolve from 'novice models' over 'explanatory models' to 'expert models' (or 'scientific models'), where the amount of information about causal relationships referring to phenomena in the real world increases from the former to the latter (Ifenthaler 2006). It is, however, important to note that expert models are not considered the desired aim of learning in any case. Due to the complexity of expert models, ad-hoc decisions based on perceived situations become more difficult and the perceived 'usefulness' of the mental models degrades (Ifenthaler 2006). In most cases, explanatory models are perceived as 'most useful', as they contain all information necessary to correctly judge a given situation (Ifenthaler 2006).

Depending on the situation, explanatory models may be rather simple or complex and contain less or more information, making them either more similar to a novice or an expert model. In terms of Articulation Work, expert models are hardly ever necessary, as they would require the individual to fully comprehend the entire work situation including the contributions and rationales of all other participants. In most situations, it is sufficient to develop an explanatory model of one's role in the overall work process and the interfaces to immediate co-workers. Elaborate explanatory models reduce the perceived complexity of work situations and thus enable focusing on the actual productive cooperative work.

Mental models evolve through experience in real world situations. Whenever an individual is confronted with perceptions that cannot be assimilated by existing schemes or be explained by current mental models, these models evolve and accommodate to the new perceptions (cf. Fig. 1.3). The goal of accommodation is to enable adequate action in situations similar to the one just perceived.

Mental model change requires recognizing the lack of adequacy of one's mental model and the opportunity and willingness to reflect on and adapt the mental model. In collaborative work settings, mental model change might not be restricted to a single person, but might require that all actors are involved in the work process in the reflection and change process. The willingness of changing a mental model that has been recognized to be inadequate by an individual can be assumed (Weick et al. 2005) (not imposing any assumptions about the quality of the change). Still, having the opportunity to adapt a mental model by gathering the required input and being able to retrieve it in an adequate form, can be an issue (ibid.). Furthermore, in collaborative settings, the willingness of other actors to change their mental models must not be assumed. If they do not perceive the environmental setting to be 'problematic' Strauss (1988), inquiries for change are usually met with resistance (Ifenthaler et al. 2007).

The challenges outlined above can be met with explicit activities dedicated to articulation, reflection, and alignment of individual mental models (Seel et al. 2009). Such activities need to be facilitated by providing artifacts that can serve as focal points of discussion and act as anchors for developing mutual understanding about the subject at hand (Dix and Gongora 2011). Conceptual models have been widely recognized as an appropriate mean to serve as external artifacts representing mental models (Novak 1995; Pirnay-Dummer and Lachner 2008; Chabeli 2010).

1.5 Collaborative Multi-perspective Modeling

Using collaborative conceptual modeling activities for creating a shared understanding about organizational phenomena has already been discussed extensively in prior research. Recently, research in the area of conceptual modeling has recognized that the added value of collaborative modeling not only is generated via the resulting models, but also by creating common ground about the modeled process for the involved people (Hoppenbrouwers et al. 2005). Research has started to examine how these modeling processes can be facilitated to support the evolution of common ground (Hoppenbrouwers and Rouwette 2012). In this line of research, several efforts have been made to qualitatively describe the effects occurring in such modeling sessions (Rittgen 2007; Seeber et al. 2012). The modeling process is considered to be a series of negotiation acts, with the model being an artifact generated as an outcome. Support measures in the process of modeling consequently focus on enabling and documenting negotiation acts. The process of process modeling has also been examined from a cognitive perspective, focusing on the development of understanding on the subject of modeling for the individual modeler (Soffer et al. 2011), where the authors discuss the cognitive fit of available modeling constructs as a factor influencing the process of modeling.

In the area of conceptual modeling of work processes, the idea of enabling multiple actors to explicitly articulate their individual understanding of their work contribution in separate models and use them as the foundation for consolidation in a structured way was first proposed by Mullery (1979). The multi-perspective modeling paradigm focuses on the representation of individual work contributions in models and subsequently merges them into a common model by agreeing on the interfaces among the individual models. It explicitly specifies the model elements which are subject to alignment, distinguishing them from the model parts that remain the responsibility of the individual actors.

This approach has been picked up by Türetken and Demirörs (2011), who propose a decentralized process elicitation approach ("Plural") in which individuals describe their own work. It uses eEPC (Nüttgens and Rump 2002) as a modeling language. Plural uses tool support built upon a commercial modeling environment, which identifies inconsistencies between individual models. Front et al. (2017) adopt multi-perspective modeling in the ISEA approach ('Identification, Simulation, Evaluation, Amelioration'). Perspectives here do not exclusively refer to individual work contributions, but are understood as putting different aspects of an organization into the focus of observation (e.g., information, organization, interaction). Modeling is tightly integrated with means of simulation, which allows to evaluate the perceived correctness of the models and to alter them accordingly.

Collaborative modeling and negotiation are also promoted by the Collaborative Modeling Architecture (COMA) approach (Rittgen 2009), which focuses on providing support for articulating and consolidating models during collaborative modeling with a language-agnostic negotiation approach. The COMA tool enables actors to communicate via the software in a structured way specified by the COMA methodology. Following its negotiation-oriented approach, COMA provides guidance for model consolidation (i.e., the negotiation process), which thus makes explicit divergent views and suggestions for a common view, which is ultimately agreed upon with the support of a human facilitator.

The usefulness of multi-perspective modeling as proposed by Mullery (1979) has also been backed by results for cognitive sciences in the field of collaborative learning (Engelmann and Hesse 2010) and mutually revealing and understanding mental models (Groeben and Scheele 2000). Engelmann and Hesse (2010) show that sharing of individually created concept maps about a topic improves mutual understanding within a group and improves the group members' performance in terms of problem solving skills related to this topic. Groeben and Scheele (2000) propose to adopt a dialogical approach to create a shared understanding about mental models. They use a tailored conceptual modeling language to explicitly represent these mental models and make them a subject of dialogue that ultimately reflects the reached consensus.

Dean et al. (2000) have examined the effects of different group modeling approaches, and found that having participants work on separate parts of a single model increases individual involvement, but leads to inconsistencies that need to be resolved in a separate step. These inconsistencies can be partially prevented when using a modeling approach that is guided by a human facilitator. Similar results have been observed by Hjalmarsson et al. (2015), who conducted empirical research in the area of facilitation of business process modeling workshops. They were able to identify different facilitation styles that are characterized by different behavioral patterns of the facilitator. The appropriateness of these styles is dependent on situational factors of the modeling setting and prior modeling knowledge of the participants.

1.6 Natural Versus Techno-Centric Modeling

The involvement of process participants in modeling tasks is linked to a major challenge: they cannot be expected to have modeling skills, and might not be willing to acquire these skills (Prilla and Nolte 2012). Trying to deploy modeling languages with a strict syntax and semantics and many different symbols often leads to even more resistance, as its added value does not become immediately visible (ibid.). What process participants would prefer is describing their knowledge through representational means that are as simple as possible in terms of both syntax and semantics (Zarwin et al. 2014). Zarwin et al. (2014) refer to these preferences as *natural modeling*. This term shifts the focus of attention from the technical and formal aspects of modeling to human aspects, with the aim of making it more widely accepted. Natural modeling follows three principles:

- modeling should be based on intuitive symbols and constructs
- modeling should be collaborative, so that models can serve as vehicles of communication facilitating knowledge sharing and promoting negotiation and commonly agreed-upon decisions, and
- modeling should be flexible in a sense that the symbols do not have a predefined meaning but rather the language used should emerge dynamically based on the situation at hand

Only if the ultimate goal of a model is its technical processing, modeling support instruments need to enable modelers to work in a continuum between "natural and formal modelling", which "should be fundamentally understood as the two polarities" (Zarwin et al. 2014, p. 29) on a continuum—the degree of formal syntax and semantics a model adheres to thus can evolve over time during its design.

Much existing research on collaborative modeling focuses on natural modeling practices (although not necessarily referred to as such). Research on supporting inexperienced modelers focuses on measures to guide them through the process of creating a model without overloading them with syntactic formalism. Existing research (e.g., Santoro et al. 2010; Fahland and Weidlich 2010; Kabicher and Rinderle-Ma 2011; Lai et al.

2014) suggests that starting modeling based upon a concrete work case makes it easier for inexperienced modelers to develop an understanding of the concepts necessary to represent a work process in an abstract model.

Using a case-based approach to modeling also reduces the number of language elements necessary to depict the work process. Case-based modeling omits alternatives in a process and exception handling and thus leads to smaller models, which usually also do not require complex semantic constructs. While the number of modeling elements alone appears not to have a notable impact on the understanding of a modeling language for inexperienced modelers (Recker and Dreiling 2007), empirical evidence shows that the number of language constructs used during modeling is limited and highly dependent on the modeling objective (Muehlen and Recker 2008). When involving inexperienced modelers, it seems to be appropriate to limit the number of available language constructs a priori to those appropriate for the intended modeling perspective and targeted outcome (Genon et al. 2011; Britton and Jones 1999).

Furthermore, Herrmann and Nolte (2014) and Santoro et al. (2010) provide evidence that non-formalized information and annotations to model elements can aid the externalization process, as this does not force the modelers to express all information using the constructs of the modeling language. Some results also point at the importance of (human or automatic) facilitation and scaffolding during the model creation process (Hjalmarsson et al. 2015) and the model alignment process (Rittgen 2007), particularly for inexperienced modelers (Davies et al. 2006). In addition, procedural and structural scaffolds provided by a facilitator or an automated system may support the elaboration of incomplete models (Herrmann and Loser 2013; Hoppenbrouwers et al. 2013; Oppl 2016; Oppl and Hoppenbrouwers 2016).

1.7 Taking an Integrated Socio-technical System Perspective

The presented kernel theories have been used as the foundation for artifact development as discussed in the introduction to this section. The MTO-framework (Strohm and Ulich 1997) can be used again to

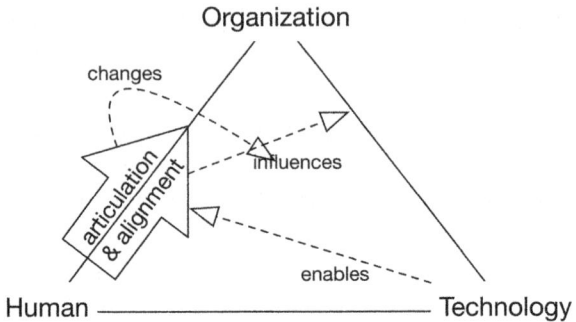

Fig. 1.4 Foci of research addressed in this book

visualize the different foci of research addressed in this book (cf. Fig. 1.4).

The main focus of the digital work design is to facilitate human actors' articulation and the alignment of their views on collaborative organizational work practices. Socio-technical artifacts are developed to enable this facilitation. In the following chapters, we examine how the deployment of such artifacts change the involved actor's perception of their work in an organizational context and how they progress to develop a shared understanding about their collaborative work. The articulation results are represented in a form that enables to influence existing enterprise architectures on both, an organizational and technical level, making use of concepts developed in the fields of business process management and information system design.

In this way, we further enrich the design space of socio-technical system design. While human resource management and work process organization from a technical perspective are understood in most cases (cf. Attewell 1992; Orlikowski 2000), we incorporate conceptual models of mental representations into socio-technical development cycles. The promoted integrated business and knowledge management perspective separates running business operations from dynamic capabilities while keeping them aligned through (i) deriving knowledge claims from existing operational procedures and (ii) either embodying accepted knowledge claims to changes in the business processing environment, or in all other cases, keep the handled knowledge claims in some living organizational design memory.

The approach gives space for development drivers in motivating and shaping cross-functional collaboration and allowing members of an organization to elaborate how operation could work across different boundaries (cf. Hsiao et al. 2012; Beane and Orlikowski 2015). Moving beyond singular dimensions of developing organizations allows suggesting a conceptual framework capturing the dynamics of social and technical work system patterns (cf. Edmondson et al. 2003; Jones 2013). It enriches the original socio-technical system paradigm (cf. Trist 1981; Mumford 2000) by explication of mental models, while keeping the assessment of system-wide implications of change and process innovation. The organization as a social subsystem of people and a technical subsystem of work process elements is linked through support instruments for continuous adaptation.

We supplement the original technical subsystem model comprising the structures, tools, and knowledge needed to perform the work with methodologically grounded technologies for handling the social system's attitudes, beliefs, and relationships between individuals and among groups. Active alignment support ensures the compatibility of individual mental models and finally that of the social and the technical subsystem. Hence, the technical and social subsystems form the entire work system when being kept adjusted to its development system (cf. Teece 2018). They require joint consideration to reflect on organizational enabling conditions and to promote people *and* technology as key drivers of development. The presented interventions and artifacts show the facilities to be encountered for stakeholder support.

References

Argyris, C., and D. Schön. 1978. *Organizational Learning: A Theory of Action Perspective*. Addison-Wesley.

Attewell, Paul. 1992. Technology Diffusion and Organizational Learning: The Case of Business Computing. *Organization Science* 3 (1): 1–19 (Informs).

Baxter, Gordon, and Ian Sommerville. 2011. Socio-Technical Systems: From Design Methods to Systems Engineering. *Interacting with Computers* 23 (1): 4–17 (Elsevier).

Beane, Matt, and Wanda J. Orlikowski. 2015. What Difference Does a Robot Make? The Material Enactment of Distributed Coordination. *Organization Science* 26 (6): 1553–1573 (Informs).

Botta-Genoulaz, Valérie, and Pierre-Alain Millet. 2006. An Investigation into the Use of ERP Systems in the Service Sector. *International Journal of Production Economics* 99 (1): 202–221 (Elsevier).

Britton, Carol, and Sara Jones. 1999. The Untrained Eye: How Languages for Software Specification Support Understanding in Untrained Users. *Human-Computer Interaction* 14 (1–2): 191–244. https://doi.org/10.1080/07370024.1999.9667269.

Chabeli, M. 2010. Concept-Mapping as a Teaching Method to Facilitate Critical Thinking in Nursing Education: A Review of the Literature. *Health SA Gesondheid* 15 (1) (Open Journals Publishing).

Convertino, Gregorio, Helena M. Mentis, Mary Beth Rosson, John M. Carroll, Aleksandra Slavkovic, and Craig H. Ganoe. 2008. Articulating Common Ground in Cooperative Work: Content and Process. In *CHI '08: Proceeding of the Twenty-Sixth Annual SIGCHI Conference on Human Factors in Computing Systems*, 1637–1646. New York: ACM. https://doi.org/10.1145/1357054.1357310.

Curtis, B., Marc I. Kellner, and Jim Over. 1992. Process Modeling. *Communications of the ACM* 35 (9): 75–90 (New York: ACM Press).

Dann, H.-D. 1992. Variation von Lege-Strukturen zur Wissensrepräsentation. In *Struktur-Lege-Verfahren als Dialog-Konsens-Methodik. Ein Zwischenfazit zur Forschungsentwicklung bei der rekonstruktiven Erhebung subjektiver Theorien*, 2–41. Münster: Aschendorff.

Davidson, Elizabeth. 2006. A Technological Frames Perspective on Information Technology and Organizational Change. *The Journal of Applied Behavioral Science* 42 (1): 23–39 (Sage Publications).

Davies, Islay, Peter Green, Michael Rosemann, Marta Indulska, and Stan Gallo. 2006. How Do Practitioners Use Conceptual Modeling in Practice? *Data & Knowledge Engineering* 58 (3): 358–380.

Dean, Douglas, Richard Orwig, and Douglas Vogel. 2000. Facilitation Methods for Collaborative Modeling Tools. *Group Decision and Negotiation* 9 (2): 109–128 (Springer).

Dix, A., and L. Gongora. 2011. Externalisation and Design. In *Proceedings of the Second Conference on Creativity and Innovation in Design*, 31–42. ACM.

Edmondson, Amy C., Ann B. Winslow, Richard M.J. Bohmer, and Gary P. Pisano. 2003. Learning How and Learning What: Effects of Tacit and Codified Knowledge on Performance Improvement Following Technology Adoption. *Decision Sciences* 34 (2): 197–224 (Wiley Online Library).

Engelmann, T., and F.W. Hesse. 2010. How Digital Concept Maps About the Collaborators' Knowledge and Information Influence Computer-Supported Collaborative Problem Solving. *International Journal of Computer-Supported Collaborative Learning* 5 (3): 299–319.

Fahland, D., and M. Weidlich. 2010. Scenario-Based Process Modeling with GRETA. In *Proceedings of the BPM 2010 Demonstration Track*, 52–57.

Feldman, Martha S., and Brian T. Pentland. 2003. Reconceptualizing Organizational Routines as a Source of Flexibility and Change. *Administrative Science Quarterly* 48 (1): 94–118 (SAGE Publications).

Firestone, J.M., and M.W. McElroy. 2003. *Key Issues in the New Knowledge Management*. Butterworth-Heinemann.

Frederiks, P.J.M., and Th.P. van der Weide. 2006. Information Modeling: The Process and the Required Competencies of Its Participants. *Data & Knowledge Engineering* 58 (1): 4–20. https://doi.org/10.1016/j.datak.2005.05.007.

Front, A., Dominique Rieu, Marco Santorum, and Fatemeh Movahedian. 2017. A Participative End-User Method for Multi-Perspective Business Process Elicitation and Improvement. *Software & Systems Modeling*, 1–24 (Springer).

Fujimura, J.H. 1987. Constructing 'Do-Able' Problems in Cancer Research: Articulating Alignment. *Social Studies of Science* 17 (2): 257–293 (SAGE Publications).

Genon, Nicolas, Patrick Heymans, and Daniel Amyot. 2011. Analysing the Cognitive Effectiveness of the BPMN 2.0 Visual Notation. In *Software Language Engineering*, Lecture Notes in Computer Science, vol. 6563, 377–396. Berlin and Heidelberg: Springer Berlin Heidelberg. https://doi.org/10.1007/978-3-642-19440-5_25.

Gerson, E.M., and Susan Leigh Star. 1986. Analyzing Due Process in the Workplace. *ACM Transactions on Information Systems (TOIS)* 4 (3): 257–270 (ACM Press).

Goncalves, Joao Carlos de A.R., Flávia Maria Santoro, and Fernanda Araujo Baiao. 2009. Business Process Mining from Group Stories. In 161–166. IEEE. https://doi.org/10.1109/CSCWD.2009.4968052.

Groeben, Norbert, and Brigitte Scheele. 2000. Dialogue-Hermeneutic Method and the "Research Program Subjective Theories". *Forum Qualitative Sozialforschung/Forum: Qualitative Social Research* 1 (2): 1–10. http://nbn-resolving.de/urn:nbn:de:0114-fqs0002105.

Haddara, Moutaz, and Ondrej Zach. 2012. ERP Systems in SMEs: An Extended Literature Review. *International Journal of Information Science* 2 (6): 106–116 (Scientific & Academic Publishing).

Helmberger, P., and S. Hoos. 1962. Cooperative Enterprise and Organization Theory. *American Journal of Agricultural Economics* 44 (2): 275.

Herrmann, Thomas, and Alexander Nolte. 2014. Combining Collaborative Modeling with Collaborative Creativity for Process Design. In *COOP 2014—Proceedings of the 11th International Conference on the Design of Cooperative Systems, 27–30 May 2014, Nice (France)*, 377–392. Cham: Springer International Publishing. https://doi.org/10.1007/978-3-319-06498-7_23.

Herrmann, T., and K.-U. Loser. 2013. Facilitating and Prompting of Collaborative Reflection of Process Models. In *Proceedings of MoRoCo@ECSCW 2013*, 17–24. ceur-ws.org.

Herrmann, Thomas, M. Hoffmann, G. Kunau, and K.U. Loser. 2002. Modelling Cooperative Work: Chances and Risks of Structuring. In *Cooperative Systems Design, a Challenge of the Mobility Age. Proceedings of COOP 2002*, 53–70. IOS Press.

Hjalmarsson, Anders, Jan C. Recker, Michael Rosemann, and Mikael Lind. 2015. Understanding the Behavior of Workshop Facilitators in Systems Analysis and Design Projects: Developing Theory from Process Modeling Projects. *Communications of the AIS* 36 (22): 421–447.

Hoppenbrouwers, Stijn, and Etiënne Rouwette. 2012. A Dialogue Game for Analysing Group Model Building: Framing Collaborative Modelling and Its Facilitation. *International Journal of Organisational Design and Engineering* 2 (1): 19–40 (Inderscience Publishers Ltd).

Hoppenbrouwers, Stijn, Henderik Alex Proper, and Theo P. van der Weide. 2005. A Fundamental View on the Process of Conceptual Modeling. In *Conceptual Modeling—ER 2005*, ed. L. Delcambre, C. Kop, H. C. Mayr, J. Mylopoulos, and O. Pastor, 128–143 (Chap. 9). Lecture Notes in Computer Science, vol. 3716. Berlin and Heidelberg: Springer Berlin Heidelberg. https://doi.org/10.1007/11568322_9.

Hoppenbrouwers, Stijn, Rob Thijssen, and Jan Vogels. 2013. Operationalizing Dialogue Games for Collaborative Modeling. In 41–48.

Hsiao, Ruey-Lin, Dun-Hou Tsai, and Ching-Fang Lee. 2012. Collaborative Knowing: The Adaptive Nature of Cross-Boundary Spanning. *Journal of Management Studies* 49 (3): 463–491 (Wiley Online Library).

Ifenthaler, D. 2006. Diagnose Lernabhängiger Veränderung Mentaler Modelle—Entwicklung Der SMD-Technologie Als Methodologisches Verfahren Zur Relationalen, Strukturellen Und Semantischen Analyse Individueller Modellkonstruktionen. University of Freiburg.

Ifenthaler, D., Pablo N. Pirnay-Dummer, and Norbert M. Seel. 2007. The Role of Cognitive Learning Strategies and Intellectual Abilities in Mental Model Building Processes. *Technology, Instruction, Cognition and Learning* 5: 353–366.

Johnson-Laird, P.N. 1981. Mental Models in Cognitive Science. *Cognitive Science* 4 (1): 71–115 (Elsevier).

Jones, Gareth R. 2013. *Organizational Theory, Design, and Change*. Upper Saddle River, NJ: Pearson.

Jonkers, Henk, Marc Lankhorst, Rene Van Buuren, Stijn Hoppenbrouwers, Marcello Bonsangue, and Leendert Van Der Torre. 2004. Concepts for Modeling Enterprise Architectures. *International Journal of Cooperative Information Systems* 13 (3): 257–287 (World Scientific).

Jonkers, Henk, Marc M. Lankhorst, Hugo W.L. ter Doest, Farhad Arbab, Hans Bosma, and Roel J. Wieringa. 2006. Enterprise Architecture: Management Tool and Blueprint for the Organisation. *Information Systems Frontiers* 8 (2): 63–66 (Springer).

Kabicher, Sonja, and Stefanie Rinderle-Ma. 2011. Human-Centered Process Engineering Based on Content Analysis and Process View Aggregation. In *Advanced Information Systems Engineering. CAiSE 2011*, ed. H. Mouratidis and C. Rolland, 467–481 (Chap. 35). Lecture Notes in Computer Science, vol. 6741. Berlin and Heidelberg: Springer Berlin Heidelberg. https://doi.org/10.1007/978-3-642-21640-4_35.

Lai, Han, Rong Peng, and Yuze Ni. 2014. A Collaborative Method for Business Process Oriented Requirements Acquisition and Refining. In *Proceedings of ICSSP 2014*, 84–93. New York: ACM Press. https://doi.org/10.1145/2600821.2600831.

Muehlen, zur Michael, and J.C. Recker. 2008. How Much Language Is Enough? Theoretical and Practical Use of the Business Process Modeling Notation. In *Advanced Information Systems Engineering. CAiSE 2008*, Lecture Notes in Computer Science, vol. 5074, ed. Z. Bellahsène and M. Léonard, 465–479. Berlin and Heidelberg: Springer Berlin Heidelberg. https://doi.org/10.1007/978-3-540-69534-9_35.

Mullery, G.P. 1979. CORE-a Method for Controlled Requirement Specification. In *ICSE '79—Proceedings of the 4th International Conference on Software Engineering*, 126–135.

Mumford, Enid. 2000. A Socio-Technical Approach to Systems Design. *Requirements Engineering* 5 (2): 125–133 (Springer).

Novak, Joseph D. 1995. Concept Mapping to Facilitate Teaching and Learning. *Prospects* 25 (1): 79–86 (Kluwer Academic Publishers). https://doi.org/10.1007/BF02334286.

Nüttgens, M., and F.J. Rump. 2002. Syntax Und Semantik Ereignisgesteuerter Prozessketten (EPK). *Promise*, 64–77.

Oppl, Stefan. 2016. Towards Scaffolding Collaborative Articulation and Alignment of Mental Models. *Procedia Computer Science* 99: 124–145. https://doi.org/10.1016/j.procs.2016.09.106.

Oppl, Stefan, and Stijn Hoppenbrouwers. 2016. Scaffolding Stakeholder-Centric Enterprise Model Articulation. In *The Practice of Enterprise Modeling*, Lecture Notes in Business Information Processing, vol. 267, 133–147. Springer International Publishing. https://doi.org/10.1007/978-3-319-48393-1_10.

Orlikowski, Wanda J. 2000. Using Technology and Constituting Structures: A Practice Lens for Studying Technology in Organizations. *Organization Science* 11 (4): 404–428 (Informs).

Orlikowski, Wanda J., and C. Suzanne Iacono. 2001. Research Commentary: Desperately Seeking the 'IT' in IT Research—A Call to Theorizing the IT Artifact. *Information Systems Research* 12 (2): 121–134 (Informs).

Pirnay-Dummer, Pablo N., and A. Lachner. 2008. Towards Model Based Knowledge Management. A New Approach to the Assessment and Development of Organizational Knowledge. In *Annual Proceedings of the AECT 2008*, ed. M. Simonson, 178–118.

Prilla, M., and Alexander Nolte. 2012. Integrating Ordinary Users Into Process Management: Towards Implementing Bottom-Up, People-Centric BPM. In *Enterprise, Business-Process and Information Systems Modeling*, 182–194. Springer.

Ragowsky, Arik, and Tomi Somers. 2002. Enterprise Resource Planning. *Journal of Management Information Systems* 19 (1): 11–15 (Taylor & Francis).

Recker, J.C., and A. Dreiling. 2007. Does It Matter Which Process Modelling Language We Teach or Use? An Experimental Study on Understanding Process Modelling Languages Without Formal Education. In *Proceedings of 18th Australasian Conference on Information Systems*, Toowoomba, Australia, pp. 356–366.

———. 2011. The Effects of Content Presentation Format and User Characteristics on Novice Developers' Understanding of Process Models. *Communications of the Association for Information Systems* 28 (6): 65–84.

Recker, J.C., and Michael Rosemann. 2009. Teaching Business Process Modelling: Experiences and Recommendations. *Communications of the Association for Information Systems* 25 (1): 32.

Rittgen, Peter. 2007. Negotiating Models. In *Advanced Information Systems Engineering*, Lecture Notes in Computer Science, vol. 4495, ed. J. Krogstie and Andreas Opdahl, 561–573. Berlin and Heidelberg: Springer Berlin Heidelberg. https://doi.org/10.1007/978-3-540-72988-4_39.

---. 2009. Collaborative Modeling of Business Processes: A Comparative Case Study. In *Proceedings of the 2009 ACM Symposium on Applied Computing*, 225–230. New York: ACM Press. https://doi.org/10.1145/1529282.1529333.

Santoro, Flávia Maria, Marcos R.S. Borges, and José A. Pino. 2010. Acquiring Knowledge on Business Processes from Stakeholders' Stories. *Advanced Engineering Informatics* 24 (2): 138–148. https://doi.org/10.1016/j.aei.2009.07.002.

Seeber, I., R. Maier, and B. Weber. 2012. CoPrA: A Process Analysis Technique to Investigate Collaboration in Groups. In *2012 45th Hawaii International Conference on System Sciences*, 363–372. IEEE.

Seel, Norbert M. 1991. *Weltwissen Und Mentale Modelle*. Göttingen u.a.: Hogrefe.

---. 2003. Model-Centered Learning and Instruction. *Technology, Instruction, Cognition and Learning* 1 (1): 59–85 (Old City Publishing).

Seel, Norbert M., D. Ifenthaler, and Pablo N. Pirnay-Dummer. 2009. Mental Models and Problem Solving: Technological Solutions for Measurement and Assessment of the Development of Expertise. In *Model-Based Approaches to Learning: Using Systems Models and Simulations to Improve Understanding and Problem Solving in Complex Domains*, Modeling and Simulations for Learning and Instruction, vol. 4, ed. P. Blumschein, W. Hung, and J. Strobel, 17–40. Sense Publishers.

Simões, David, Pedro Antunes, and Jocelyn Cranefield. 2016. Enriching Knowledge in Business Process Modelling: A Storytelling Approach. In *Innovations in Knowledge Management*, 241–267. Springer.

Soffer, P., M. Kaner, and Y. Wand. 2011. Towards Understanding the Process of Process Modeling: Theoretical and Empirical Considerations. In *International Conference on Business Process Management*, 357–369. Berlin, Heidelberg: Springer.

Soh, Christina, Siew Kien Sia, Wai Fong Boh, and May Tang. 2003. Misalignments in ERP Implementation: A Dialectic Perspective. *International Journal of Human-Computer Interaction* 16 (1): 81–100 (Taylor & Francis).

Strauss, A. 1985. Work and the Division of Labor. *The Sociological Quarterly* 26 (1): 1–19 (Blackwell Publishing Ltd).

---. 1988. The Articulation of Project Work: An Organizational Process. *The Sociological Quarterly* 29 (2): 163–178.

---. 1993. *Continual Permutations of Action*. New York: Aldine de Gruyter.

Strohm, O., and E. Ulich. 1997. *Unternehmen Arbeitspsychologisch Bewerten: Ein Mehr-Ebenen-Ansatz Unter Besonderer Berucksichtigung Von Mensch, Technik Und Organisation*. Zürich: vdf Hochschulverlag.

Teece, David J. 2018. Business Models and Dynamic Capabilities. *Long Range Planning* 51 (1): 40–49 (Elsevier).

Thome, Rainer. 1982. Wirtschaftlichkeitsrechnung in Der Informationsverarbeitung. *Zeitschrift Für Betriebswirtschaft (ZfB)* 52 (6): 555–579 (Springer).

Trist, Eric. 1981. The Evolution of Socio-Technical Systems. *Occasional Paper*, no. 2.

Türetken, Oktay, and Onur Demirörs. 2011. Plural: A Decentralized Business Process Modeling Method. *Information & Management* 48 (6): 235–247. https://doi.org/10.1016/j.im.2011.06.001.

Weick, Karl E., Kathleen M Sutcliffe, and David Obstfeld. 2005. Organizing and the Process of Sensemaking. *Organization Science* 16 (4): 409–421 (Informs).

Wenger, E. 2000. Communities of Practice and Social Learning Systems. *Organization* 7 (2): 225–246.

Zarwin, Z., M. Bjekovic, J.M. Favre, J.S. Sottet, and Erik Proper. 2014. Natural Modelling. *Journal of Object Technology* 13 (3): 4:1–36. https://doi.org/10.5381/jot.2014.13.3.a4.

Open Access This chapter is licensed under the terms of the Creative Commons Attribution 4.0 International License (http://creativecommons.org/licenses/by/4.0/), which permits use, sharing, adaptation, distribution and reproduction in any medium or format, as long as you give appropriate credit to the original author(s) and the source, provide a link to the Creative Commons licence and indicate if changes were made.

The images or other third party material in this chapter are included in the chapter's Creative Commons licence, unless indicated otherwise in a credit line to the material. If material is not included in the chapter's Creative Commons licence and your intended use is not permitted by statutory regulation or exceeds the permitted use, you will need to obtain permission directly from the copyright holder.

2

Elicitation Requirements

This chapter discusses the elicitation in work process design and its requirements on socio-technical support instruments. It provides the conceptual underpinnings of the articulation and alignment processes occurring during work process elicitation, drawing from different disciplines such as social psychology, cognitive sciences, knowledge management, and computer-supported collaborative work. We finally offer a theory-based synthesis of the concepts developed in these areas to inform and reflect on the methods' design in the following chapters.

Although a thorough acquisition of work knowledge is almost never readily available for development, requirements can be identified on how information could be articulated and aligned for further processing, both, in terms of elicitation, and representation, as well as inherent conditions and support. Much of the adjacent methodological and technological requirements are not documented—they reside in the minds of experienced developers or stakeholders concerned with organizational design. Although requirements for system design need to be elicited or drawn out, the methodology on how to thoroughly identify the stakeholder capabilities, needs, risks, and assumptions associated with a given work setting, business, or project is unclear in most cases.

In the following, we start with an individual perspective on elicitation and call for role awareness in this process, as work processes can be distinguished at least by functional roles individual actors need to take in order to achieve business objectives. Understanding one's own role(s) lays the ground to adopt various perspectives on work procedures and consider context relevant for role-specific behavior. The resulting situatedness enables reflecting on the scope of work tasks and re-shaping organizational structures in collaborative settings. In order to handle complex situations, a systems-of-system perspective could help. Bringing intangible or implicit knowledge to the surface and to represent it qualifies for aligning mental models on existing work procedures and behaviors in a comprehensive way. It facilitates co-creating future work settings, in particular, taking into account the continuous penetration of digital systems into work task accomplishment.

2.1 Setting the Stage—Awareness on Roles and Their Management

Traditionally, the preparation for elicitation is the first step. It aims towards a comprehensive and an accurate understanding of the work situation and the needs of involved stakeholders. During the elicitation process, an analyst's understanding of the work needs helps in scoping and selecting proper stakeholders and elicitation techniques. Hence, stakeholders need to get actively engaged in articulation and alignment. Stakeholders here are understood as any persons that are directly or indirectly affected by a work process or engage in it. This may include customers/end users, suppliers, the project manager, quality assurance, regulators, business partners, operational support, domain subject matter experts, and implementation specialists.

A facilitator needs to recruit appropriate stakeholders based on the intended project or scope of activities. After a facilitator has identified and recruited relevant stakeholders, before method(s) by which elicitation can be performed, it is advisable to create awareness on roles and role identities, in particular due to the proliferation of digital media and their social media capabilities:

> New communication technologies have freed interaction from the requirements of physical copresence; these technologies have expanded the array of generalized others contributing to the construction of the self. Several research foci emerge from this development: the substance of 'I', 'me', and the generalized other in a milieu void of place, the establishment of 'communities of the mind', and the negotiation of copresent and cyberspace identities. (Cerulo 1997, p. 386)

Consequently, not only at the workplace, but also in all of today's societal communities, stakeholders have had to learn dealing with a variety of roles. They can present themselves differently based on who they are talking to and what an interaction is about (cf. Castells 1997). When using content management systems or social media to share their experiences with work processes, they act in a certain role. The role is based on technological affordances and immediate context. Roles may either be described in certain profiles using registration wizards, or recorded along the interaction, for example, documenting paths in business information systems.

The first case might be obvious for role design and presenting oneself, whereas the latter most of us become aware of once receiving own behavior data, for example, when having searched for information and receiving proposals referring to our search pattern. Hence, role design and management have become increasingly important when multiple situation elements occur in some concerted manner. Consider, for instance, searching for information on a product in an online catalogue. The user could be a novice in product management or customer service. It could also be an experienced product manager or a barely skilled customer agent. Role types occur along various dimensions and domains, such as level of skill with respect to features or technologies, and expertise in a domain. They need to be recognized for articulating and aligning work knowledge, in particular when involving multiple technical communities, as our exemplified user may also ask questions in a product forum in which authors address novice workers (cf. Ellison et al. 2006).

If stakeholders are more aware about their content creation and usage as well as communication acts, their role in interaction becomes more transparent to them, and they are able to articulate their knowledge in a more reflected way. It has been observed that people react to situations based on context rather than fixed behavior patterns (cf. Meyrowitz 1986). In our example, all three items, that is, the level of competence in product handling, searching with descriptors and meta-data, and interactive navigation have to be considered in their mutual context.

In an information-based—and yet more important, in a knowledge-based—work environment, roles are functional entities based on the stakeholder identities, evolving over time (Castells 1997). Their management goes beyond traditional presentation formats, such as yellow page entries or personal web pages, as stakeholders are acting in various roles in dynamically changing (virtual) communities (cf. Jensen Schau and Gilly 2003). Virtual communities in the knowledge age society are groups of people connected via social and knowledge media. They engage in knowledge creation, documentation, sharing, collective use, and distribution.

Community members take the role of content providers, explorers, and respondents. They may change these roles dynamically, driven by their personal identities triggering their behavior (cf. Montague 2013; Ackerman et al. 2017). Such static descriptions of the Self are more structured than blogs and information boards, presuming ongoing interactions among community members (Robinson 2007). However, in virtual communities, goal-oriented interaction forms the awareness of its members and finally, their individual activities (Ellison et al. 2006).

Meyrowitz (1986) has already observed that social media tend to blur the lines between 'front stage' (what should be visible) and 'back stage' behaviors (what currently is not, but potentially should, become visible to others). Consequently, facilitators and analysts need to look at dealing with the 'front' and 'back' stage dynamically. Bridges are the features of new media, in particular, when operating under the control of stakeholders. Context thus becomes paramount in a virtual community, and the role of management within it (Ferscha et al. 2004a). However, self-regulated role management seems to be a challenging task. Jarvis (2009) and Jarvis and Watts (2012) indicate that role management is a learning task, as becoming a self in society, both mind and self are socially learned phenomena. It has to deal with informed learning activities and might include conflicting individual and social interests.

Although roles can be part of various contexts, they constitute the appearance of individual actors. Even when related to learning how to

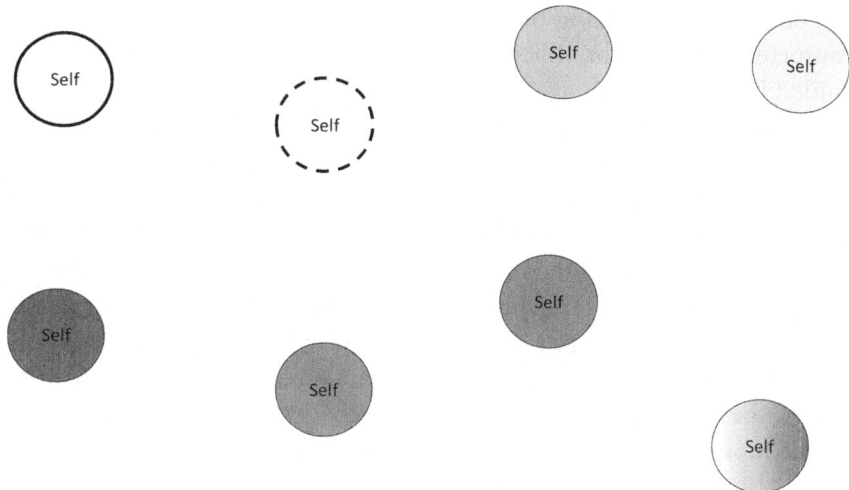

Fig. 2.1 Awareness on roles

manage various roles, their set up is relevant to how stakeholders get involved in work knowledge elicitation. Consequently, articulation support requires features for stakeholders to make roles transparent, if not build capacity to manage them in a reflected and structured way.

Figure 2.1 visualizes setting the stage in terms of identifying one's role. *Self* denotes an actor who has a certain role. Different Selfs are represented by different colors or tones. The roles are pictured by the surrounding circles. As shown in the figure, actors can play different roles—the Self with the white background has two roles, which is denoted by different outlines of the circle.

2.2 Situation Awareness

As already mentioned above, the development of organizations, and thus socio-technical systems, is increasingly driven by its members. Hence, stakeholders need to spend socio-cognitive effort when articulating and aligning knowledge about their work. Role- and task-specific behavior of stakeholders is framed by its triggers, such as individual intention, and its expected effects or outcome. This framing can be done on arbitrary levels of granularity, depending on a stakeholder's perspective and/or level of competence or insight. Elicitation, modeling, and probing should be guided by direct recall, avoiding errors, incompatibilities, and inconsistencies grounded in articulation and representation (Harman et al. 2015).

Such recalls require insight into the situations that stakeholders face or are part of when accomplishing tasks. They lay the ground for situated articulation and allow framing activities with situation-specific information, both on triggering them, and on effectuation (cf. Gross et al. 2005). Triggers can be events (externally set) or intentions (stakeholder-specific) in combination with input data to be processed, whereas effectuation is represented through output in terms of data or states, and the outcome, that is, the (intended) effect of a certain activity or a set of actions.

The most authentic articulation and representation of situations can be assumed to stem from stakeholders experiencing these situations. In self-contained articulation settings, stakeholders do not have to rely on information provided by analysts, in contrast to settings involving external people, such as for interviewing, where it cannot be assumed that analysts or facilitators are familiar with the field (Parsaye and Chignell 1988). Moreover, stakeholders, in particular experts, when asked explicitly, forget to mention tasks they assume to be widely known, or have difficulties explaining what they do when not actually doing it at the same time (Grosskopf et al. 2010). Knowledge is thus inseparable from doing (cf. Brey 2005).

Putting situated cognition theory in the context of representation, generated models in a natural and intuitive way potentially have greater accuracy than what could traditionally be achieved with common acquisition and analysis techniques (cf. Harman et al. 2015). Reducing the requirement of involving external people enables a wider scope of self-organizing work, as many more stakeholders can participate in organizational change and development.

An underlying concept in this context seems to be 'agency.' According to Himma (2009),

> [the] idea of agency is conceptually associated with the idea of being capable of doing something that counts as an act or action. As a conceptual matter, X is an agent if and only if X is capable of performing action; breathing is something we do, but it does not count as an action. Typing these words is an action, and it is in virtue of my ability to do this kind of thing

that, as a conceptual matter, I am an agent. ... Agents are not merely capable of performing acts; they inevitably perform them (in the relevant sense). ... The very concept of agency presupposes that agents are conscious. (p. 19)

Reflecting this understanding reveals the way of involvement in a situation when humans are acting or interacting. It underpins the requirement to devote design effort to human issues to the same extent developers spend for technical ones. The recognition of user modeling can be considered such an endeavor (cf. Brusilovsky and Cooper 2002). Situatedness is awareness about its world, comprising communities, organizations, societies, or other contingent systems of systems, and its capability to induce changes on it (cf. Campos et al. 2009).

The essence of situation awareness lies in the monitoring of various entities and the relations that occur among them. Since the properties of relations, unlike the properties of objects, are not directly measurable, one needs to have some background knowledge (such as ontologies and rules) to specify how to derive the existence and meaning of particular relations. (Matheus et al. 2005)

Consequently, system development, concerning cognition, organizations, social or technological systems, should be driven by different systemic perspectives and lead to architectures allowing dynamic changes (cf. Rolland et al. 1999). Situatedness of development processes is a key issue in software and method engineering communities (cf. Barwise and Perry 1981). Prescriptions, from either the user interaction or the task handling perspective, need to be adapted to the situation at hand, allowing for systems dynamics in the course of task or interaction processes (Christian Stary 2017a).

According to findings in cognitive science, actors (there referred to as 'agents') are considered as *embodied* and interactively *situated* in worlds (Dobbyn and Stuart 2003). When analyzing the meanings attached to these terms a set of conditions for situatedness and embodiment can be derived, based on the conclusive assumption that external representational schemes are required for adaptation. While virtual agents in virtual worlds are considered neither situated nor embodied, awareness of evolving goals, various modalities for interaction and task accomplishment procedures could lead to a rich repertoire of interactions (cf. Gross et al. 2005).

Embedded actors could develop individual points of view, relative to their starting position work spaces, and have a capacity to develop a dedicated interaction space. None of these capabilities are possible without representation of work activities. They can either rely on engineering work flows, as for example, in Business Process Management (Weske 2010), or on engineering of cognitive support, such as model-based approaches (cf. Christian Stary 2000). The latter need to relate to cognitive constructs (cf. Eberle et al. 2011). Thereby, mutual relationships between user properties and interaction styles can be captured in terms of cognitive characteristics. In addition, rules for dynamically tuning task accomplishment and interaction can be kept in dedicated representation schemes, such as adaptation models.

The problem with this type of context information is that it cannot be encoded with standardized approaches, such as BPMN (Business Process Modeling Notation; www.bpmn.org). While an expert may be able to explain the rationale for work activities, these normative representations do not convey required context to information (cf. Brown et al. 1989). Hence, additional effort is required to provide adequate context information, for example, through apprenticeships (Lave 1988). From this, theories of explicit memory, sometimes referred to as tacit knowledge, have emerged, as knowledge cannot easily be conveyed to other people. To retrieve this information, it is easiest to use a simulation-based approach for memory recall (Rubin 2006).

For articulating context when capturing role- or task-specific work knowledge, activity-relevant information can be framed in a structured way (cf. Christian Stary 2017b). As shown in Fig. 2.2, a tripartite approach could consist of:

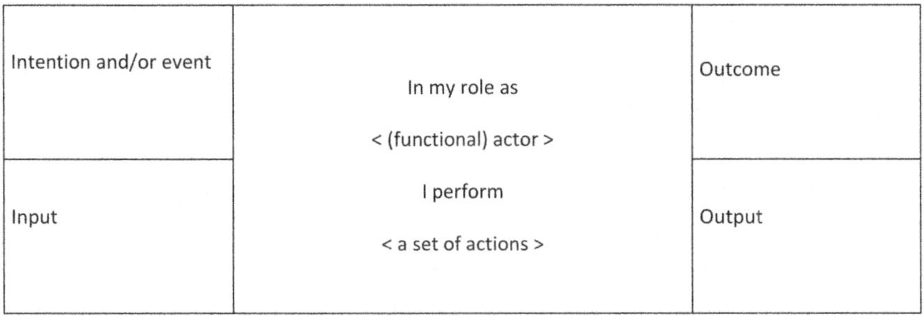

Intention and/or event	In my role as < (functional) actor >	Outcome
Input	I perform < a set of actions >	Output

Fig. 2.2 The articulation scheme containing trigger, role-specific activity, and effect

1. *Trigger and incoming information*: Hereby we distinguish pragmatically and semantically relevant information (context) from syntactic structure (input). At least the context should be given when a task chain is started.
2. *Functional processing information*: It specifies not only the function in terms of activities to be set, but rather the role in which a work task is performed. In this way, the context can be represented more accurately compared with purely functional specifications.
3. *Effect and deliverables*: Again, we distinguish pragmatically and semantically relevant information (outcome denoting the effect of using a feature) from the syntactic structure (output). At least some outcome should be generated once a work task chain is completed.

For each task-relevant behavior, a separate representation could be generated by stakeholders in the course of eliciting work knowledge (cf. Christian Stary 2017d).

Framing of role-specific actions by triggering and effectuating behavior allows for scoping actor behaviors, as the following example demonstrates. A service provider in the field of software development has a stakeholder in the functional role of a Customer Service Agent who articulates how a product claim from a customer is framed. The input is a product claim, for example, when a product does not meet a customer requirement. The intention is to help the concerned customer, until he/she is satisfied. The output of this activity is either a hint about how the requirement has already been met, or a change request for product development, in case it could not be met so far (Fig. 2.3).

From a work process perspective, this representation constitutes a particular actor with behavior. Although in the course of articulation, the

Help customer	In my role as Customer Service Agent I handle a product claim.	Customer is satisfied
Product claim		Hint to change request

Fig. 2.3 Customer service actor behavior handling customer product claims

Innovate product	In my role as Customer Service Agent I re-formulated	Product changes
Product claim from customer		Idea ticket
	The product claim Towards innovative featuring	

Fig. 2.4 Scoping another actor behavior—Idea Provider

functional role (Customer Service Agent) provides an intuitive entry point, the label could more accurately read 'product claim customer handling', as it is very likely that the work agenda of the Customer Service Agent comprises additional actions.

In case the Customer Service Agent reports in constructive way and has an idea for innovating the product based on product claims or customer requests, the articulation scheme enables switching the role in that context. Figure 2.4 shows a coherent representation for that case.

The consequences for work process modeling are substantial, since handling a product claim as a 'Customer Service Agent' shapes an actor taking a functional role, communicating with the customer, and product department. A particular role—'Idea Provider'—allows not only in reducing the complexity when the workplace of a service agent is described, but rather enables developing a product improvement or organizational learning procedure that could serve as a pattern across organizational units or domains.

The latter model could serve as input for the change manager to implement product innovation processes after the proposal has been collectively reflected on. For a complete task chain, and thus business process specification, each output of an activity needs to correspond to an input of an adjacent activity.

Procedural requirements. When framing role- or task-specific behavior in the way described above, contextual representations need to be set up along a procedure allowing to articulate intentions. Grice (1969) has already investigated the relationship between meaning and intention of utterers. From Böhm (1997)'s research, we can conclude that meaning constitutes sense-making for humans, as it needs to be seen intertwined with the functional context of a person and the goals this person is trying to achieve individually (ibid., p. 69).

Sheeran (2002) has studied possible gaps between behavior and intention. Looking for psychological variables to 'bridge' possible intention–behavior gaps, the author's meta-analysis of meta-analyses has led to a conceptualization of intention–behavior discrepancies. Four groups of variables, namely behavior type, intention type, properties of intention, and cognitive and personality variables, could be clustered as they moderate intention–behavior relations. Once behavior specifications contain a task description according to individual mental models, any verbalization of intention respects the stakeholder's personality and the cognitive model of a situation. As the intention type is not essential when articulating triggers of actions, each stakeholder can describe the way he/she perceives it in the intentional context of the action (set) at hand.

Hug et al. (2012) have referred to intentions in the context of process engineering. Rather than detailing how to facilitate stakeholder articulation with respect to intentional behavior, "the intentional level is used to guide engineers through IS [Information Systems] processes by dynamic choices. Each time an intention is achieved the model suggests the next steps that can be enacted and new ways to achieve them. The resulting IS development process is adaptive and flexible as it is dynamically constructed" (ibid., p. 204). As we will see below, for establishing intentional fit of activities, this input is valuable.

The presented sample scheme should illustrate how behavior could be captured in a context-rich way when articulating knowledge on work tasks. The scheme frames activities by triggers (incoming side) and intended effects (outgoing side). As such, activities can be contextualized with situation-specific information.

Figure 2.5 visualizes situation-awareness of actors in specific roles. The Self denotes an actor who plays a certain role in a certain situation. The role is pictured by the surrounding circle, whereas the situation context is denoted by a dotted cloud symbol. As shown in the figure, actors can not only play different roles, but also act in a certain role in different situations—the Self with the white background has two roles (denoted by

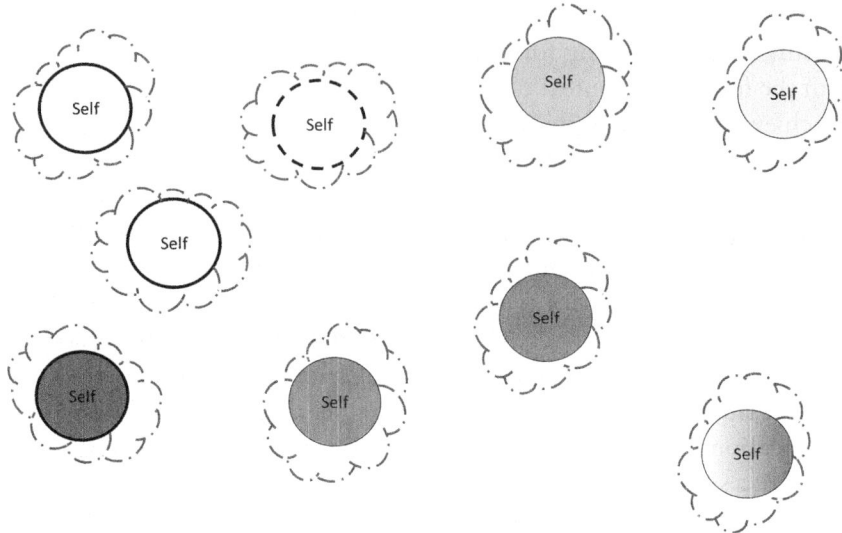

Fig. 2.5 Situation awareness

different shapes of the circle), with one role (the one with the solid circle in the figure) considered relevant for two different situations.

2.3 Conceptual Understanding of Complex Systems

The advent of digital transformation invading all societal and economic systems requires a re-consideration of the generative nature of sociotechnical system design. In particular, this transformation needs be studied with regard to how links are continuously explored and accelerated between existing as well new value spaces (Bounfour 2016). The accelerated production of relations does not only substantiate system thinking (cf. Senge 1990; Senge and Sterman 1992), but also characterizes the fundamental nature of digital production systems—relations are considered essential drivers for creating value in digital spaces (Bounfour 2016). We need to delineate their nature, as being transactional, organic, or semi-organic, since they lead to deep changes in the way we 'produce', and finally affect business models and power relations of organizations and societies (ibid.).

As digital transformations are complex, some scholars have already called for a system science approach to deal with these challenges (Flood and Carson 2013). Thereby, traditional cognitive or top-down approaches to regulate or control dynamic processes are seen as a 'last resort' (Colander and Kupers 2014). Evolving complex systems bear systemic challenges (ibid.), which are wicked due to their social or cultural nature and incomplete, contradictory, interconnected, and changing requirements that are often difficult to recognize. Bringing together complexity and wicked problem theories to understand how individual organizations and change agents can better influence large system change, Waddock et al. (2015) developed a respective framework. It integrates wicked problems and complexity theories to cope with large systems interventions while taking the perspective of individual change agents. Although the authors concluded their study that change agents in organizations can enhance their influence and use the power of system dynamics to support positive action for sustainable change, they recognized that effective large-scale change still has limited theoretical understanding.

Consequently, not only do we need to put forward theoretical understanding of change management by positioning the organization in the context of a broader system, but we also need to define its role in creating change based on articulation of individual stakeholders (cf. Senge's perspective on a learning organization requiring learning members of that organization). Individually informed articulation (e.g., on principles for acting) is likely to facilitate addressing the nature of wicked problems by setting informed relations between individual systems and the large systems where they are embedded.

Accepting the wickedness of challenges and complex problems, we need to shed light on the relation of individuals as change agents and their relations to organizations and society in transformational change. These transformations can be substantial and lead to emerging individual and social behaviors due to that change. Research reveals the essential role of individuals when structuring situated cognitive transformation processes (Kihlstrom 2013):

Evocation, selection, and manipulation all change the environment through overt behavior—either the behavior of the person him- or herself, or that of other people. In each case, someone does something overtly that changes the objective character of the environment—that is, changes the environment for everyone in it, not just for the person itself. But these three modes do not exhaust the effects of the person on the environment. People also engage in *covert* mental activities that alter their *mental representations* of their subjective environment—that is, the environment as they privately experience it. As opposed to behavioral manipulation, cognitive transformation does not act directly on the objective environment—the environment as it would be described in the third person by an objective observer and experienced by everyone in it. Rather, transformation acts on the *subjective* environment. Through cognitive transformations, people can change their internal, mental representations of the external physical and social environment—perceiving it differently, categorizing it differently, giving it a different meaning than before. In cognitive transformation, the objective features of the environment remain intact—they have not been altered through evocation, selection, and manipulation. Rather, the cognitive transformation has altered the environment *for that person* only. The environment is unchanged for everyone else—unless and until the cognitive transformation leads the person to engage in selective and manipulative behavior that, as described earlier, will change the environment for everyone in it. (Kihlstrom 2013, p. 798)

We cannot foresee how the various systems will act, and deal with traditional mechanisms to organize and control. We need to assume anarchic patterns, questioning traditional authority or other controlling systems.

One way to deal with the social dimensions of organizations and the resulting dynamics of systems involving embodied stakeholders is to take a Complex Adaptive Systems (CAS) perspective. According to Chan (2001), CAS started in US to oppose the European 'natural science' tradition in the area of cybernetics and systems. Although CAS theory shares the subject of general properties of complex systems across traditional disciplinary boundaries (like in cybernetics and systems), it relies on computer simulations as a research tool (as pointed out by Holland in 1992 initially (Holland 1992)), and considers less integrated or 'organized' systems, such as ecologies, in contrast to organisms, machines, or enterprises. Many artificial systems are characterized by apparently complex behaviors due to often non-linear spatio-temporal interactions among a large number of component systems at different levels of organization; they have been termed Complex Adaptive Systems (CAS).

CAS are dynamic systems able to adapt in and evolve with a changing environment. It is important to realize that there is no separation between a system and its environment in the idea that a system always adapts to a changing environment. Rather, the concept to be examined is that of a system closely linked with all other related systems making up an ecosystem. Within such a context, change needs to be seen in terms of co-evolution with all other related systems, rather than as adaptation to a separate and distinct environment (Chan 2001, p. 2). CAS have several constituent properties (ibid., p. 3ff):

- *Distributed control*: There is no single centralized control mechanism that governs system behavior. Although the interrelationships between elements of the system produce coherence, the overall behavior usually cannot be explained merely as the sum of individual parts.
- *Connectivity*: A system does not only consist of relations between its elements, but also of relations with its environment. Consequently, a decision or action by one part within a system influences all other related parts.
- *Co-evolution*: With co-evolution, elements in a system can change based on their interactions with one another and with the environment. Additionally, patterns of behavior can change over time.
- *Sensitive dependence on initial conditions*: CAS are sensitive due to their dependence on initial conditions. Changes in the input characteristics or rules are not correlated in a linear fashion with outcomes. Small changes can have a surprisingly profound impact on overall behavior, or vice-versa, a huge upset to the system may not affect it. ... This means the end of scientific certainty, which is a property of 'simple' systems (e.g., the ones used for electric lights, motors, and electronic devices). Consequently, socio-technical systems are fundamentally unpredictable in their behavior. Long-term prediction and control are therefore believed to not be possible in complex systems.

- *Emergent order:* Complexity in CAS refers to the potential for emergent behavior in complex and unpredictable phenomena. Once systems are not in equilibrium they tend to create different structures and new patterns of relationships. CAS function best when they combine order and chaos in an appropriate measure—this phenomenon has been termed Far from Equilibrium. CAS in their dynamics combine order and chaos, and thus, stability and instability, competition and cooperation, order and disorder—being termed the State of Paradox.

A complex socio-technical system is a group of different types of elements (i.e., related nodes of a network), existing far from equilibrium, when forming interdependent, dynamic evolutionary networks that are sensitive dependent and fractionally organized (Fichter et al. 2010). Taking a CAS perspective requires system thinking in terms of networked but modular elements acting in parallel (Holland 2006). In socio-technical settings, these elements can be individuals, technical systems or their features. Understood as CAS, they form and use internal models to anticipate the future, basing current actions on expected outcomes. It is this attribute that distinguishes CAS from other kinds of complex systems; it is also this attribute that makes the emergent behavior of CAS intricate and difficult to understand (Holland 1992, p. 24).

According to CAS theory, in CAS settings each element sends and receives signals in parallel, as the setting is constituted by each element's interactions with other elements. Actions are triggered upon other elements' signals. In this way, each element also adapts and thus, evolves through changes over time. Self-regulation and self-management have become crucial assets in dynamically changing socio-technical settings, such as organizations (Allee 2009; Firestone and McElroy 2003). Self-organization of concerned stakeholders as system elements is considered key in handling requirements for change. However, for self-organization to happen, stakeholders need to have access to relevant information of a situation. Since the behavior of autonomous stakeholders cannot be predicted, a structured process is required to guide behavior management according to the understanding of stakeholders and their capabilities to change their situation individually (Allee 2009; Christian Stary 2014).

From the interaction of the individual system elements arises some kind of global property or pattern, something that could not have been predicted from understanding each particular element (Chan 2001). A typical emergent phenomenon is a social media momentum stemming from the interaction of the users when deciding upon a certain behavior, such as spontaneous meetings (Ferscha et al. 2004b). Global properties result from the aggregate behavior of individual elements. Although it is still an open question how to apply CAS to engineering systems with emergent behavior (Holland 1992), in case of socio-technical system design pre-programmed behavior is a challenging task, as humans may change behavioral structures in response to external or internal stimuli. As such, stakeholders in these systems (self-)organize evolvement and adapt to a changing environment, usually generating more complexity in the process.

System-of-Systems (SoS) thinking is considered an effective way of handling CAS, in particular when developing complex artifacts in a structured way (Jamshidi 2008). According to Institute of Electrical and Electronics Engineers (IEEE's) Reliability Society, a system is "a group of interacting elements (or subsystems) having an internal structure which links them into a unified whole. The boundary of a system is to be defined, as well as the nature of the internal structure linking its elements (physical, logical, etc.). Its essential properties are autonomy, coherence, permanence, and organization" (IEEE-Reliability Society Technical Committee on Systems of Systems 2014). A System-of-Systems (SoS) is a system that involves several systems "that are operated independently but have to share the same space and somehow cooperate" (ibid., p. 2).

As such, they have several properties in common: operational and managerial independence, geographical distribution, emergent behavior, evolutionary development, and heterogeneity of constituent systems (ibid.). These properties affect setting the boundaries of SoS and the internal behavior of SoS, and thus, influence methodological SoS developments (Jaradat et al. 2014, p. 206). SoS are distinct with respect to:

1. autonomy where constituent systems within SoS can operate and function independently and the capabilities of the SoS depends on this autonomy
2. belonging (integration), which implies that the constituent systems and their parts have the option to integrate to enable SoS capabilities
3. connectivity between components and their environment
4. diversity (different perspectives and functions)
5. emergence (foreseen or unexpected) (ibid.)

Several structures and categorization schemes have been used when considering complex systems as System-of Systems, ranging from close coupling (systems within systems) to loose coupling (assemblage of system). They constitute embodied systems cooperating in an interoperable way (Chris Stary and Wachholder 2016; Christian Stary 2017c; Weichhart et al. 2018), allowing for the autonomous behavior of each system while contributing through collaboration with other systems, in order to achieve the objective of the networked systems (SoS) (Maier 2005).

Referring to structural and dynamic complexity, structural complexity derives from (i) heterogeneity of components across different technological domains due to increased integration among systems and (ii) scale and dimensionality of connectivity through a large number of components (nodes) highly interconnected by dependences and interdependences. Dynamic complexity manifests through the emergence of (unexpected) system behavior in response to changes in the environmental and operational conditions of its components (IEEE-Reliability Society Technical Committee on Systems of Systems 2014).

A typical technical SoS example is contextualized apps available on a smartphone. Each of them can be considered as a system. When adjusting them along a workflow, for example, to raise alert and guide a patient to the doctor, in case certain thresholds with respect to medical conditions are reached for a specific user, several of these systems, such as the blood pressure app, calendar app, and navigation app, need to be coordinated and aligned for personal healthcare, updating the task manager of the involved users. In this case, the smartphone serves as an SoS carrier, supporting the patient-oriented redesign of the workflow, and thus, the SoS structure. The apps of the smartphone can still be used stand-alone,

Elicitation Requirements

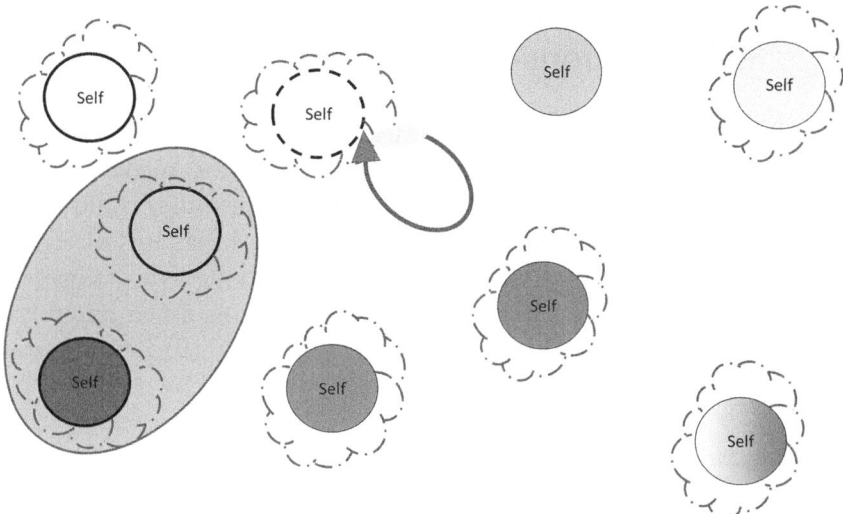

Fig. 2.6 Conceptual understanding of complex systems

while the smartphone serves as a communication infrastructure and provider of networked healthcare-relevant subsystems. It is the latter property that qualifies the smartphone as a carrier of an SoS.

When we project this concept on understanding complex organization of work, actors can become aware of their capability to act autonomously while at the same time being part of a bigger whole, namely the business organization (or even of several organizations). Figure 2.6 visualizes awareness of actors of being part of a complex systems, in this case a System-of-Systems, in their specific roles. Again, Self denotes an actor who plays a certain role (pictured by the surrounding circle) in a certain situation (denoted by a dotted cloud symbol). As shown in the figure, actors need to become aware of which System-of-Systems they are part of (they can be part of various Systems-of-Systems). In the shown case, the Self with the white background is part of a System-of-Systems consisting of two systems where the considered Self is in one role part of system one, whereas the other roles with gray backgrounds constitute the other, larger system. The second role of the Self with the white background is not part of the currently considered System-of-Systems (but might be part of other systems, which are currently out of scope for the actor reflection on being part of a complex system).

2.4 Creating a Reflective Practice for Situations-to-Be

Articulation and alignment of knowledge on work processes can be directed towards reflecting work procedures (i) as they worked in the past, (ii) as they are performed now, (iii) as well as how they could work in the future. It might depend on the current work patterns of actors whose perspective is taken. However, with respect to the style of organizing work and handling work processes, Dewey distinguished impulsive and routine from reflective action (cf. Dewey 1910, 1933), since any professional behavior can have three flavors:

- Impulsive action is based on trial and error.
- Routine action is based largely on authority and tradition.
- Reflective action is based on "the active, persistent and careful consideration of any belief or supposed form of knowledge in the light of the grounds that support it" (Dewey 1933, p. 9).

Dewey explains reflective thinking as a 'chain' not only involving "a sequence of ideas but a *con*-sequence" of thoughts (Dewey 1933, p. 4). In his understanding, acting in open-mindedness and responsibility are consequences of reflective thinking, both facilitating developing commitment to tasks and opening for new ideas.

Schön's Reflective Practitioner approach deepens insights in reflection activities when aiming at professional capabilities to handle complex and unpredictable problems of actual practice with confidence, skill, and care (Schön 1984). A professional practitioner "can think while acting and thus respond to the uncertainty, uniqueness, and conflict involved in the situations in which professionals practice" (Adler 1991). As such, propositional knowledge is tightly coupled with know-how when instantiated in solving knowledge-intense tasks. Hence, it is the knowledge by acquaintance enabling confidence and care tackling even complex problems, which in turn requires know-how and propositional knowledge to perform tasks in a skilled way in those situations. Unique or surprising situations are handled through reframing and finding new solutions ("reflection-in-action"). This process is

1. a conscious one, though not necessarily articulated in words
2. a critiquing one, as it leads to questions and re-structuring
3. immediately significant for action (most important) (cf. Schön 1987, p. 29)

When reviewing actions in the past rather than in-situ, "reflection-on-action" (Schön 1987) leads to evaluating already experienced situations. In case it has consequences for future action (as understood by Dewey), this reflection is transformative. Methodologically, personal narratives and autobiographies have turned out to facilitate self-exploration, in particular looking beyond or behind professional activities, such as social conditions. They allow a more comprehensive personal picture, and consequently unwrapping existing forms of Gestalt and reframing.

An andragogical premise to self-managed (co-)creation assumes the nature and characteristics of actors as maturing persons moving their self-concepts from dependencies from surrounding systems towards self-directedness and autonomy in an evolving world. While experience forms the richest resource for development, readiness to act in accordance with an aligned Self is a prerequisite for (co-)creation, thus, linking task accomplishment to social behavior and endeavor (Böhm 1997).

An agogic (i.e., learning-) and situation-aware mind-set asserts that an actor's time perspective changes from postponed application of experiences and knowledge to immediacy of application and accordingly, orientation to acting shifts from subject-centered activities to focused interaction in co-creative settings (Bronfenbrenner 1981). In social settings of this kind, several agogic principles apply:

- Activities are set in accordance with the needs of participating actors under the given conditions and capabilities to act.
- Each actor has certain resources that are not only the starting point for but also the subject of design activities. These resources are accepted to be limited.
- Actors determine their way and pace of developments, as development needs to in balanced with the current conditions. Both, active participation and retreat are part of development processes.

It is the latter principle that is of crucial importance for triggering individual development and bringing it to life in a co-creative setting.

Agogic actors need to embody (Rogers 1951; Pörtner 2008), and thus self-manage

- Empathy as sensitive understanding of others
- Appreciation of another personality without preconditioning acceptance and respect
- Congruence meaning the authenticity and coherence of one's person and behavior

The first two behaviors are based on the flow from surrounding systems to the Self, whereas congruence is decisive in making visible individual values and their attributes to other systems, and thus, part of the surrounding system. Authenticity refers to meeting a person 'as a person', to the equal of a person, experiencing a situation with the entire spectrum of channels (perceived impulses, feelings, impression, etc.). Coherence includes judging in how far or at what point in time the individual space can be shared with others, that is, becoming visible in an outer space. An essential part of congruence is that all participating actors have the same, transparent understanding of a co-creative system, including pre-set conditions and irreversible process design, for example, normative or role-specific behavior (Spindler and Stary 2017).

Motschnig-Pitrik and Nykl (2001) argued "that problem solving within an individual's context is particularly effective, since it most closely matches the living, sensing, and experience of this individual and has the highest potential for disposition and reuse of the individual's experience" (p. 275). Agogic at the workplace—here referred as work-agogy—(see Fig. 2.7) indicates sensing crucial to cognitive intentional acts, to be captured by in-depth asking:

- WHAT IS? What did you see, hear, smell, taste, feel? What happened, when and how? Can you describe it in detail?
- WHAT SHOULD BE? Which perspective, which sense do you see? What needs to be achieved? Which priorities do you want to set? What do you want exactly? And why? Which state satisfies you?
- WHY? Which meaning do the observations have for you? Which relations do you recognize? What do you reckon? How can you explain that? What are your conclusions?

Elicitation Requirements

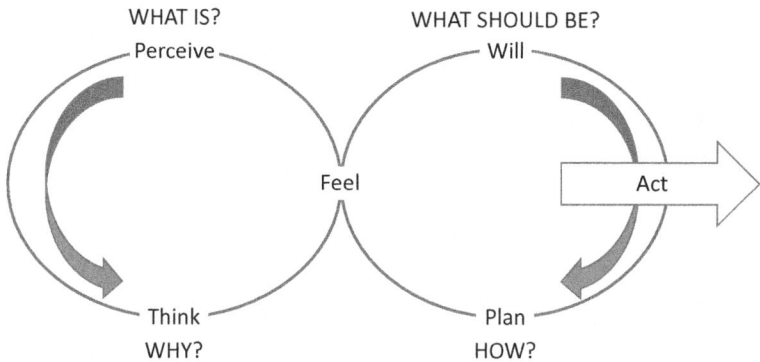

Fig. 2.7 Work-agogy (according to Arbeitsagogik.ch)

- HOW? How to proceed? Which means shall be used? Which tactics shall we chose? What is to be done? Who does what, with what, whom, when, and how?

As indicated in Fig. 2.7, work-agogy in the context of work processes captures the rationale of doing in terms of perceiving a situation and cognitive reflection of perceived information, as some pre-processor to doing, guided by intention and planned action. According to that model, various subsystems are involved in preparing actions through reflecting outer-space information and bringing action from inner space processing to become visible for others in the outer space.

According to Rogers (1961), a facilitating social atmosphere is required for understanding and acceptance of the individual to develop ('grow'). It will then "will become more similar to the person he would like to be; will be more self-directing and self-confident; will become more of a person, more unique and more self-expressive; will be more understanding, more acceptant of others; will be able to cope with the problems of life more adequately and more comfortably" (Rogers 1961, pp. 37–38). In this, the inner space of a person can become part of the outer inner space, for example, through his/her understanding the role, as required for co-creating the organization of work.

Figure 2.8 visualizes the results of developing a reflective practice for situations-to-be. We refer to the actors (represented by individual Selfs in various roles [capturing their inner space] and involved in specific situations),

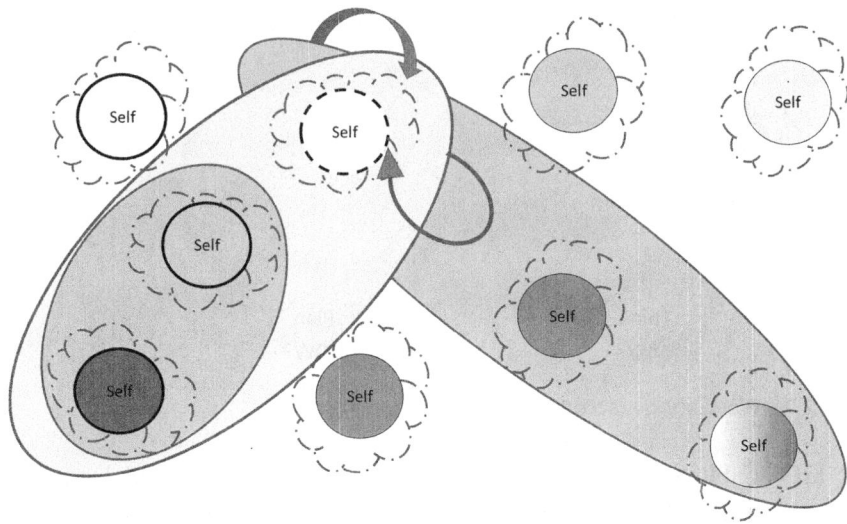

Fig. 2.8 Creating a reflective practice for situations-to-be

being part of a complex systems in terms of a System-of-Systems. Both, the situation and the System-of-Systems represent the outer space of an actor. As indicated by the upper arrow on top of the Systems-of-Systems in the figure, actors need to develop an understanding of novel system constellations. Potential scenarios need to be evaluated, like in the shown case adding actors with gray background as part of an additional System-of-Systems (depicted from the middle to the lower right) consisting of three systems, where the considered Self (white background) is in the role of potentially becoming part of two Systems of Systems, leading to an enriched overall system.

2.5 Focusing While Utilizing Multiple Perspectives

Individual introspection into personal views on one's work by means of externalization can be considered a prerequisite for the development of common views on work and organizational improvement, respectively. The role of the individual in this context has not only been an issue in organizational research (e.g., Sachs 1995; Suchman 1995), but has also been addressed regarding the learning aspects for both individuals and groups. Theories originating in cognitive sciences offer an explanatory approach for how individual perceptions and pictures and (organizational) reality are mutually influenced. One of these is 'mental models' (cf. Johnson-Laird 1981; Ford et al. 1991), as they are considered to explain the foundation of thought processes. Whenever humans are confronted with situations in which they should act, they create an explanatory model in their mind. The contents of this model are based on individual perception of the situation, previous experiences, and personal values.

In organizational settings, mental models also guide an individual's way of interacting with others. This includes decisions on when to explicitly cooperate, with whom to cooperate, in which way, when to expect input from others, and when to deliver results to others. In order to interact successfully, the individual mental models have to fit each other. Mental models are purely cognitive constructs and are per definition inaccessible to others. In order to align mental models, the involved individuals first have to make their mental models visible to others. In many situations, verbal expressions may not lead to sufficient visibility required for successful alignment. When the work setting is perceived as complex or when unexpected contingencies arise, more explicit representations of mental models are needed (Russell et al. 1993; Klein et al. 2006).

Explicit representations of mental models are called 'externalizations'. In collaborative work, externalization is necessary to provide people with a common ground for sharing and negotiation of different views. Shared views in turn change individual mental models. In this way, a common understanding of interaction emerges. Externalization can be supported methodologically and by using tools (Pirnay-Dummer 2006, see also, Ifenthaler (2006) for an overview of established techniques in this field).

Structure elaboration techniques are an effective means to create physical representations of mental models (Dann 1992). In a moderated process (the dialogue-hermeneutic method), the participants create a graphical representation of their mental models by placing labeled cards on a modeling surface. Subsequently, they relate each other using associations. Dann (1992) has stressed the importance of the immediacy of representation in the structuring process. This immediacy is attained by the physical creation of the model. Participants immediately refer to a physical representation rather than abstract items. They create and modify the model in a dialogue-based way until reaching consensus about what is represented. Mental models of individuals are externalized, questioned, and can be modified at the same time. The procedure ends once all participants feel comfortable with the result.

Structure elaboration techniques are highly sophisticated approaches with respect to the specification of both, the methodology and the instruments to be used. However, their suitability for the externalization of mental models has already been evaluated empirically (Groeben and Scheele 2000; Ifenthaler 2006). Some researchers (e.g., Dann 1992) have suggested that structure elaboration techniques should always be adapted to the case at hand, for example, in terms of prescribed modeling elements or methodology. Presumably, such an adaptation could be necessary when used for externalization.

Due to its minimalist approach to semantics and syntax, concept mapping (Novak and Canas 2006) is widely used to elaborate on structures. These maps contain mutually linked nodes corresponding to (mental) concepts. In contrast to other structure elaboration technologies, concept mapping does not explicitly aim at creating consensus of how to interpret the externalization among the involved individuals. In concept mapping, concepts are collected directly during structuring, which allows for immediate, contextualized specification of new aspects of the model. Concept maps also support defining concept classes (such as 'persons', 'tasks' etc.) for additional (hierarchical) structuring and do not give any constraints on which or how many classes to use.

As such, the concept mapping approach is considered to be suitable for externalization of mental models (Pirnay-Dummer 2006). In the course of mapping, constructs are arranged according to an issue of interest, for example, individual organization of work (Oppl 2006). The constructs are named and structured by associating them. In this way, a contextual specification is established. Such mappings have already been applied in structured domains, such as mathematics, allowing for individually arranging domain content (Brinkmann 2003), or for generating meaningful representations from scratch according to individual mental models (Coffey and Hoffman 2003). While for the first setting, the focus of mapping lies on the arrangement of previously known elements, the latter requires an open space to identify, name, and arrange content.

Some of the existing tools for structure elaboration, do not only provide support for the articulation process itself, but also allow assessing the quality of representations, for example, based on metrics derived from graph-theory for concept maps (Ruiz-Primo and Shavelson 1996). Other tool approaches offer a tight integration with the computer desktop environment and enable links to digital resources (see for concept maps, Canas et al. 2004). In particular, concept maps seem to have potential for usage in daily work, as they can be integrated into and consulted from existing (computer-supported) workflows.

At the center of articulation in the course of knowledge elicitation is the ability to learn about mental models. Structure elaboration in terms of mapping mental constructs to diagrammatic expressions has already turned out to be useful to generate ideas, to design a structure, such as organization of work, to communicate ideas, and to aid learning by explicitly integrating new and old knowledge. By communicating diagrammatic representations, such as concept maps, misunderstandings can be avoided (Ausubel 2000), a prerequisite for shared reflection and collective knowledge creation.

Although the format of representing articulated knowledge may be open with respect to syntax and semantics, as in the case of structure elaboration, elicitation of work knowledge can profit from a fundamental perspective on human work. It can be directed towards information or communication, as different strategies of organizing knowledge are related to them (F. Fuchs-Kittowski and Fuchs-Kittowski 2007): formalization, codification, personalization, and socialization. In particular, the latter is of importance for alignment and shared understanding—see Table 2.1 (according to F. Fuchs-Kittowski and Fuchs-Kittowski 2007).

Finally, eliciting knowledge is influenced by the individual perception and representation of work practices (Bossen 2017). For instance, the description of a scheduling procedure for consultation of medical experts in an outpatient clinic is likely to differ whether one asks the patient, administrative staff, or the medical experts. Hence, the challenge of elicitation in this context is grounded in the role- or task-specific perspective

Table 2.1 Managing elicited knowledge (according to and translated from F. Fuchs-Kittowski and Fuchs-Kittowski 2007)

	Information-oriented elicitation		Communication-oriented elicitation	
Strategy Focus	*Formalization* Individual provision of codified knowledge	*Codification* Exchange of explicit and codified knowledge	*Personalization* Exchange of knowledge between actors	*Socialization* Interaction within community
Knowledge	Independent of people	Independent of people	Depending on people	Social product
Objective	Active control and information provision	Reuse of (codified) knowledge	Situation-aware generation of knowledge	Development of common knowledge
Task	Anticipation of existing knowledge and knowledge demands	Codification of existing knowledge	Access to existing expert knowledge	Provision of working conditions
Focus	Technology Formalized, schema-like	Technology Iterative, schema-like	People complex, non-schema-like	Community Creative (problem solving)
Supported activity	Automation systems	Content Management Systems	Competence Management Systems	Interaction/cooperative systems
Type of knowledge management				
Application samples	AI systems, Workflow-Management Systems, Information Logistics	Document/Content-Management, Intranet	Skill-Management, Expert Yellow Pages	Groupware, Communityware, Social Software (Wiki, Blog)

the stakeholder tasks. It lays ground to sociological theories, such as Strauss' theory of action.

> He opposed representations of action as if concerning *an* act (singular) with a beginning and an end by one actor following a set course of action. This linear and 'rationalistic'—in the sense of producing a simplifying and rationalizing depiction—can be contrasted to an interactional model: Looking closer, acts come forward as involving multiple steps in which emergent circumstances and the interaction with other actors have to be monitored by the actor, who has to adjust her actions to the contingencies arising in an ongoing manner, and which results in 'an act' as requiring efforts of aligning, coordinating, monitoring and being more convoluted than in former the linear representation. (Bossen 2017, p. 79; Strauss 1993)

Bossen (2017, p. 79f) concludes that "representations of practices should then not be made too rashly and should build on detailed empirical knowledge: Streamlining work into linear, rational models entails the risk of ignoring or forgetting central features of the apparent mess of work. Further, since no description of a phenomenon can capture all its aspects, but will highlight some and push others to the background, the act of representing requires making choices of what to make visible" (Suchman 1995).

Figure 2.9 visualizes the situation where different stakeholders pursue different interests in various situations, and may take different perspectives upon work practices in their mental models, including the interaction with actors in specific roles. In the figure, Self denotes an actor who plays a certain role in a certain situation on which he/she has a certain perspective according to individual perception of the corresponding work. As shown in the figure, each actor has a certain perspective which might overlap with others or not (represented by stars in the figure). The perception can depend on the role and situation of an actor, as shown by the Self with the white background in the figure. The lower right part of the figure shows a constellation of overlap as perspectives can be shared and include the interaction beyond plain information exchange.

Once articulation of work knowledge makes visible the multiple perspectives on work due to the individual mental models of tasks or roles, the design and structuring of work can be enriched by parameters determining the quality, and final success of operating a business.

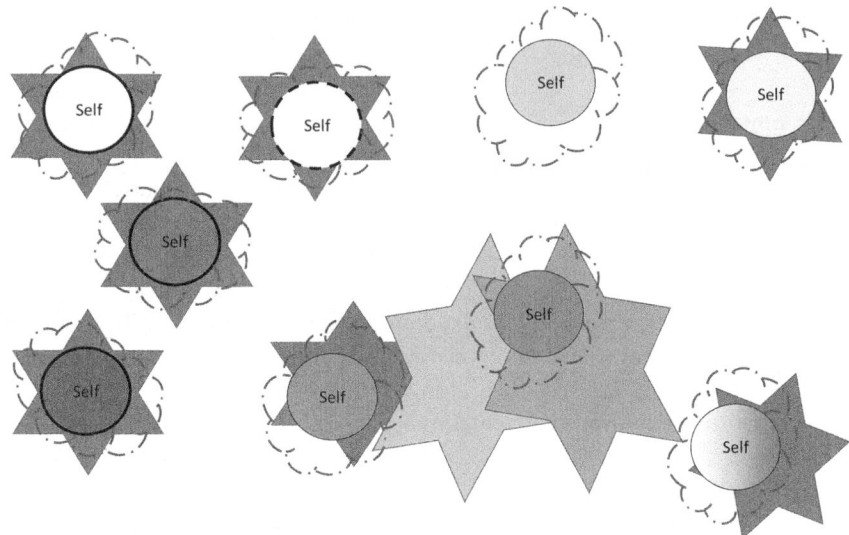

Fig. 2.9 Focusing while utilizing multiple perspectives

2.6 Articulating Intangible Assets

"How people work is one of the best kept secrets in America." The (location-independent) validity of this statement by Wellman (cited in Suchman 1995) has been underlined in various contexts, for example, by Polanyi (1958) when referring to 'ineffable knowledge' that does not allow workers to reflect about their work without becoming conscious about work structures. Nonaka and Takeuchi (1995) even referred to the problems caused by changing those structures.

Strauss has pointed out the importance of *Articulation Work* (Strauss 1985) in that context. This term is dichotomous and has always to be considered in both of its meanings: *Articulation* Work is talking about one's work in order to be able to work together with others. Articulation *Work* is an integral part of work in general, particularly in the sense that it takes effort to realize it. Articulation Work is considered as a conceptual complement to 'Production Work', that is, the work dedicated to achieve organizational goals (Fujimura 1987).

Most of the time, Articulation Work happens implicitly (Strauss 1988), that is, none of the involved participants consciously and actively communicates his/her view on his/her work. However, a common understanding is created by simply working together. This phenomenon corresponds to the phenomenon of *socialization* described in Nonaka and Takeuchi's Socialization, Externalization, Combination, Internalization (SECI)-model (Nonaka and Takeuchi 1995).

Similar findings have resulted from studies in work and cognitive psychology. It has been shown that an essential part of the user's task-relevant knowledge is tacit (i.e., unconscious). Knowledge either becomes tacit through automation of work procedures, that is, formerly explicit knowledge lapses into the unconscious and, by that, becomes tacit (Hacker 1998), or the tacit knowledge is acquired through implicit learning, that is, task-relevant knowledge is learned without awareness through personal experience and practical examples, similar to a master–apprentice relationship (Neuweg 2004).

In a variety of professions that rely on complex problem-solving capabilities and creativity like law, medicine, sales, teaching, or management, tacit knowledge is considered as a crucial factor for success (Sternberg et al. 1999). In particular, it plays a central role in dealing with critical, that is, non-routine situations at work (Büssing et al. 2002). The main characteristic of the tacit dimension of knowledge is that it is difficult to communicate and formalize (Nonaka and Takeuchi 1995; Polanyi 1966). Consequently, tacit knowledge is difficult to capture with traditional task elicitation methods like questionnaires, surveys, structured interviews, or analyses of existing documentations. The task analyst simply does not know what kind of questions to ask (Beyer and Holtzblatt 1997). When eliciting user-task information, developers, therefore, have to deal with the tacit dimension that indwells work procedures.

As in established and routine task settings, workers are not always conscious of how and why they act in a certain way, problems that might occur once established work practices need to be adapted (Gasser 1986; Gerson and Star 1986). However, the term 'established work practices' is ambiguous. Strauss (1988) and Fujimura (1987) distinguish routine work form problematic work, the latter increasing the need for Articulation Work. Regarding routine work, Strauss however states that *one man's routine of work is made up of the emergencies of other people*

(Hughes 1971 cited in Strauss 1993, p. 43). According to this understanding, established work practices are only those procedures where all involved people are able to routinely handle the steps required to complete the work.

Consequently, established work practices can turn into problematic situations anytime. Introducing new people or changes in the working environment can lead to unforeseeable contingencies that require Articulation Work to be resolved. Changes in the working environment that cause the established work practice to break down can be as simple as printers running out of paper (Bendifallah and Scacchi 1987). This is a contingency that can be resolved rather quickly and simply. There are, however, situations that require more effort to be resolved (ibid.).

According to Strauss, *explicit* Articulation Work (in contrast to implicit one) (Strauss 1988) becomes increasingly important, the more complex and problematic a work situation is perceived by the people involved ("Problematic interactions involve 'thought', or when more than one interactant is involved then also 'discussion'. An important aspect of problematic action can also be 'debate'—disagreement over issues or resolutions" (Strauss 1993, p. 43)).

Since there is still a strong tendency towards standardization and explicit definition of work routines (cf. business process modeling Scheer 2003), workers are considered more and more (error-prone) system elements from a socio-technical system. As such, their individual influence has to be reduced as far as possible. While this view has facilitated the development of mankind during the last centuries, it has clearly reached its limits, according to studies on work transformation (Sachs 1995).

Today's complex business environments require skills, which have not been considered important anymore for frontline workers. In settings, where exception handling might become the standard process, automated execution of workflows using human manpower does not work anymore. Much of what in former times has been regarded routine work (or operations) is fully automated today. Humans get in charge mostly when something goes wrong or cannot be decided based on a set of predefined rules. When people in such cases do not consciously know what is going on in their work environment—when their work is a secret to them—they experience troubles. In today's business settings consciousness of work practices is required increasingly, as

- the demand for and to develop further skills needs to be identified (Hampson and Junor 2005)
- work practices and interfaces have to be negotiated in collaborative work settings (Strauss 1988)
- work processes needs to be improved continuously (Caetano et al. 2005)
- exceptions need to be tackled in a straightforward way (Gerson and Star 1986), and
- work practices need to be communicated to others for support (Herrmann et al. 2004)

The common prerequisite of all these settings is individual awareness about how work is done, in which context it happens, which goals are to be reached by which skills. Sachs (1995) suggests taking an alternative view on work, regarding not only organizational tasks, but also the given human-activity-centered aspects, the context of work and its understanding by human beings, as they are highly relevant for economic success.

According to Strauss (1988), explicit Articulation Work aims at unveiling these issues and making them communicable to others. It enables people to externalize their individual views on work, to reflect upon it, and to present. A means to support explicit Articulation Work is using representations of work as a basis and facilitator for externalization (Suchman 1995) ("A map or other representational device is a piece of craftwork, crafted in the interest of making something visible. Things are made visible so that they can be seen, talked about, and potentially manipulated," ibid.). Representations of work in terms of Suchman (ibid.) "(…) are interpretations in the service of particular interests and purposes, created by actors specifically positioned with respect to the work represented."

In this respect, it doesn't matter, "(…) whether (these representations are) created from within the work practices represented or in the context of externally-based design initiatives (…)" (ibid.). Following Suchman, representations of work can either be a result of work or describe work from a bird's eye view (with people stepping out of the system to describe it)—or both. In terms of explicit Articulation Work, representations from a bird's eye view are the results of articulation. Actual results of work might serve as a basis for explicit articulation and facilitate it, but they are not in the focus of this work.

Representations from a bird's eye view can be codified in different forms. A common form is to use textual descriptions of work (Kyng 1995). Textual codification allows capturing work with the whole expressional power of natural language. Reflection about and communication of the structure of work, however, is better facilitated by diagrammatical representations or graphical models (Hahn and Kim 1999). Models have proven to serve as mediators and boundary objects for people communicating about their work (Boland and Tenkasi 1995, cited in Krogstie et al. 2006).

Models are built using modeling languages using a syntactically fixed and semantically predefined set of symbols. These constraints are necessary for further processing, but appear to hinder the modeling process itself (Jørgensen 2004). Most modeling languages force modelers to use representational schemes that do not necessarily correspond to their individual understanding of work (Oppl 2018). This mismatch often leads to situations where the modeling language is inappropriate to express what people consider relevant—"Indeed, I would go so far as to claim that constraining practitioners during early design to use some fixed notation with a fixed semantics would slow them down, by forcing them to pay more attention to the limitations of the notation than to the details of their problem" (Goguen 1993).

For support of explicit Articulation Work, it has to be assured that all aspects of work considered relevant by people can be expressed by the modeling language (Oppl 2016). Moreover, modeling requires the recognition of relevant real-world phenomena, to abstract and conceptualize them, and to represent them with the means of the modeling language. These are non-trivial tasks, which might be very challenging—if not overstraining—for people inexperienced in modeling (Goguen 1993). Articulation Work, however, has to be performed by everybody involved in the work process (Strauss 1988), also—and especially—frontline workers, who very rarely have experience in modeling (Oppl 2017).

Figure 2.10 visualizes the recognition of intangible assets, both, on the level of individual actors, and the collective layer. As a prerequisite for designing situations-to-be, actors need to reveal and communicate information that influence their perception, thinking, and doing—they need to engage in explicit Articulation Work. In the figure, the actors are rep-

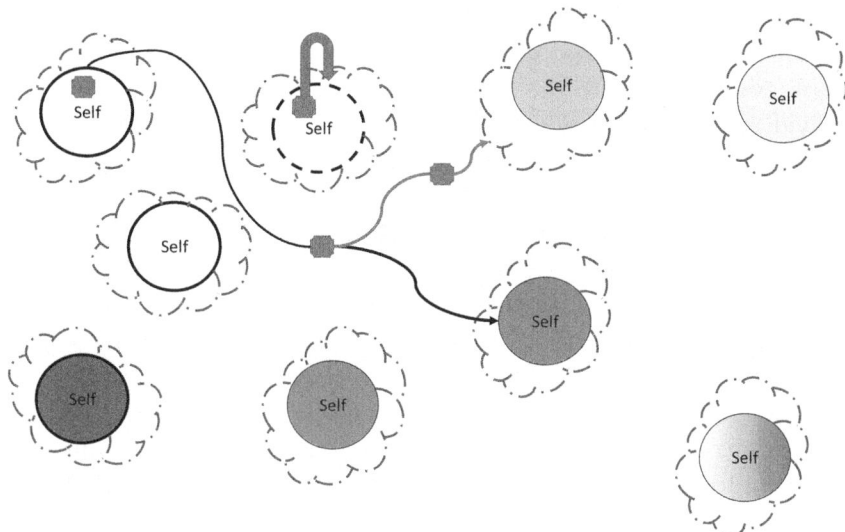

Fig. 2.10 Articulating intangible assets

resented by their individual Selfs taking various roles and being involved in various specific situations. They might have blind spots, indicated by the big dots in the figure worth being elicited and evaluated in terms of implications for themselves, and when interacting with others (as indicated by the links between Selfs).

2.7 Engage in Alignment for Collective Intelligence

Herrmann et al. (2002) have shown that workers should not only be able to describe their *particular* view on the assigned work tasks, but also co-construct a common understanding of collaborative work tasks. Such type of participation facilitates technology development, even when different paths to accomplish a certain task are followed by individual workers. Empirical results from work psychology, too, give evidence that there are many alternative efficient and effective procedures when users have freedom in their task accomplishment procedure (Ulich 1994).

Hence, when dealing with different users and different individual perception of tasks and task accomplishment procedures, elicitation techniques should support the elicitation of both, idiographic (i.e., the individual user's perception of the task) and co-constructive (i.e., common aspects of user groups in task perception) user-task information. In order to achieve this objective, the elicitation, as well as the representation of user-tasks has to be context-sensitive (Mirel 2004). However, from the method perspective successful elicitation should avoid the influence of representational structures to cognition, in particular, when capturing the tacit dimension, value both, individual differences and commonalties of user-work information (Hemmecke and Stary 2006).

Although articulation can be guided by modeling, thus leading to representations of work knowledge, developing a shared understanding of such manifestations should be considered a learning process (Seel 2003). Those processes are most successful when the gap between mental models and representations can be kept minimal. In her extensive empirical work, Maria Montessori has identified several cornerstones for successful knowledge creation and acquisition to that respect (cf. Montessori 2005; Ludwig et al. 2002):

- Both have to be *tuned to individual types of stakeholders*. Learning should be an individualized process that might also occur in group settings.
- Acquiring and creating knowledge are oriented towards *individual acting*. Stakeholders should acquire competence and skills directly working with subjects or manipulating content.
- Knowledge creation should be under the *control* of the stakeholders, including setting the stage for sensible learning phases (which are essential for understanding).
- The acquisition and creation of knowledge should *lead to and be built on visible structures* (inner structure requires external structure) with a maximum degree of freedom to act individually and express mental models accurately.
- Knowledge creation should be *based on some material or pre-structured content* to direct the attention of individuals.

- Creating and acquiring knowledge should occur *in a comprehensive, but focused way* (in-depth concentration on the subject of acquisition). Subject-specific elements should be complemented by transformation tasks. For instance, business process modeling using event-driven process chains in ARIS (Architektur integrierter Informationssysteme; Architecture of Integrated Information Systems) (Scheer 2003) should be complemented by UML (Unified Modeling Language)-models, since the latter provide an additional, object-oriented perspective on process-model elements.
- Active acquisition should be *observed* by coaches, providing intervention on demand. Such a setting allows for misconceptions, faulty or misleading procedures, for example, caused by opinion leaders in group settings.

Maria Montessori's observation let her conclude that any learning process should be facilitated by allowing stakeholders to manipulate objects in a self-managed way. This process should be implemented in a well-prepared environment. This environment is shared with the mentor and/or peers, for sharing experience, guidance, and help. However, the acquisition of knowledge is the responsibility of each stakeholder. The role of the stakeholder is to handle the material according to inherent properties of the content and few inputs provided by a facilitator. In the ideal case, the prepared environment guides the stakeholder to domain-specific properties and tasks that can be accomplished in a self-managed way using the manipulative elements of the environment—a strategy technological instruments aim to follow (Zuckerman et al. 2005).

The tasks that are traditionally performed in Montessori-oriented settings start on a straightforward level and become increasingly complex:

1. *Structuring* (Ordering) elements: Montessori considers (mathematical) structuring the training in exact thinking. She has recognized the domain-specific grouping of elements, the correct assignment of phenomena, and the multi-dimensional capturing of things in the world substantial for further acquisition processes. Exact working in natural sciences, however, requires the combination of motor- and sensor experience.
2. *Communication* of models or concepts and transformation processes by means of language. The verbal handling and the semantically cor-

rect application of domain ontologies are at the center of knowledge acquisition and creation. Language has to be materialized and embodied in cognition.
3. *Cosmic education* through comprehensive and symbolic application of knowledge. Montessori's constructionist approach envisions learning to occur in and lead to a well-organized 'home' with harmonized arrangements and objects that can be found according to their scope of use.

For Maria Montessori, the exploration of the environment and self-managed handling of content elements is the key to comprehensive and holistic understanding. Stakeholders should (re)construct knowledge in an environment prepared accordingly. The environment has to contain the means for self-education. It has to contain activating objects of interest for sharing, acquiring, or creating knowledge, rather than isolated pieces of information or objects without indication of their usage.

Facilitators should motivate the acquisition, facilitate the acquisition and transfer process, and resolve conflicts. They serve as mediators between content elements and individuals in the environment. Understanding focusses on content elements and their interactive handling.

In case digital work should enrich human perceptual capabilities, metaphors could help when constructing socio-technical work spaces (cf. Turkle 1998, p. 291; Oppl and Stary 2011b). Thereby, humans do not interact as a separate part of the socio-technical environment, they are part of it. This phenomenon is also termed immersion. Immersion facilitates active participation in processes (rather than consumption of visual information) through manipulation of objects (Oppl 2006).

Given immersion, another factor moves also to the center of interest: the capability to share experiences and to interact in a common context even over large distances. It is the idea of structural and dynamic networking. Focusing on networking and context-sensitive interaction allows for more than the reproduction of predefined sequences of interaction with a limited set of features. It allows for exploration, self-management, and social process support. In this way, they support human-centered concept developments, for instance to move forward from 'simple' training mechanisms in the sense of reproducing activities and facts in a predefined domain towards collaborative knowledge exploration in an open space.

With respect to content, Norman and Spohrer (1996) have found out that high quality material in general should provide a high degree of confidence in their (i) usefulness, (ii) interest (which is particularly in line with Maria Montessori—see above), and (iii) effectiveness. They have elaborated their principles of 'learner-centered education' in terms of individual engagement, effectiveness, and viability. Engagement means collaboration with highly motivated learners in the course of education. It is enabled through "rapid, compelling interaction, and feedback" (ibid., p. 26). Effectiveness, in the sense of Norman and Spohrer, denotes the depth of understanding and the skills students acquire. The viability addresses the seriousness of the problems tackled, the relevance of the topics, and the accuracy of tools for the process of knowledge creation and representation.

One way to meet these objectives in virtual settings or augmented environments has been to recognize the multiple dimensions of knowledge sharing and creation and to tackle them explicitly. For instance, Resnick et al. (1996) have observed: "Educational technology has too heavily emphasized the equivalent of stereos and CDs and not emphasized computational pianos enough" (ibid., p. 42). The researchers' goal was to develop computational construction kit development "enabling people to express themselves in increasingly ever-more complex ways, deepening their relationships with new domains of knowledge" (ibid., p. 42).

The theory of constructional design focuses on a constructionist approach to individual knowledge acquisition. Constructional design of content is a type of meta-design (designing for designers) to support learners in their own design activities and thus leading to hands-on experience in construction. Papert (1993) argues for a constructionist approach to learning: In design-based learning, things that people design (such as Lego® constructions) "serve as external shadows of the designer's internal mental models. These external creations provide an opportunity for people to reflect on—and then revise and extend—their internal models of the world" (Resnick et al. 1996, p. 42).

Engagement, as demanded by Norman and Spohrer, needs to be implemented through something more than learning-by-doing, since, in contrast to learning-by-doing little attention has been given to the "general principles governing the kinds of 'doing' most conductive to learning" (Resnick et al. 1996, p. 42). Two general principles should guide the design of activities binding individuals to an object: personal and epistemological connection. They have been defined as follows:

- *Personal connections.* Constructions kits and activities should connect to users' interests, passions, and experiences. The point is not simply to make the activities more 'motivating'. When activities involve objects and actions that are familiar, users can draw on their previous knowledge, connecting new ideas to their pre-existing intuitions.
- *Epistemological connections.* Construction kits and activities should connect to important domains of knowledge—and, more significantly, encourage new way of thinking (end even new ways of thinking about thinking). A well-designed construction kit makes certain ideas and ways of thinking particularly salient, so that users are likely to connect with those ideas in a natural way in the process of designing and creating. (Resnick et al. 1996, p. 42)

Materials enabling rich learning experience should provide both types of connections. Two ways of implementations have been pursued: enrichment of existing objects and virtualizing the core material. In the 'Things That Think' initiative (MIT's Media Lab), everyday objects should embed computational capabilities, not only to accomplish particular tasks more cheaply or easily or intelligently, but to enable people to think about things in new ways (Weiser 1991). One solution was programmable bricks. Structures and mechanisms have been developed using programmable Lego®-bricks for car and castles building *including* behaviors. Typical creations are: real animals, step-trackers, science experiments, and smart rooms. The program is stored in the brick after a download from the PC. Actually, a brick is a very personal computer. In this way, a strong *personal connection* is established, since the brick is part of the learners' culture and life. The bricks allow to compare artificial with natural beings (e.g., robots and animals) as well as to understand complex systems' behavior, for example, feedback strategies. In that way, an *epistemological connection* can be set up.

Narrative-based, Immersive, Constructionist/Collaborative Environments (NICE's) underlying theoretical framework "combines constructivist educational theory with ideas that emphasize the importance of collaborative learning and narrative development" (Roussos et al. 1997, p. 62). Constructivist pedagogy is one "by which learners actively construct and interrelate knowledge and ideas" (ibid.). These findings lead us to the conclusion that the more objects are available in a concrete form and way, and the more focused communication occurs, the more effectively (and efficiently) knowledge-creation and sharing can be supported (Oppl and Stary 2011a, 2014).

The involvement of individuals seems to play a central role for knowledge acquisition and throughout the process of creating mutual understanding, redefining the role of developers: "The process of constructional design is not a simple matter of 'programming in' the right type of connections" (Resnick et al. 1996, p. 49), since behavior is not predictable by developers. "Developers of design-oriented learning environments need to adopt a relaxed sense of 'control' " (ibid.) in the sense of creating 'spaces' for *possible* activities and experiences rather than limiting the interaction space (which, again, is in line with Montessori). However, developers have to make those spaces dense with personal and epistemological connections. Then, there will be defined regions, both appealing and intellectually interesting (as demanded by Montessori or Norman and Spohrer).

Understanding immersion in the sketched sense of *individual and social engagement* in knowledge creation and sharing processes enables more than scanning and retrieving information. Both, constructionist and constructivist acquisition support the personal and epistemological connection of individuals to subjects.

From the perspective of socio-technical design of digitized work systems with such engaging environments for articulation and representation, the emotional side has to receive attention, equal to social and cognitive aspects of knowledge creation and sharing. Hedonic qualities address the matter of emotion and pleasure when persons interact with artifacts. For interactive systems, they have become a matter of competitiveness (Subramanya and Yi 2007). The factors contributing to a rich and satisfying user experience include interactions "that are natural, intuitive, simple, pleasant, easy to remember, and adaptive to individuals' idiosyncrasies" (ibid., p. 114). Millard et al. (1999) have shown joy of using an artifact might increase the quality of work significantly. Several dimensions have been identified for design taking into account user experience:

- *Devices*: Factors related to this dimension comprise the use of colors for display, and touch-sensitive screens.
- *Communication and social interaction*: Relevant issues to that respect are the provision of a (virtual) vicinity, feelings of personal touch, gestures, and differentiated communication based on relationship to persons.
- *Application*: Pleasing user interaction is based on a minimal feature list, non-intrusive media (e.g., hands-free usage of mobile devices), personalization of content, and the combination of stimuli or multi-modality.

Although there is a long tradition in handling user properties and individual differences in human–computer interaction (Egan 1988), only few engineering practices tackle them in connection to design. The current practice taking into account multiple perspectives focuses on model-driven development (Gruhn et al. 2007; Petrasch and Meimberg 2006). It enforces an implementation-independent representation of interactive systems, relying on diagrammatic representations to reflect a status-quo and exchange design ideas. The models allow a structured procedure, due to the mutually tuned representation of content—a demand that has also been uttered in the context of structured knowledge creation and sharing, with respect to learning resources (Kurzel et al. 2003).

Figure 2.11 visualizes Selfs actively involved in sharing and re-arranging information they have been revealing through the activities described in the previous subsections, such as externalizing intangible assets. Following the reflective practice for situations-to-be, actors in their various roles and involved in specific situations need to align their interactional understanding, in order to proceed with developing their organization of work. As indicated by the three clouds, aligned situations emerge in the course of alignment, based on a common understanding of articulated knowledge.

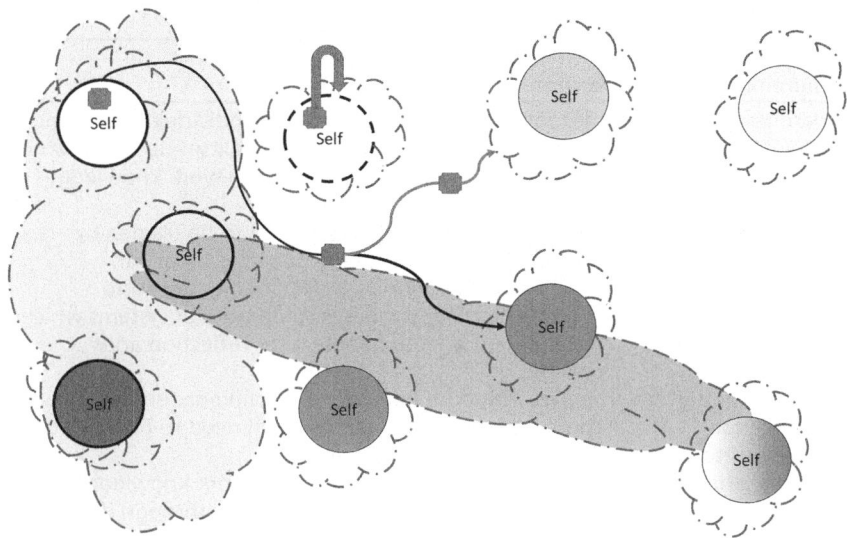

Fig. 2.11 Engage in alignment for collective intelligence

2.8 Synthesis

In Table 2.2, we give an overview of the requirements collected from the various disciplines and approaches. They have been detailed in the previous sections. We synthesize their meaning for each of the requirements. It becomes evident that the starting point is the individual Self of each actor which is challenged to open up for developing awareness, if not in-depth understanding, of

- roles taken by the actor
- context given by situations the actor perceives to be relevant
- complex systems the actor is part of
- reflecting on past, present, and future scenarios of work the actor participates in
- how to focus while taking different perspective on work processes
- intangible work assets provided and required by the actor
- consolidating actor-specific work knowledge when aiming for collective intelligence

Table 2.2 Summary of elicitation requirements

Elicitation requirement	Description
Awareness on role(s) and their management	Roles constitute the appearance of individual actors and can be part of various contexts. Their set up is relevant to how stakeholders get involved in work knowledge elicitation.
Situation Awareness	Role- or task-specific activities need to be framed by information of the situation an actor is part of.
Conceptual understanding of complex systems	Networked and continuous development of socio-technical settings increases complexity of systems which requires concepts to handle it for reflection and change.
Creating a reflective practice for situations-to-be	Theories influence mental model building, either consciously or unconsciously. Both need to be tackled for articulating the future.
Focusing while utilizing multiple perspectives	Determining the target of eliciting work knowledge becomes more focused when looking through different glasses on work.
Articulating intangible assets	Elicitation has to tackle both, explicit and implicit knowledge on work, in order to achieve a complete picture of the relevant work situation.
Engage in alignment for collective intelligence	Being part of a system plays a crucial role in externalizing knowledge, as one is the observer who needs to observe him/herself while being an integral part of a work organization. Of particular importance is intelligibility and purposeful involvement when one's implicit knowledge is codified to be understood by other stakeholders.

The respective individual reflection processes lay the ground for the development of collective intelligence, which frames the articulation alignment activities, which eventually lead to embodiment into work processes and finally, business operation.

From a procedural perspective, elicitation requires

1. A preparation of the setting, actors, and instruments. It includes the scope or universe of discourse, such as a business case, a motivating articulation environment including graspable material, and actors willing to learn both, express their mental models, and engage in co-creative reflection and generation processes

2. Situation-sensitive articulation features as different people externalize knowledge on roles and work tasks differently
3. Facilitation encouraging stakeholders to look beyond well-established boundaries and patterns, and deal with high complexity of work situations and organizational structures
4. Representational alignment as a consolidated representation serves as a baseline for documentation and further development
5. Organizational alignment once elicited knowledge should be embodied in the workspaces of an organization

We will use this table and procedural cornerstones to put the results of the next sections into the context of elicitation requirements. Methodological approaches to articulation and alignment of mental models as well as corresponding tool support can be considered with respect to these requirements. They allow appraising the results concerning their effectiveness and usefulness in dynamic work practices in digitalized work settings.

References

Ackerman, Mark, Michael Prilla, Christian Stary, Thomas Herrmann, and Sean Goggins. 2017. *Designing Healthcare That Works: A Sociotechnical Approach*. Academic Press.

Adler, Susan. 1991. The Reflective Practitioner and the Curriculum of Teacher Education. *Journal of Education for Teaching* 17 (2): 139–150 (Taylor & Francis).

Allee, Verna. 2009. Value-Creating Networks: Organizational Issues and Challenges. *The Learning Organization* 16 (6): 427–442 (Emerald Group Publishing Limited).

Ausubel, J. 2000. *The Acquisition and Retention of Knowledge: A Cognitive View*. Kluwer.

Barwise, Jon, and John Perry. 1981. Situations and Attitudes. *The Journal of Philosophy* 78 (11): 668–691 (JSTOR).

Bendifallah, S., and W. Scacchi. 1987. Understanding Software Maintenance Work. *IEEE Transactions on Software Engineering* 13 (3): 311–323.

Beyer, H., and K. Holtzblatt. 1997. *Contextual Design: Defining Customer-Centered Systems*. Morgan Kaufmann.

Boland, Richard J., Jr., and Ramkrishnan V. Tenkasi. 1995. Perspective Making and Perspective Taking in Communities of Knowing. *Organization Science* 6 (4): 350–372 (Informs).

Bossen, Claus. 2017. Socio-Technical Betwixtness: Design Rationales for Health Care IT. In *Designing Healthcare That Works*, 77–94. Elsevier.

Bounfour, Ahmed. 2016. *Digital Futures, Digital Transformation: From Lean Production to Acceluction*. Cham: Springer International Publishing.

Böhm, Winfried. 1997. *Entwürfe Zu Einer Pädagogik Der Person: Gesammelte Aufsätze*. Julius Klinkhardt.

Brey, Philip. 2005. The Epistemology and Ontology of Human-Computer Interaction. *Minds and Machines* 15 (3–4): 383–398 (Springer).

Brinkmann, Astrid. 2003. Graphical Knowledge Display—Mind Mapping and Concept Mapping as Efficient Tools in Mathematics Education. *Mathematics Education Review* 16 (4): 35–48.

Bronfenbrenner, U. 1981. *Die Ökologie Der Menschlichen Entwicklung*. Klett-Cotta.

Brown, John Seely, Allan Collins, and Paul Duguid. 1989. Situated Cognition and the Culture of Learning. *Educational Researcher* 18 (1): 32–42 (Thousand Oaks, CA: Sage Publications).

Brusilovsky, P., and D.W. Cooper. 2002, January. Domain, Task, and User Models for an Adaptive Hypermedia Performance Support System. In *Proceedings of the 7th International Conference on Intelligent User Interfaces*, 23–30. ACM.

Büssing, A., B. Herbig, and T. Ewert. 2002. Implizites Wissen und erfahrungsgeleitetes arbeitshandeln: Entwicklung einer Methode zur Explikation in der Krankenpflege [Implicit Knowledge and Experience Guided Working: Development of a Method of Explication in Nursing]. *Zeitschrift für Arbeits- und Organisationspsychologie* 46 (1): 2–21.

Caetano, A., A.R. Silva, and Jose Tribolet. 2005. Using Roles and Business Objects to Model and Understand Business Processes. In *Proceedings of the 2005 ACM Symposium on Applied Computing*, 1308–1313. New York: ACM Press.

Campos, Jordi, Maite López-Sánchez, Juan Antonia Rodríguez-Aguilar, and Marc Esteva. 2009. Formalising Situatedness and Adaptation in Electronic Institutions. In *Coordination, Organizations, Institutions and Norms in Agent Systems IV*, 126–139. Springer.

Canas, A.J., G. Hill, R. Carff, N. Suri, J. Lott, T. Eskridge, G. Gómez, M. Arroyo, and R. Carvajal. 2004. CmapTools: A Knowledge Modeling and Sharing

Environment. In *Concept Maps: Theory, Methodology, Technology, Proceedings of the 1st International Conference on Concept Mapping*. Pamplona, Spain: Universidad Púbica De Navarra.

Castells, Manuel. 1997. *Power of Identity: The Information Age: Economy, Society, and Culture*. Blackwell Publishers, Inc.

Cerulo, Karen A. 1997. Identity Construction: New Issues, New Directions. *Annual Review of Sociology* 23 (1): 385–409 (Annual Reviews).

Chan, S. 2001, October. Complex Adaptive Systems. In *ESD. 83 Research Seminar in Engineering Systems*, vol. 31, 1–9.

Coffey, John W., and Robert R. Hoffman. 2003. Knowledge Modeling for the Preservation of Institutional Memory. *Journal of Knowledge Management* 7 (3): 38–52 (MCB UP Ltd).

Colander, David, and Roland Kupers. 2014. *Complexity and the Art of Public Policy: Solving Society's Problems From the Bottom Up*. Princeton University Press.

Dann, H.D. 1992. Variation Von Lege-Strukturen Zur Wissensrepräsentation. In *Struktur-Lege-Verfahren Als Dialog-Konsens-Methodik*, Arbeiten Zur Sozialwissenschaftlichen Psychologie, vol. 25, ed. Brigitte Scheele, 2–41. Aschendorff.

Dewey, John. 1910. *Educational Essays*. Bath: Cedric Chivers.

———. 1933. *How We Think: A Restatement of the Reflective Thinking to the Educative Process*. Heath.

Dobbyn, Chris, and Susan Stuart. 2003. The Self as an Embedded Agent. *Minds and Machines* 13 (2): 187–201 (Springer).

Eberle, Peter, Christian Schwarzinger, and Christian Stary. 2011. User Modelling and Cognitive User Support: Towards Structured Development. *Universal Access in the Information Society* 10 (3): 275–293 (Springer).

Egan, Dennis E. 1988. Individual Differences in Human-Computer Interaction. In *Handbook of Human-Computer Interaction*, 543–568. Elsevier.

Ellison, Nicole, Rebecca Heino, and Jennifer Gibbs. 2006. Managing Impressions Online: Self-Presentation Processes in the Online Dating Environment. *Journal of Computer-Mediated Communication* 11 (2): 415–441 (Oxford, UK: Oxford University Press).

Ferscha, Alois, Clemens Holzmann, and Stefan Oppl. 2004a. Context Awareness for Group Interaction Support. In *Proceedings of the Second International Workshop on Mobility Management & Wireless Access Protocols*, 88–97. New York: ACM. https://doi.org/10.1145/1023783.1023801.

———. 2004b. Team Awareness in Personalised Learning Environments. In *Proceedings of MLEARN 2004*, Bracciano, Italy, pp. 67–72.

Fichter, Lynn S., E.J. Pyle, and S.J. Whitmeyer. 2010. Expanding Evolutionary Theory Beyond Darwinism with Elaborating, Self-Organizing, and Fractionating Complex Evolutionary Systems. *Journal of Geoscience Education* 58 (2): 58–64.

Firestone, J.M., and M.W. McElroy. 2003. *Key Issues in the New Knowledge Management*. Butterworth-Heinemann.

Flood, Robert L., and Ewart R. Carson. 2013. *Dealing with Complexity: An Introduction to the Theory and Application of Systems Science*. Springer Science & Business Media.

Ford, Kenneth M., Frederick E. Petry, Jack R. Adams-Webber, and Paul J. Chang. 1991. An Approach to Knowledge Acquisition Based on the Structure of Personal Construct Systems. *IEEE Transactions on Knowledge and Data Engineering* 3 (1): 78–88 (IEEE).

Fuchs-Kittowski, Frank, and Klaus Fuchs-Kittowski. 2007. Knowledge Management between Provision and Generation of Knowledge. In *Knowledge Management: Innovation, Technology and Cultures*, 165–175. World Scientific.

Fujimura, J.H. 1987. Constructing 'Do-Able' Problems in Cancer Research: Articulating Alignment. *Social Studies of Science* 17 (2): 257–293 (Sage Publications).

Gasser, L. 1986. The Integration of Computing and Routine Work. *ACM Transactions on Office Information Systems* 4 (3): 205–225 (ACM Press).

Gerson, E.M., and Susan Leigh Star. 1986. Analyzing Due Process in the Workplace. *ACM Transactions on Information Systems (TOIS)* 4 (3): 257–270 (ACM Press).

Goguen, J. 1993. On Notation. Department of Computer Science and Engineering, University of California at San Diego.

Grice, H. Paul. 1969. Utterer's Meaning and Intention. *The Philosophical Review* 78 (2): 147–177 (JSTOR).

Groeben, N., and B. Scheele. 2000. Dialog-Konsens-Methodik im Forschungsprogramm Subjektive Theorien. *Forum Qualitative Sozialforschung/Forum: Qualitative Social Research* 1 (2). Deutschland.

Gross, Tom, Chris Stary, and Alex Totter. 2005. User-Centered Awareness in Computer-Supported Cooperative Work-Systems: Structured Embedding of Findings from Social Sciences. *International Journal of Human-Computer Interaction* 18 (3): 323–360 (Taylor & Francis).

Grosskopf, A., J. Edelman, and Mathias Weske. 2010. Tangible Business Process Modeling—Methodology and Experiment Design. In *Proceedings of Business Process Management Workshops (BPM 2009)*, pp. 489–500.

Gruhn, Volker, Daniel Pieper, and Carsten Röttgers. 2007. *MDA®: Effektives Software-Engineering Mit UML2® Und EclipseTM*. Springer-Verlag.

Hacker, Winfried. 1998. *Allgemeine Arbeitspsychologie: Psychische Regulation Von Arbeitstätigkeiten*. H. Huber.

Hahn, J., and J. Kim. 1999. Why Are Some Diagrams Easier to Work with? Effects of Diagrammatic Representation on the Cognitive Integration Process of Systems Analysis and Design. *ACM Transactions on Computer-Human Interaction (TOCHI)* 6 (3): 181–213 (New York: ACM Press).

Hampson, Ian, and Anne Junor. 2005. Invisible Work, Invisible Skills: Interactive Customer Service as Articulation Work. *New Technology, Work & Employment* 20 (2): 166–181.

Harman, Joel, Ross Brown, Udo Kannengiesser, Nils Meyer, and Thomas Rothschädl. 2015. Model as You Do: Engaging an S-BPM Vendor on Process Modelling in 3D Virtual Worlds. In *S-BPM in the Wild*, ed. A. Fleischmann, W. Schmidt, and C. Stary, 113–133. Cham: Springer.

Hemmecke, J., and C. Stary. 2006. The Tacit Dimension of User Tasks: Elicitation and Contextual Representation. In *International Workshop on Task Models and Diagrams for User Interface Design*, 308–323. Berlin, Heidelberg: Springer.

Herrmann, Thomas, G. Kunau, K.U. Loser, and N. Menold. 2004. Socio-Technical Walkthrough: Designing Technology Along Work Processes. In *Proceedings of the Eighth Conference on Participatory Design: Artful Integration: Interweaving Media, Materials and Practices*, 132–141. New York: ACM Press.

Herrmann, Thomas, M. Hoffmann, G. Kunau, and K.U. Loser. 2002. Modelling Cooperative Work: Chances and Risks of Structuring. In *Proceedings of COOP 2002: Cooperative Systems Design, a Challenge of the Mobility Age*, 53–70. IOS Press.

Himma, Kenneth Einar. 2009. Artificial Agency, Consciousness, and the Criteria for Moral Agency: What Properties Must an Artificial Agent Have to Be a Moral Agent? *Ethics and Information Technology* 11 (1): 19–29 (Springer).

Holland, John H. 1992. Complex Adaptive Systems. *Daedalus* 121 (1): 17–30 (JSTOR).

———. 2006. Studying Complex Adaptive Systems. *Journal of Systems Science and Complexity* 19 (1): 1–8 (Springer).

Hug, C., R. Deneckère, and C. Salinesi 2012, May. Map-TBS: Map Process Enactment Traces and Analysis. In *2012 Sixth International Conference on Research Challenges in Information Science (RCIS)*, 1–6. IEEE.

Hughes, F. 1971. *The Sociological Eye*. Aldine de Gruyter.

IEEE-Reliability Society Technical Committee on Systems of Systems. 2014. Systems of Systems Whitepaper.

Ifenthaler, D. 2006. Diagnose Lernabhängiger Veränderung Mentaler Modelle— Entwicklung Der SMD-Technologie Als Methodologisches Verfahren Zur Relationalen, Strukturellen Und Semantischen Analyse Individueller Modellkonstruktionen. University of Freiburg.

Jamshidi, Mo. 2008. *System of Systems Engineering*. Hoboken, NJ: John Wiley & Sons, Inc.

Jaradat, Raed M, Charles B. Keating, and Joseph M. Bradley. 2014. A Histogram Analysis for System of Systems. *International Journal of System of Systems Engineering* 5 (3): 193–227 (Inderscience Publishers Ltd).

Jarvis, P. 2009. Learning to Be a Person in Society. Learning to Be Me. W: K. Illeris (Red.), *Contemporary Theories of Learning. Learning Theorists… in Their Own Words*. London and New York: Routledge.

Jarvis, Peter, and Mary Watts. 2012. *The Routledge International Handbook of Learning*. Routledge.

Jensen Schau, Hope, and Mary C. Gilly. 2003. We Are What We Post? Self-Presentation in Personal Web Space. *Journal of Consumer Research* 30 (3): 385–404 (The University of Chicago Press).

Johnson-Laird, P.N. 1981. Mental Models in Cognitive Science. *Cognitive Science* 4 (1): 71–115 (Elsevier).

Jørgensen, H.D. 2004. Interactive Process Models. PhD thesis, Department of Computer and Information Sciences, Norwegian University of Science and Technology Trondheim, Norway.

Kihlstrom, John F. 2013. The Person-Situation Interaction. In *The Oxford Handbook of Social Cognition*, ed. Donal E. Carlston, 768–805. Oxford: Oxford University Press.

Klein, G., B. Moon, and R.R. Hoffman. 2006. Making Sense of Sensemaking 1: Alternative Perspectives. *IEEE Intelligent Systems* 21 (4): 70–73 (IEEE Educational Activities Department, Piscataway, NJ).

Krogstie, J., Guttorm Sindre, and H.D. Jørgensen. 2006. Process Models Representing Knowledge for Action: A Revised Quality Framework. *European Journal of Information Systems* 15 (1): 91–102. https://doi.org/10.1057/palgrave.ejis.3000598.

Kurzel, Frank, Jill Slay, and Kim Hagenus. 2003. Personalising the Learning Environment. In *Proceedings of Informing Science and Information Technology Education 2003 (InSITE)*.

Kyng, M. 1995. Making Representations Work. *Communications of the ACM* 38 (9): 46–55 (New York: ACM Press).

Lave, J. 1988. *The Culture of Acquisition and the Practice of Understanding* (Tech. Rep. No. 88-0007). Palo Alto, CA: Institute for Research on Learning.

Ludwig, H., Ch. Fischer, and R. Fischer. 2002. *Montessori-Pädagogik in Deutschland. Rückblick—Aktualität—Zukunftsperspektiven: 40 Jahre Montessori-Vereinigung E.v.* LIT Verlag.

Maier, M.W. (2005). Research Challenges for Systems-of-Systems. In *2005 IEEE International Conference on Systems, Man and Cybernetics*, vol. 4, 3149–3154. IEEE.

Matheus, C.J., M.M. Kokar, K. Baclawski, and J.J. Letkowski. 2005, November. An Application of Semantic Web Technologies to Situation Awareness. In *International Semantic Web Conference*, 944–958. Berlin, Heidelberg: Springer.

Meyrowitz, Joshua. 1986. *No Sense of Place: The Impact of Electronic Media on Social Behavior*. Oxford University Press.

Millard, N., L. Hole, and S. Crowle. 1999, August. Smiling Through: Motivation at the User Interface. In *Proceedings of HCI International (the 8th International Conference on Human-Computer Interaction) on Human-Computer Interaction: Ergonomics and User Interfaces-Volume I—Volume I*, 824–828. L. Erlbaum Associates Inc. ISO 690.

Mirel, Barbara. 2004. *Interaction Design for Complex Problem Solving: Developing Useful and Usable Software*. Morgan Kaufmann.

Montague, Gerard P. 2013. *Who Am I? Who Is She?: A Naturalistic, Holistic, Somatic Approach to Personal Identity*. Walter de Gruyter.

Montessori, M. 2005. *The Montessori Method*. Kessinger Publishing.

Motschnig-Pitrik, R., and L. Nykl. 2001, July. The Role and Modeling of Context in a Cognitive Model of Rogers' Person-Centred Approach. In *International and Interdisciplinary Conference on Modeling and Using Context*, 275–289. Berlin, Heidelberg: Springer.

Neuweg, Georg Hans. 2004. Tacit Knowing and Implicit Learning. In *European Perspectives on Learning at Work: The Acquisition of Work Process Knowledge*, Cedefob Reference Series. Luxemburg: Office for Official Publications for the European Communities.

Nonaka, I., and H. Takeuchi. 1995. *The Knowledge-Creating Company: How Japanese Companies Create the Dynamics of Innovation*. Oxford University Press.

Norman, Donald A., and James C. Spohrer. 1996. Learner-Centered Education. *Communications of the ACM* 39 (4): 24–27.

Novak, Joseph D., and A.J. Canas. 2006. *The Theory Underlying Concept Maps and How to Construct Them*. Florida Institute for Human and Machine Cognition.

Oppl, Stefan. 2006. Towards Intuitive Work Modeling with a Tangible Collaboration Interface Approach. In *Proceedings of WETICE '06*, June. IEEE Press, 400–405. https://doi.org/10.1109/WETICE.2006.71.

———. 2016. Articulation of Work Process Models for Organizational Alignment and Informed Information System Design. *Information & Management* 53 (5): 591–608. https://doi.org/10.1016/j.im.2016.01.004.

———. 2017. Evaluation of Collaborative Modeling Processes for Knowledge Articulation and Alignment. *Information Systems and E-Business Management* 15 (3): 717–749 (Springer Berlin Heidelberg). https://doi.org/10.1007/s10257-016-0324-9.

———. 2018. Which Concepts Do Inexperienced Modelers Use to Model Work?—An Exploratory Study. In *Proceedings of MKWI 2018*.

Oppl, Stefan, and C. Stary. 2011a. Effects of a Tabletop Interface on the Co-Construction of Concept Maps. In *Proceedings of the 13th IFIP TC13 Conference on Human-Computer Interaction (INTERACT 2011)*, 443–460. Berlin and Heidelberg: Springer. https://doi.org/10.1007/978-3-642-23765-2_31.

———. 2011b. Towards Informed Metaphor Selection for TUIs. In *Proceedings of the 3rd ACM SIGCHI Symposium on Engineering Interactive Computing Systems (EICS 2011)*, ed. F. Paternò, 16–33 (Chap. 2). Communications in Computer and Information Science, vol. 213. Berlin and Heidelberg: ACM Press. https://doi.org/10.1007/978-3-642-23471-2_2.

———. 2014. Facilitating Shared Understanding of Work Situations Using a Tangible Tabletop Interface. *Behaviour & Information Technology* 33 (6): 619–635. https://doi.org/10.1080/0144929X.2013.833293.

Papert, Seymour. 1993. *The Children's Machine: Rethinking School in the Age of the Computer*. Basis Books.

Parsaye, Kamran, and Mark Chignell. 1988. *Expert Systems for Experts*. New York: Wiley.

Petrasch, R., and O. Meimberg. 2006. *Model-Driven Architecture. Eine Praxisgerechte Einführung in Die MDA*. Heidelberg: dpunkt.

Pirnay-Dummer, Pablo N. 2006. Expertise Und Modellbildung—MITOCAR. University of Freiburg.

Polanyi, M. 1958. *Personal Knowledge: Towards a Post-Critical Philosophy*. Chicago: University of Chicago Press.

———. 1966. *The Tacit Dimension*. Doubleday & Co.

Pörtner, Marlis. 2008. *Ernstnehmen-Zutrauen-Verstehen: Personzentrierte Haltung Im Umgang Mit Geistig Behinderten Und Pflegebedürftigen Menschen*. Klett-Cotta.

Resnick, Mitchel, Amy Bruckman, and Fred Martin. 1996. Pianos Not Stereos: Creating Computational Construction Kits. *Interactions* 3 (5): 40–50 (ACM).

Robinson, Laura. 2007. The Cyberself: The Self-Ing Project Goes Online, Symbolic Interaction in the Digital Age. *New Media & Society* 9 (1): 93–110 (Thousand Oaks, CA: Sage Publications).

Rogers, Carl R. 1951. *Client-Centered Therapy: Its Current Practice, Implications, and Theory.* Boston, MA: Houghton Mifflin.

———. 1961. *On Becoming a Person: A Therapist's Point of View of Psychotherapy.* Alemar.

Rolland, Colette, Naveen Prakash, and Adolphe Benjamen. 1999. A Multi-Model View of Process Modelling. *Requirements Engineering* 4 (4): 169–187 (Springer).

Roussos, Maria, Andrew E. Johnson, Jason Leigh, Christina A. Vasilakis, Craig R. Barnes, and Thomas G. Moher. 1997. NICE: Combining Constructionism, Narrative and Collaboration in a Virtual Learning Environment. *Computer Graphics—New York—Association for Computing Machinery* 31: 62–63 (ACM Association for Computing Machinery).

Rubin, David C. 2006. The Basic-Systems Model of Episodic Memory. *Perspectives on Psychological Science* 1 (4): 277–311 (Los Angeles, CA: SAGE Publications).

Ruiz-Primo, M.A., and R.J. Shavelson. 1996. Problems and Issues in the Use of Concept Maps in Science Assessment. *Journal of Research in Science Teaching* 33 (6): 569–600.

Russell, D.M., M.J. Stefik, P. Pirolli, and S.K. Card. 1993. The Cost Structure of Sensemaking. In *Proceedings of the SIGCHI Conference on Human Factors in Computing Systems.* New York, pp. 269–276.

Sachs, Patricia. 1995. Transforming Work: Collaboration, Learning, and Design. *Communications of the ACM* 38 (9): 36–44 (New York: ACM Press).

Scheer, August Wilhelm. 2003. *ARIS—Business Process Modeling.* 3rd ed. Springer.

Schön, D. 1984. *The Reflective Practitioner: How Professionals Think in Action.* Basic Books.

Schön, Donald A. 1987. *Educating the Reflective Practitioner.* San Francisco: Jossey-Bass.

Seel, Norbert M. 2003. Model-Centered Learning and Instruction. *Technology, Instruction, Cognition and Learning* 1 (1): 59–85 (Old City Publishing).

Senge, P.M. 1990. *The Fifth Discipline: The Art and Practice of the Learning Organization.* Doubleday/Currency.

Senge, Peter M., and John D. Sterman. 1992. Systems Thinking and Organizational Learning: Acting Locally and Thinking Globally in the

Organization of the Future. *European Journal of Operational Research* 59 (1): 137–150 (Elsevier).

Sheeran, Paschal. 2002. Intention—Behavior Relations: A Conceptual and Empirical Review. *European Review of Social Psychology* 12 (1): 1–36 (Taylor & Francis).

Spindler, M., and Christian Stary. 2017. SoS: Anarchy Active Inner Spacing for Co-Creating Future Outer Space. *Challenging Organisations and Society* 6 (1): 1013–1054.

Stary, C., and D. Wachholder. 2016. System-of-Systems Support—A Bigraph Approach to Interoperability and Emergent Behavior. *Data & Knowledge Engineering* 105: 155–172.

Stary, Christian. 2000. TADEUS: Seamless Development of Task-Based and User-Oriented Interfaces. *IEEE Transactions on Systems, Man, and Cybernetics- Part A: Systems and Humans* 30 (5): 509–525 (IEEE).

———. 2014. Non-Disruptive Knowledge and Business Processing in Knowledge Life Cycles—Aligning Value Network Analysis to Process Management. *Journal of Knowledge Management* 18 (4): 651–686 (Emerald Group Publishing Limited).

———. 2017a. Interactive Articulation and Probing of Processes: Capturing Intention and Outcome for Coherent Workplace Design. In *The Future Information Society: Social and Technological Problems*, 241–257. World Scientific.

———. 2017b. Requirements Elicitation and Specification Using the S-BPM Paradigm. *REFSQ Workshops Proceedings, Workshop on Continuous Requirements Engineering*, 1796.

———. 2017c. System-of-Systems Design Thinking on Behavior. *Systems* 5 (1): 3 (Multidisciplinary Digital Publishing Institute).

———. 2017d. Walking on 2 Legs: 3D-Structured Method Alignment in Project Management. In *Interactivity, Game Creation, Design, Learning, and Innovation*, 33–42. Springer.

Sternberg, Robert J., Joseph A. Horvath, and others. 1999. *Tacit Knowledge in Professional Practice: Researcher and Practitioner Perspectives*. Psychology Press.

Strauss, A. 1985. Work and the Division of Labor. *The Sociological Quarterly* 26 (1): 1–19 (Blackwell Publishing Ltd).

———. 1988. The Articulation of Project Work: An Organizational Process. *The Sociological Quarterly* 29 (2): 163–178.

———. 1993. *Continual Permutations of Action*. New York: Aldine de Gruyter.

Subramanya, S.R., and Byung K. Yi. 2007. Enhancing the User Experience in Mobile Phones. *IEEE Computer* 40 (12): 114–117.

Suchman, L. 1995. Making Work Visible. *Communications of the ACM* 38 (9): 56–64 (New York: ACM Press).
Turkle, Sh. 1998. *Leben Im Netz. Identitäten in Zeiten Des Internet.* Reinbeck bei Hamburg: Rowohlt.
Ulich, E. 1994. Arbeitspsychologie. 4., Überarbeitete Und Erweiterte Auflage, 1998. Stuttgart: Schäffer-Poeschel.
Waddock, Sandra, Greta M. Meszoely, Steve Waddell, and Domenico Dentoni. 2015. The Complexity of Wicked Problems in Large Scale Change. *Journal of Organizational Change Management* 28 (6): 993–1012 (Emerald Group Publishing Limited).
Weichhart, Georg, Christian Stary, and François Vernadat. 2018. Enterprise Modelling for Interoperable and Knowledge-Based Enterprises. *International Journal of Production Research* 56 (8): 2818–2840 (Taylor & Francis).
Weiser, M. 1991. The Computer for the 21st Century. *Scientific American* 265: 94–104.
Weske, Mathias. 2010. *Business Process Management: Concepts, Languages, Architectures.* Springer.
Zuckerman, Oren, S. Arida, and M. Resnick. 2005. Extending Tangible Interfaces for Education: Digital Montessori-Inspired Manipulatives. In *Proceedings of the SIGCHI Conference on Human Factors in Computing Systems (CHI)*, 859–868. New York: ACM Press.

Open Access This chapter is licensed under the terms of the Creative Commons Attribution 4.0 International License (http://creativecommons.org/licenses/by/4.0/), which permits use, sharing, adaptation, distribution and reproduction in any medium or format, as long as you give appropriate credit to the original author(s) and the source, provide a link to the Creative Commons licence and indicate if changes were made.

The images or other third party material in this chapter are included in the chapter's Creative Commons licence, unless indicated otherwise in a credit line to the material. If material is not included in the chapter's Creative Commons licence and your intended use is not permitted by statutory regulation or exceeds the permitted use, you will need to obtain permission directly from the copyright holder.

3

Value-Oriented Articulation

When the articulation of values in the context of business operation and managing organizations is brought up, stakeholders often start referring to the value of business assets, for example, IT. Such a perspective is driven by a focus on business enablers and resources required to generate valuable assets for the market. And most decision makers look at those elements from a 'requirements engineering' perspective to deliver products and services. This understanding is grounded in Michael Porter's concept of value chain and its analysis that helps in apprehending how an enterprise creates valuable elements through a set of core (like sales) and support activities. Both are assumed to contribute to the sustainable existence of the producing organization in competitive and continuously changing environments, based on products or services for which customers are creating revenues.

Products and services are produced along business processes. These are composed of functional activities transforming incoming goods and information through a series of cross-functional steps in the course of business operation. Such an approach to business analysis considers value creation to reside in the design and execution of work processes (rather than the processed or created assets) that leads to a result for customers or consumers. Although value created in this way has a tangible component,

© The Author(s) 2019
S. Oppl, C. Stary, *Designing Digital Work*,
https://doi.org/10.1007/978-3-030-12259-1_3

for example, effort to be spent for production, its second component, the intangible part, such as distance to customer needs, is of equal importance. However, capturing both requires explication and representation of stakeholder knowledge of work processes and their structure. The more the stakeholders know of business processes, the higher are the chances of promising improvement ideas stemming from business operation.

The challenge is now to effectively develop techniques to bring up opportunities to organize work in a way that it creates intangible and tangible value. Based on existing work processes, potential arrangements of operational structures need to be articulated by the concerned stakeholders. Knowing how to express what an organization knows in terms of structuring work to achieve business objectives reveals development opportunities without anticipating prospective operational structures. In this chapter, we introduce three different foci of articulation:

- Stakeholders identify and refine their role identity in the context of collaborative task accomplishment
- Stakeholders explicate information needs and supplies when accomplishing functional tasks
- Stakeholders elaborate on collaboration with others by specifying transactions between functional roles

Although each of the approaches finally ends up with revealing and documenting essential process elements in terms of stakeholder roles, activities, and relationships, they differ in terms of their means, externalize, and represent knowledge. In this way, particular aspects when articulating knowledge can move to the center of interest. We start out with subject-oriented articulation support allowing to shape the understanding of a role (termed subject) through natural language expressions and identifying interaction patterns when accomplishing business-relevant tasks. We proceed to demonstrate a card-based structural elaboration approach for developing interaction patterns starting from a functional role perspective and progressing to an overall interactional perspective, in order to capture relevant business operations. We finally show an approach aimed at a detailed understanding of interactions in terms of formal and informal relations between stakeholder roles, when completing the chapter, with an account on value networks.

3.1 Shaping Role Identities Through Contextual Behavior Articulation

The approach introduced in this chapter aims to utilize human language skills when stakeholders describe their operational behavior at work. They also allow taking into account to capture the interaction with other stakeholders. The underlying framework for representation, subject orientation, is explained with its ontological background based on Fleischmann et al. (2012). Sample applications demonstrate the practical benefits of the approach. They cumulate in the execution of behavior representations that facilitate process development in terms of seamless round-trip engineering. Behavior patterns can be deployed dynamically when operational knowledge needs to be adapted or when aiming to transform organizations in a non-disruptive way, for example, in the course of digitalization projects.

3.1.1 Start Simple, Using Natural Language

When following natural language sentences, stakeholders can describe their behavior in terms of contextual activities in specific situations, that is, framing activities through some active role and affected objects. Thereby, we can use several constituent elements of sentences: subject, predicate, and object, referring to WHO is DOING WHAT (some activity) handling WHICH OBJECT. In case the person articulating knowledge is the addressed, WHO, he/she can describe actions and objects in a straightforward way. Besides involving only a single actor, descriptions created in this way should be easily understood due to the tripartite structure, and thus being used without further transformation when communicating work-articulated information to other stakeholders.

When using subject-oriented representations, we can utilize this information structure as models. Following the structure of natural language sentences, processes executed by digital systems can also be expressed. However, the articulation needs to start like any development of an information system or digital artifact in a socio-technical system with identifying a specific scope or universe of discourse. It is that part of the observed reality that is supposed to be supported by an information system or technological artifact.

The identified scope determines a so-called universe of discourse (i.e., the space or field of concern) in which structural qualities and behavioral elements have a certain meaning for those using them. Typically, stakeholders refer to work situations, such as handling business cases that become part of subject-oriented representations. These models include the interactions of behavioral entities (humans or technological artifacts) occurring in a work environment. It is this kind of information that qualifies subject-oriented models to be executed without further transformation, as they contain the control flow required for processing specified activities in a certain sequence.

In the course of articulation, model elements are considered either essential or complementary. The latter are grouped around the essential elements, and trigger modeling processes (Scholz and Holl 1999) embodied in various existing modeling paradigms. Typical paradigms are functional ones leading to control or data flow diagrams, data-oriented ones leading to data models such as Entity Relationship diagrams, and object-oriented ones using modeling languages such as UML (Unified Modeling Language). Likewise, subject-oriented articulation follows notational conventions, namely those that lead to subject-oriented models. Thereby, stakeholders identify roles or small sets of tasks using notations or modeling languages like the ones mentioned above. Subject orientation allows representing parts of the observed reality in terms of natural language sentence structures. Hence, these models can be used for any other representation or modeling approach universal use, due to the familiarity of natural language in daily communication, and the availability of a structural semantics for sentences, comprising subject, predicate, and object.

Since the use of natural language does not prevent misunderstandings, this simplified sentence semantics should help to initially clarify roles, activities, and concerned objects of work before engaging in more structured forms of representation. It might require some exercise to strictly apply it; however, it aims to deliver more complete descriptions of situations compared to purely functional descriptions of workplaces. The structural sentence semantics of natural language 'subject-predicate-object' corresponds to subject-oriented modeling, in several ways:

- A *subject* is the starting point for describing a situation or events.
- An activity is denoted by a *predicate*.
- An activity concerns an (abstract) *object*.

The distinction between essential and supplementary aspects can be kept for natural language articulation, since humans also tend to use passive sentences in case they do not take into consideration any particular actor explicitly. Such sentences could convey events or specific contextual information of situations. For precise representation, however, each activity has to be assigned to a specific subject (actor). In behavior models, acting roles, for example, the employees are distinguished from predicates defining the activities of acting roles, and objects denoting the purpose of these activities. In the course of accomplishing their tasks, they receive work inputs and pass on results. Hence, we consider interaction and communication, either direct or indirect, to be an essential activity of acting roles for subject-oriented articulation and representation.

We introduce the subject-oriented articulation approach using a common work situation: Employees have to apply for going on holidays or taking days off. It allows us to demonstrate the fundamental and supplementary aspects of the sentence structure 'subject-predicate-object'. Figure 3.1 shows the natural language description of the respective work procedure.

Holiday application procedure:
An employee fills in a holiday application form. He/She puts in a start and end date of his/her planned vacations. The responsible manager checks the application and informs the employee about his/her decision; the holiday request might be rejected or approved. In case of approval the holiday data are sent to the human resource department (HR) which updates the days-off file.

Fig. 3.1 Natural language description of an application procedure for vacation (released under a Creative Commons Attribution 4.0 International License (CC BY 4.0))

As long as modeling is focused on activities, predicates are essential, whereas subjects and objects are considered as complementary elements. However, in subject-oriented terms, the subject and activities are essential, as they constitute the core concepts to representing processes in this approach. A subject scopes a specific set of send-, receive-, and do-activities (Fleischmann et al. 2012).

3.1.2 Roles As Semantic and Pragmatic Entities

When applying subject-oriented articulation to picture reality and represent situations according to its essential elements, some properties of subject systems can be identified. They finally guide the articulation of behavior:

- *Being in the World*: Identifying a subject means bringing a self-contained entity to life—it is a behavior encapsulation of an active entity, and also subject to the 'world' (i.e., identified universe of discourse). The latter results from the fact that a subject can be addressed (only) by other, existing subjects of the world. Consequently, being a subject *in* the world also means being subject *to* the world.
- *Subjects are social and private at the same time*: Exchanging messages is interaction via send and receive pairs. Hence, subjects are open for message passing, either for being informed or for further handling and delivering a business object. However, how they process incoming messages and produce output remains encapsulated in the (internal) behavior description. In this way, subjects align individuals with communities—they allow stakeholders having a cognitive identity while behaving as a social being.
- *Subjects are dynamic entities while keeping the outer structure stable*: They can change their internal behavior while remaining a stable communication partner. In this way, self-organizing communities can be represented. It increases flexibility of structures, even when changing their manifest form. New gadgets can take over new responsibilities, such as calendar, meeting, cinema proposal, or sensor systems, just to name a few, replacing or encapsulating existing behavior patterns.
- *Subjects make the world more concrete* due to their nature of being a boundary object. Such a boundary object can be communicated

among stakeholders and thus, understood by people with different backgrounds (Arias and Fischer 2000). Subject representations can be read in natural language using active sentences. This property ensures some understanding and allows active participation of all stakeholders, even when requiring some self-discipline to use active sentence and complete natural language expressions to describe situations. It brings the approach to integrated thinking and acting of stakeholders, as proposed by Heidegger (cf. (Han 2015), p. 53).

- *Subject orientation scales* due to the decentralized management mechanism. It enables setting up and configuring a large number of actors or systems. The latter is of particular importance in networked settings. Thereby, subjects correspond to autonomous agents, not only being capable to implement certain task behaviors, but also to monitor the status of other elements or systems. For instance, in safety-critical settings, such monitoring and supervision services may be a requirement.
- *Subjects are part of choreography.* In this way, lifecycle activities of certain systems or elements can become part of continuous development without endangering ongoing operations of networked actors. As long as the communication interface remains, internal subject behavior can be replaced and modified.
- *Subject-oriented representations allow for problem- and domain-specific abstraction.* This feature provides uniform addressable interfaces for resource control and management.

Overall, a subject-oriented representation of any setting can come close to the 'reality' as perceived and pictured by humans, both in terms of its elements as behavioral entities including their set of activities and interactions, and in terms of its description, as natural language can directly be used conveying the meaning encoded in work processes.

3.1.3 Acting in a Specific Role—Pragmatic Modeling

The semantics of a situation and activities of embodied actors refer to the pragmatic aspects of a situation and thus, influence the pragmatic quality of a representation or behavior model.

Pragmatic quality is the correspondence between the model and the audience's interpretation of the model and has one goal, comprehension, meaning that the model has been understood. Means to increase pragmatic quality include not only executability, animation, and simulation but also more advanced techniques like model transformations, model filtering to present model abstractions from several viewpoints, model translation, and explanation generation. (Krogstie et al. 2006, p. 94, according to Stamper 1996)

In the following subsections, we start out with the general perspective on the world as perceived by stakeholders from a subject-oriented view and proceed with constructing work models based on a role-specific behavior understanding (cf. Fleischmann and Stary 2012).

3.1.3.1 The World As Network of Roles

Articulating the world in a subject-oriented way means trying to represent each observation in terms of networked active elements termed subjects, assumed to act in parallel (Fleischmann et al. 2012). Since each of those actors or subjects can be described in terms of its behavior and has the capability to exchange messages, a federated choreographic ecosystem is established: Federation means a form or single unit, within which each actor or subject or organization keeps some internal autonomy. This form or single unit identifies the perceived part of the world that is considered relevant to describe a specific situation. It sets up the universe of discourse or context space for representation and action.

Keeping some internal autonomy at some point requires being more concrete: The 'some' is dedicated to the level of abstraction considered representative for the stakeholders or modelers, both, with respect to functional or technical activities, and interaction or communication with other subjects.

Choreographic ecosystem refers to recognizing concurrent, however, synchronized processes and activities

- in a community of interacting elements and their environment
- when considered as networked or interconnected system

According to this perspective, ecosystems operate as autonomous, concurrent behaviors of distributed subsystems or actors. A subject is a behavioral role assumed by some entity that is capable of performing actions. The entity can be a human, a piece of software, a machine (e.g., a robot), a device (e.g., a sensor), or a combination of these, such as intelligent sensor systems.

Since subjects represent systems with a uniform structure, they can be used to define federated systems or System-of-Systems (SoS) (Jamshidi 2008). SoS have as essential properties "autonomy, coherence, permanence, and organization" (ibid., p. 1) and are constituted "by many components interacting in a network structure," with most often physically and functionally heterogeneous components. For instance, education support systems comprise social media and content management systems for learning support. SoS subjects can execute local actions that do not involve interacting with other subjects (e.g., a clock providing the time in an office), and communicative actions that are concerned with exchanging messages between subjects, that is, sending and receiving messages, for example, triggering ringing a tone (Stary and Wachholder 2015; Stary 2017).

3.1.3.2 Articulation by Stepwise Behavior Abstractions

Subjects exchange messages and use operations on objects. For the holiday application, the behavior articulation starts with the identification of the actors or roles involved in the process (Bach 2000), that is, the subjects, and the messages they exchange. Actors drive a process. In order to coordinate and tune their activities, actors have to communicate and use suitable tools. Figure 3.2 shows the subjects involved in the holiday appli-

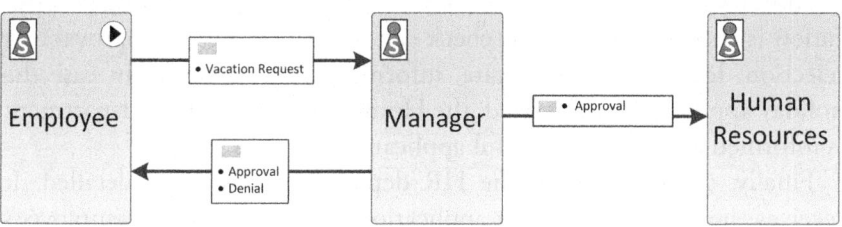

Fig. 3.2 Subject identification for the holiday application process, providing subjects and their interaction

cation process, the exchanged messages naming the transferred information. In this way, depending on the activities of the subjects, all predicates required for task completion are identified step-by-step (including the required data).

While sending messages, data is transmitted. For instance, the holiday message application sent by the employee to the manager contains the start and end of applied holidays. 'Send messages' or 'message transfer' does not imply implementation details of the underlying mechanisms for interaction. The holiday application might be transferred in carbon copy, by email, or using a web application accessible by the employee and the manager. The terms refer to a logical action rather than concrete implementations, for example, messaging systems used for data exchange.

Figure 3.2 shows only the interaction structure of a process. The first refinement concerns the sequences of interactions, that is, the behavior of each subject has to be specified. Figure 3.3 details the employee behavior, namely the sequence of sending and receiving messages and performing activities. The initial state is marked. In this state, the employee fills in a holiday application form. Upon completion, the employee's state switches to the next state via the transition 'holiday application completed'. This state is a sending state. In this state, the holiday application is sent to the manager. After successfully sending the message, the employee reaches the state 'answer of manager' waiting for approval or rejection. This state is a receiving state. In case of rejection, the process terminates. In case of approval, the holidays can be taken as applied for. Upon return of the employee, the holiday application process also terminates.

The behavior of the manager is complementary to the employee's. The messages sent by employee are received by the manager and vice versa. Figure 3.4 shows the behavior of the manager. The manager is on hold for the holiday application of the employee. Upon receipt, the holiday application is checked (state). This check can result in either an approval or a rejection, leading to either state, informing the employee. In case the holiday application is approved, the Human Resources (HR) department is informed about the successful application.

Finally, the behavior of the HR department has to be detailed. It receives the approved holiday application and puts it to the employee's days-off record, without further activities (process completion) (Fig. 3.5).

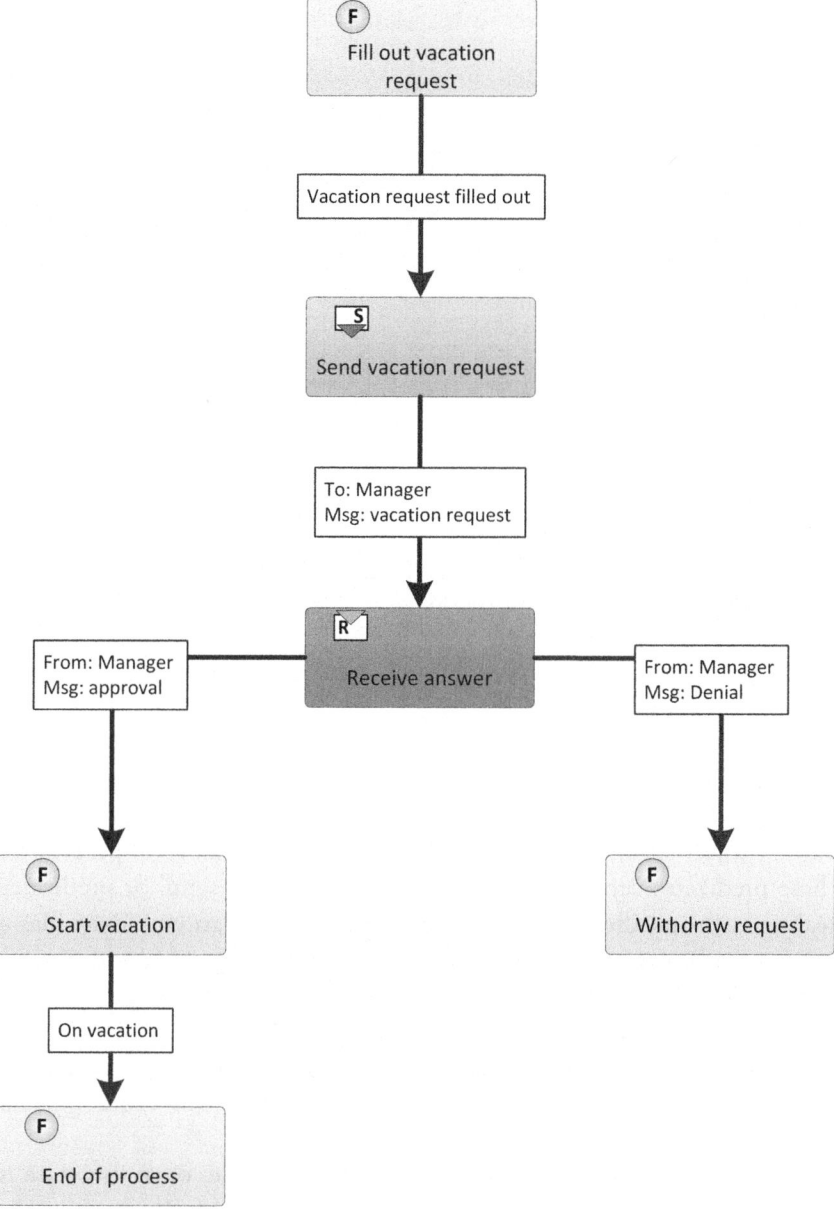

Fig. 3.3 Employee behavior in holiday application process

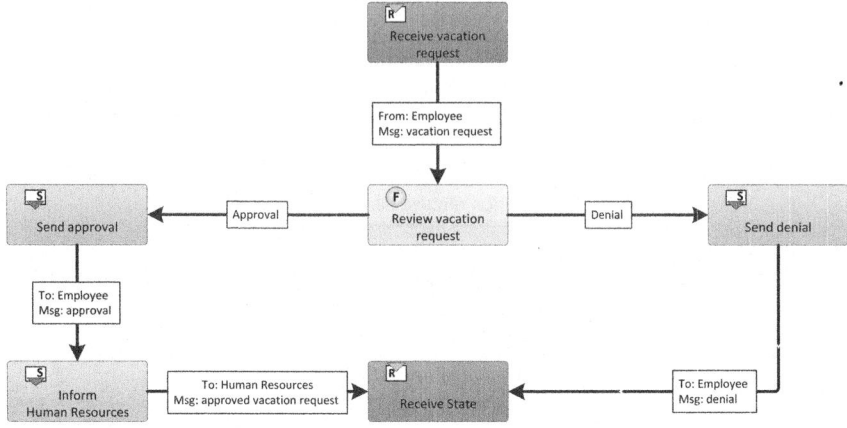

Fig. 3.4 Manager's behavior in holiday application process

So far, we have modeled:

- the subjects involved in a process
- interactions they are part of
- the data they send or receive through each interaction, and
- behavior of each subject

The description of a subject defines the sequence of sending and receiving messages, or the processing of internal functions, respectively. In this way, a subject specification contains the pushing sequence of predicates. These predicates can be the standard predicates like 'send' or predicates dealing with specific objects, such as required when an employee files a holiday application form (see Fig. 3.3). Consequently, each node (state) and transition has to be assigned an operation. The implementation of that operation does not matter at that stage, since it can be handled by object specifications. As we abstract from implementation details, it seems suitable to replace the term operation by the more general term service.

A service is assigned to an internal functional node, once this state is reached, the assigned service is triggered and processed. The end conditions correspond to links leaving the internal functional node.

Value-Oriented Articulation

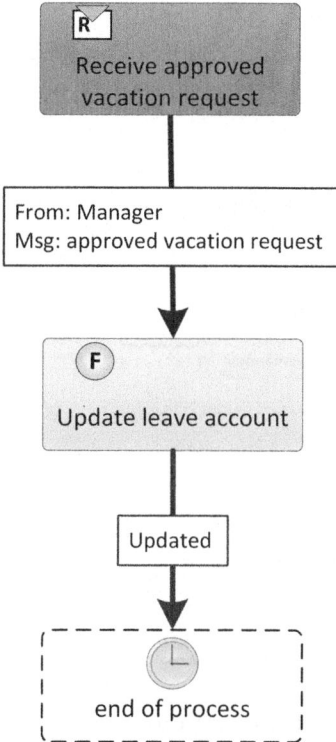

Fig. 3.5 HR department behavior in holiday application process

Each result link of a sending node (state) is assigned to a named service. Before sending, this service is triggered to identify the content or parameter of a message. The service determines the values of the message parameters transferred by the message. Analogously, each output link of a receiving node (state) is also assigned to a named service. When accepting a message in this state that service is triggered to identify the parameter of the received message. The service determines the values of the parameters transferred by the message and provides them for further processing.

These services are used to assign a certain meaning to each step in a subject. Services allow defining the predicates used in a subject. All of those are triggered in a synchronous way, that is, a subject only reaches its

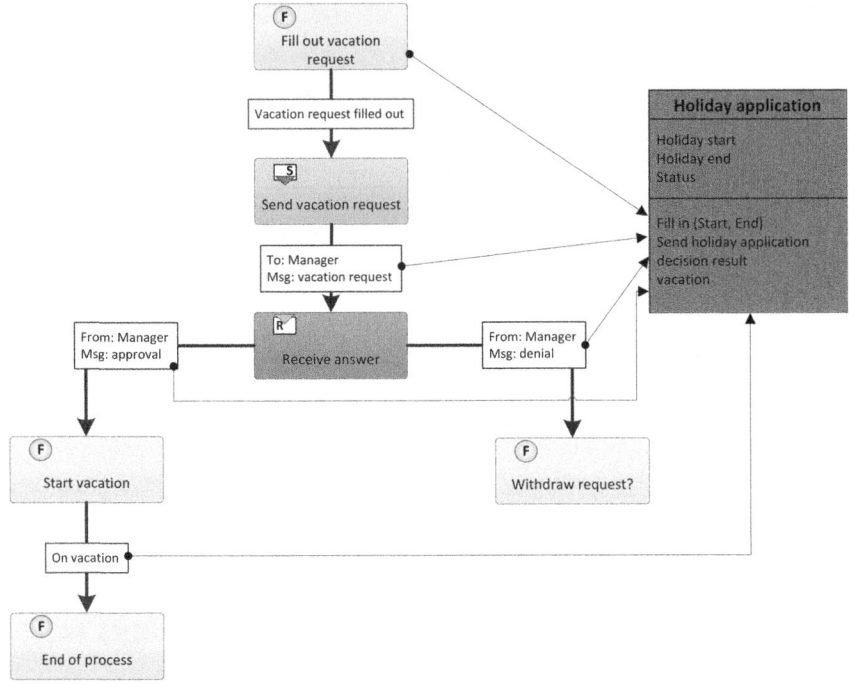

Fig. 3.6 A subject with predicates and objects

subsequent state once all triggered services have been completed. Figure 3.6 shows how the predicates of a subject are defined by means of objects.

3.1.4 Conclusive Summary

Natural language is a valid starting point for articulation and behavior representation. Structured natural language sentences can serve as a fundamental means of articulation. When using the introduced subject-oriented scheme, stakeholders recognize actors as the starting point for modeling, allowing for rich context representation of functional behavior (Brocke et al., 2015, 2016). The representation scheme ensures coherence, both, in terms of flow of control, and the addressed data. Consequently, stakeholders can benefit from specifications that contain contextual and operational information, such as social interactions, cooperation, and collaboration aspects (Neubauer and Stary 2017).

3.2 Sorting Out: Cards As Carrier of Functions and Interaction

While subject-oriented models provide a natural language-oriented form of representation, the act of modeling itself still requires the people engaged in modeling to step out of their own work context and adopt a bird's eye view on their work. This is cumbersome for inexperienced modelers, who usually have a spatio-temporally contextualized understanding of their work contributions and are mainly used to talk about single cases (i.e., instances) of their work processes. Abstracting from these instances and adopting a more generic view is a fundamental skill when engaging in modeling activities (Frederiks and van der Weide 2006). It, however, cannot be assumed to be fully developed for all modeling participants. Appropriate forms of representations and scaffolds can thus help to mitigate deficiencies in this area and allow including people without prior modeling experiences in work articulation and design activities (Oppl et al. 2017). In this light, we here introduce a method based on structure elaboration techniques (Groeben and Scheele 2000) that scaffolds the articulation process and still leads to models that represent both, the functional and interactional aspects of work processes.

Research on facilitating lay modeling focuses on measures to guide inexperienced modelers through the process of creating a model without overloading them with syntactic formalism and complex modeling constructs. Existing research (Santoro et al. 2010; Fahland and Weidlich 2010; Kabicher and Rinderle-Ma 2011; Lai et al. 2014) suggests that starting modeling based upon a concrete work case facilitates developing an understanding of the necessary concepts for inexperienced modelers when describing a work process in an abstract conceptual model. Using a case-based approach to modeling also reduces the number of language elements necessary to depict the work process.

For example, case-based models do not require decision constructs or elements for exception handling. While the number of modeling elements alone appears not to have a notable impact on the understanding of a modeling language for inexperienced modelers (Recker and Dreiling 2007), empirical evidence shows that the number of elements actually used during modeling is limited and highly dependent on the modeling objective (Muehlen and Recker 2008). When involving inexperienced modelers, it seems to be appropriate to limit the number of available modeling elements a priori to those appropriate for the intended modeling perspective and targeted outcome (Genon et al. 2011; Britton and Jones 1999). For modeling organizational work, the modeling perspective is oriented towards the work of actors and their interactions within an organization. The targeted outcome is reaching common ground on the work process for non-expert modelers.

Furthermore, Herrmann and Nolte (2014) and Santoro et al. (2010) provide evidence that non-formalized information and annotations to model elements can aid the externalization process. However, they do not force the modelers to express all information using the constructs of the modeling language. Some results also point at the importance of (human or automatic) facilitation and scaffolding during the model creation process (Hjalmarsson et al. 2015) and the model alignment process (Rittgen 2009), particularly for inexperienced modelers (Davies et al. 2006). Recent research indicates that procedural and structural scaffolds provided by a facilitator or an automated system may support the refinement of incomplete models (Oppl and Hoppenbrouwers 2016; Oppl 2016).

Summarizing, the following properties of a modeling approach support collaborative modeling by inexperienced modelers: (1) starting with case-based development of process models, (2) offering a constrained set of modeling constructs with semantics focused on the modeling objective, (3) enabling informal annotations of model elements (i.e., not adhering to formal modeling syntax), and (4) offering procedural and structural scaffolds for model creation and alignment.

3.2.1 Articulation Concepts

Models of work processes that should express the collaborative aspects of work need to provide semantic constructs to represent who is involved in the work process, which activities are performed by the involved entities, and what information or artifacts are exchanged by them. These elements describe the coordinative aspects as well as the operative aspects of work and thus, can be considered the minimal set of conceptual elements necessary to describe collaborative work (Fjuk and Dirckinck-Holmfeld 1997). This assumption has been backed by the development of business process modeling languages over the last few years, where the focus has shifted from functional approaches (e.g., Event-driven Process Chains (EPCs); Nüttgens and Rump 2002) to approaches that structure process descriptions along the involved entities and explicitly allow them to express their interaction (e.g., BPMN (Business Process Modeling Notation); White and Miers 2008 or S-BPM (Subject-oriented Business Process Management); Fleischmann et al. 2012).

The mentioned interaction-oriented modeling languages are designed to describe complex business processes, covering all their variants and potential exceptions. The modeling constructs introduced to handle this complexity, however, are not required for the articulation approach proposed here (Oppl 2018). Starting articulation with a case-based narrative approach avoids the need for control-flow constructs beyond describing sequences of activities and interaction with others. This reduces the number of modeling elements to make modeling easier for non-expert modelers. Based on empirical data collected on practitioners' use of BPMN 2.0, Muehlen and Recker (2008) show that for interaction-oriented modeling of organizational work processes, at most the following constructs are used: *Task* and *sequence flow* to indicate what is to be done in which sequence, *pools* to indicate who is doing what, *message flows* to couple the process parts in the pools, and *events* indicating the *start* and *end* of the process. Abstracting from BPMN notation, the modeling language proposed here consequently consists of the following three modeling elements (cf. Fig. 3.7):

- *WHO–element*: representing actors, roles, or organizational entities (exact semantics depending on the level of abstraction individually chosen for modeling) (➜ 'pools' in BPMN or 'subjects' in subject-oriented modeling)
- *WHAT–element*: representing activities (➜ 'tasks' in BPMN or 'states' in subject-oriented modeling)

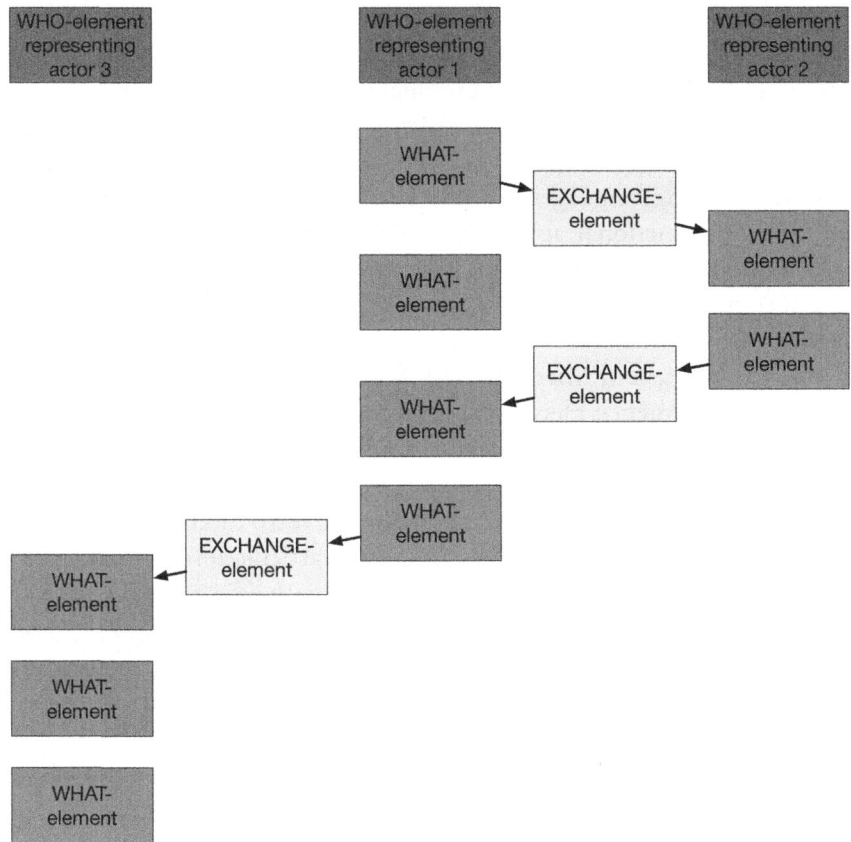

Fig. 3.7 Elements of the card-based modeling language

- *EXCHANGE-element*: describing exchange of information or artifacts among WHO-elements (exact semantics depending on designator for element) (→ 'message flow' in BPMN or 'messages' in subject-oriented modeling)

These elements are put into mutual relationship by spatially arranging them as follows (cf. Fig. 3.7):

- Each WHAT-element is assigned to a WHO-element by placing it on an imaginative straight line originating from the WHO-element

(→ assignment of 'tasks' to 'pools' in BPMN or definition of a subject's internal behavior in subject-oriented modeling)
- Causality between WHAT-elements is expressed by their order on the line starting with the one that is placed nearest to the WHO-element (→ 'sequence flow', 'start event', 'end event', in BPMN or refinement of a subject's internal behavior in subject-oriented modeling)
- EXCHANGE-elements are placed between the lines of the communicating WHO-elements and are causally related in the stream of WHAT-elements by spatial arrangement, explicitly adding connecting arrows from the activity in which or after which the exchange is triggered and to the activity that receives or is triggered by the exchange (→ 'message flow' in BPMN or definition of the interaction among subjects in subject-oriented modeling)

As shown above, the proposed language covers the elements used for interaction-oriented modeling for organizational work processes as identified by Muehlen and Recker (2008) and can be unambiguously mapped to formal business process modeling languages such as BPMN or subject-oriented process models. The number of elements has to be reduced and assigned clearly distinguishable semantics in order to meet the articulation needs of inexperienced modelers (Genon et al. 2011).

3.2.2 Articulation Process

The following spatial layout is used for the different elements described above to create a consistent form of model representation (Oppl 2015):

- WHO-items are placed on the upper border of the modeling surface, and indicate the role represented by the actor and those roles with which the modeler is perceived to interact directly.
- WHAT-items are placed below the WHO-item representing the role of the actor, and describe the actor's own activities. Their sequence indicates causal and/or temporal relationships.
- EXCHANGE-items are placed below the WHO-items of the other roles. They indicate expected exchange of information or artifacts.

Their spatial arrangement indicates the causal and/or temporal relationship to the stream of WHAT-items:

- EXCHANGE-items placed slightly above a WHAT-item indicate expected incoming information or artifacts. In case of ambiguity, this relationship can be made explicit by drawing an arrow connecting the EXCHANGE-item with the WHAT item requiring this input.
- EXCHANGE-items placed slightly below a WHAT-item indicate offered outgoing information or artifacts. In case of ambiguity, this relationship can be made explicit by drawing an arrow connecting the WHAT-item producing this output with the EXCHANGE-item.

Figure 3.8 shows the three individually articulated models for the sample process. WHO-items are represented in blue, WHAT-items are red, and EXCHANGE-items are yellow. As an example, the model of actor 2 is described in narrative form in the following: the *secretary* perceives that he has to interact with his *colleague* and his *boss* to complete his role in the process. He expects to receive a *completed application* from the *colleague* to be able to start his contribution. He *checks for conflicts* with other submitted or already confirmed applications. The *checked application* is then forwarded to the *boss*. The secretary proceeds, as soon as he receives the *confirmed application* back from the *boss*. He then *files the application* and forwards the *confirmation* to his *colleague*.

Figure 3.8 also shows semantic differences between the models on the level of WHO-elements (e.g., 'boss' vs. 'manager') and on the level of

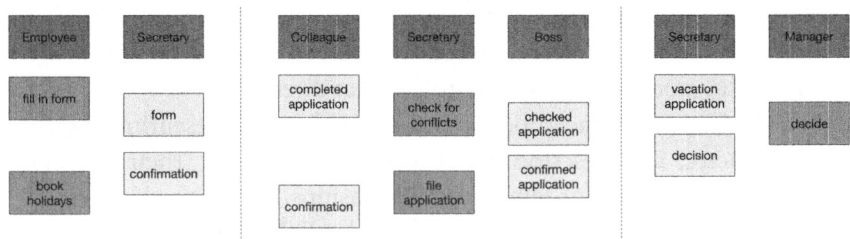

Fig. 3.8 Sample result of individual articulation

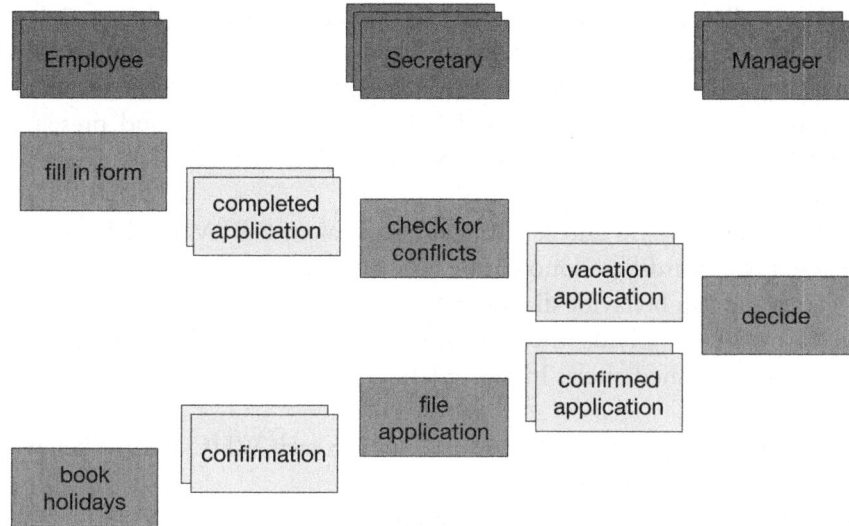

Fig. 3.9 Result of collaborative consolidation

EXCHANGE-elements (e.g., 'form' vs. 'completed application' or 'decision" vs. "confirmed application"). These differences reflect different perceptions of the work process. They are addressed in the next phase, where the individual models are consolidated into a commonly agreed-upon model. This process of consolidation is described in Chap. 4. It results in an interaction-centric model of the perceived overall process of the articulated work case as shown in Fig. 3.9.

3.2.3 Mapping to Subject-Oriented Models

The modeling approach described above has been designed to lead to models that are transformable to models created with role-aware, communication-oriented business process modeling languages such as S-BPM (Fleischmann et al. 2012) or BPMN (White and Miers 2008). The mapping from the card-based model to the target S-BPM business process model is homomorphic (i.e., fully represents the structure of the case-based model in the target S-BPM model). By applying specific transformation rules, the S-BPM model is syntactically correct. Syntactic correctness allows to further process the model with tools designed for S-BPM (cf. Krenn and Stary 2016; Oppl and Rothschädl 2014). The mapping rules are described in the following. Figure 3.10 shows a mapping from a card-based model to an S-BPM model and presents examples of the application of rules given below at the locations of the dashed-outline circles.

Syntactically valid and semantically equivalent S-BPM models can be derived from card-based models by applying the following set of rules:

Creating the behavior diagrams for each identified WHO-element is performed by applying the following rules in the given sequence (cf. numbers in dashed circles in Fig. 3.10):

1. WHO-items map to S-BPM subjects. For each WHO-item, a behavior diagram is created.
2. WHAT-items map to S-BPM function states. For each WHAT-item, an S-BPM function state of the same name is created in the according S-BPM subject. The according S-BPM subject is identified by tracing the imaginary line running vertically through the activity card up to the upper border of the model, where the heading WHO-item corresponds to the according S-BPM subject.
3. Causal relationships between S-BPM function states are identified in the original model by tracing the imaginary line running from heading WHO-item vertically down through the WHAT-items. Two vertically adjacent WHAT-items map to an S-BPM state transition from an S-BPM function state mapping to the upper WHAT-item to the S-BPM function state mapping to the lower WHAT-item.
4. The top-most WHAT-item placed below a WHO-item, maps to the S-BPM start function state of the according S-BPM subject.
5. The lower-most WHAT-item placed below a WHO-item, maps to the S-BPM end function state of the according S-BPM subject, except if the WHAT-item is the origin of a connection to an EXCHANGE-item (see next rule).
6. EXCHANGE-items connected to a WHAT-item by a directed connection originating from the WHAT-item are mapped to an S-BPM send state in the S-BPM subject mapping to the WHO-item to which the WHAT-item belongs. The S-BPM send state is inserted after the

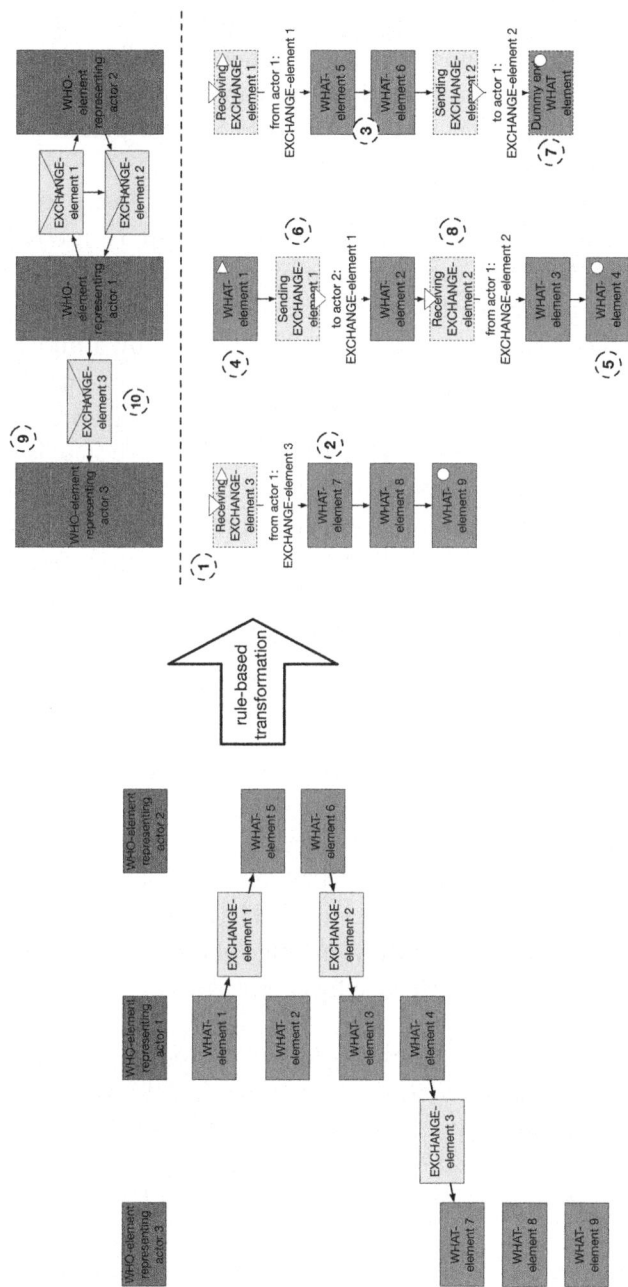

Fig. 3.10 Transformation from card-based to S-BPM model

S-BPM function state representing the originating WHAT-item. The S-BPM send state is named 'sending <name of EXCHANGE-item>'. The S-BPM send state is connected with an outgoing state transition to the S-BPM state that maps to the WHAT-item placed below the originating WHAT-item.
7. If the originating WHAT-item is the last element in its sequence of WHAT-items, an additional S-BPM function state is inserted in the according S-BPM subject as a dummy end function state (as send states cannot terminate the internal behavior of an S-BPM subject).
8. EXCHANGE-items connected to a WHAT-item by a directed connection originating from the EXCHANGE-item are mapped to a S-BPM receive state in the S-BPM subject mapping to the WHO-item the WHAT-item belongs to. The S-BPM receive state is inserted before the S-BPM function state representing the targeted WHAT-item. The S-BPM receive state is named 'receiving <name of EXCHANGE-item>'. The S-BPM receive state is connected with an incoming state transition to the S-BPM state that maps to the WHAT-item placed above the targeted WHAT-item.

The subject interaction diagram is created based on the following two rules:

1. WHO-items map to S-BPM subjects. For each WHO-item, an S-BPM subject of the same name is created.
2. EXCHANGE-items map to S-BPM message elements. For each EXCHANGE-item connecting two WHAT-items assigned to two different WHO-items, an S-BPM message of the same name is created between the according S-BPM subject elements.

The application of these rules introduces additional elements to the S-BPM model, which were not present in the original card-based model. Rules 6 and 8 add send- and receive-states to the S-BPM model. These model elements are only contained implicitly in the card-based model and are derived from the connection points between EXCHANGE- and WHAT-elements. Rule 7 introduces a dummy function state for internal behaviors that would end with a send state. This is necessary as in S-BPM models, the outgoing message information is not attached to the send-state itself, but to the following transition.

Conditional execution of process parts in internal behavior is not considered during transformation, as the card-based models due to their nature as a case-based approach do not support modeling of decisions at all. This semantic limitation is addressed after transformation to a subject-oriented model by refining it via simulated enactment (cf. Oppl 2017). This approach is described in Chap. 5.

3.3 On the Go: Capturing Functions and Interactions While Working

The capturing approaches described above allow for work knowledge representation even by inexperienced modelers and thus enable stakeholder involvement in work design processes. Knowledge capturing, however, does not always necessarily start in dedicated modeling sessions, but might already be triggered during the work process itself. This enables the capture of undistorted models of the actual flows of work that might not be obtainable during ex-post reflective modeling sessions. In-work modeling activities, thus, can provide the basis for follow-up dedicated articulation sessions, but are inherently disruptive to the actual work process (Hoppenbrouwers et al. 2018). We thus propose to mitigate the potential negative impacts of modeling while working by using an instrument that supports process knowledge capturing based on the thinking-aloud method (Van Someren et al. 1994).

Such an instrument needs to be minimally invasive and allow instant capturing and processing of work knowledge. As the articulation process is actor-centric, the instrument in particular needs to be self-contained, in order to encapsulate behavior specific to a subject, and be individual, since each stakeholder should be able to express his/her way of accomplishing tasks. In order to capture both, interactions and functions performed by stakeholders and the systems they work with, it needs to enable encoding technical activities the same way as sending and receiving messages, since they are considered to be of equivalent importance for representing role- or task-specific behavior.

We therefore devise a scheme that closely resembles the representation presented in the former section, but is even more focused in terms of semantics. Actors only distinguish between individually performed activities and interactions with others. These distinct modes of operation during a work process are identified, noted down, and put on a stack. The stack (in reverse order) represents the sequence flow of the activities and interactions of a single stakeholder in a particular instance of a work process. This information can be used as an input for card-based modeling as described in the last section and so provide the foundation for fully specified subject-oriented process models.

Modeling can be performed using physical cards or a tablet application. The latter can further reduce the effort for capturing by providing scaffolds and ad-hoc checks of the consistency of the captured information (Lerchner and Stary 2016). In both cases, stakeholders start articulating by moving a yellow or green card on the heap to the left (cf. Fig. 3.11, 'A'). A card represents a step of a work procedure. Its specifica-

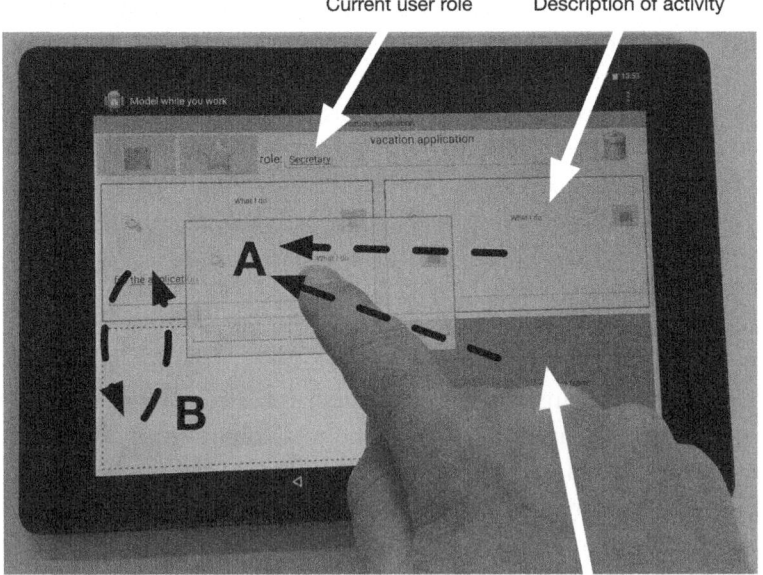

Fig. 3.11 Process capturing

Value-Oriented Articulation 109

tion requires the input of the respective data, in particular, the activity performed (yellow card), or the information to be exchanged, the communication partner, and the direction of interaction (green card).

For each step, an actor needs to decide whether the activity to be modeled is a direct manipulation task, such as calculating data in a customer order form, or is part of an interaction with other actors, for example, contacting a Customer Service Agent to provide further order details before being able to calculate the data.

After providing the data for either card, it needs be moved to 'A', in order to generate a stack of activities. A card can be moved interactively by touching it on the display in the area indicated with the red dot. In case a different card should have been moved to 'A', all the other cards can be put aside, namely moved to 'B', until the intended position is reached in stack 'A'. In this way, re-arranging a set of already captured process steps in the heap is enabled preserving the relevant order.

In this way, a first round of reflection can be supported. This feature becomes relevant, once the introspection of an actor reveals either he/she feels more confident when changing the originally modeled sequence, or once he or she has been forced to perform a certain sequence of steps due to external interference rather than his/her original intention.

In the tablet app, several process models (i.e., stacks) can be edited in parallel, as each user can switch between active modeling sessions by activating another process model. The double arrow symbol on top of the screen indicates that capability.

As context plays a crucial role for representing the behavior of stakeholders, the app enables the storage of context, both, for each activity represented by a card, and for each process model. Both can be enriched with text, audio, images, or video information. Context information may capture background information or additional data for decision making. The input of context information is enabled by the box symbol. It is displayed empty in case no context information has been provided so far, otherwise, it reflects the status as being non-empty.

In order to reduce the effort filling the cards with the required information, the tablet app provides a 'favorite'-function. It enables users to select cards with prepared content from previous modeling sessions for the stack of yellow or green cards on the right side on the main screen.

Once moved to the corresponding stack on the right hand side of the screen, the selected card can be further edited. This functionality is intended for capturing routine tasks and routine communications and works across processes.

3.4 Capturing Tangibles and Intangible Exchange Relationships

We have now presented different approaches to articulation of work knowledge using an actor-oriented and interaction-centric form of representation. Once articulation of work knowledge refers to stakeholders and their patterns of individual and collaborative behavior, we could take a closer look on the type of collaboration in which stakeholders are involved. A fine-grained understanding of interaction patterns could lead to better alignment of functional roles and their encoded work procedures. In the following, we review an articulation approach aiming to reveal both, formal and informal relations between stakeholder roles. They are captured as part of a value network.

3.4.1 Organizations As Transactional Networks of Roles

When introducing Value Network Analysis (VNA), Allee (2008) aimed at developing organizations or networks of organizations beyond the traditional value chain as mentioned in the introduction of this section. Traditional value-chain models represent a linear, if not mechanistic view of business and its operation. Complex constellations of values, however, require analyzing business relationships taking into account the role of knowledge and intangible value exchange as a foundation for value creation. Value exchange needs to be analyzed before changing business transactions in practice. In particular, complex relationships require pre-processing from a value-based perspective, as they influence effectiveness and efficiency, and possible friction in operational processes (ibid.).

VNA is meant to be a development instrument beyond engineering, as it aims to understand organizational dynamics, and thus to govern structural knowledge from a value-seeking perspective, for individuals and the organization as a whole. However, it is based on several fundamental principles and assumptions (Allee 1997, 2002, 2008; Allee et al. 2015):

- Participants of an organization and organizationally relevant network actors participate in a value network by converting what they know, both individually and collectively, into tangible and intangible value that they contribute to the network, and thus to the organization.
- Participants accrue value from their participation by converting value inputs into positive increases of their tangible and intangible assets, in ways that will allow them to continue producing value outputs in the future.
- In such a network, each participant contributes and receives value in ways that sustain both their own success and the success of the value network as a whole. This mutual dependency is a condition sine qua non. Once active participants either withdraw or are expelled, the overall system becomes unstable and may collapse, and need to reconfigure.
- Value networks require trusting relationships and a high level of integrity and transparency on the part of all participants. Then, insights can be gained into interactions by identifying and analyzing not only the patterns of exchange, but rather the impact of value transactions, exchanges, and flows, and thus, the dynamics of creating and leveraging value.
- A single transaction is only meaningful in relation to the system as a whole. It is set by role carriers who utilize incoming deliverables from other role carriers (inputs) and can assess their value, and they realize value which is manifest by generating output.

As network actors—in roles relevant for business—are responsible for handling their relations to others, the organization itself needs to be conceptualized as highly dynamic complex setting. In the following, we detail the underlying concept and methodological approach.

VNA builds upon organizations as self-adapting complex systems. These systems are modeled from that perspective by

1. Identifying patterns of interactions representing tangible and intangible relations between network actor roles
2. Describing these patterns in a structured way, recognizing

    (a) Sources and sinks of information exchanged between network actor roles
    (b) The impact of received information and objects
    (c) The capabilities of produced and delivered assets

3. Elaborating critical processes or exchanges and thus, proposing changes, from both a cognitive perspective and the flow of energy and matter

In line with the living systems perspective, VNA assumes that the basic pattern of organizing business is that of a network of tangible and intangibles exchanges. Tangible exchanges correspond to flows of energy and matter. Intangible exchanges, such as knowledge, point to cognitive processes. Describing a specific set of participating network actors and exchanges allows a detailed description of the structure of any specific organization or a network of organizations.

Although VNA considers as fundamental activity the act of exchange, it goes beyond traditional economic understanding of network actor interactions. Exchange includes goods, services, and revenue, but considers the transaction between network actors also as a representation of organizational intelligence, thus as a cognitive interaction process. Transactions ensure successful task accomplishment and business through cognitively reflected exchanges of information and knowledge sharing, opening pathways for informed decision making. Hence, exchanges do not only have value per se, but also encode the currently available collective intelligence, finally determining the current economic success.

3.4.2 Tangible and Intangible Transactions

Since in VNA knowledge and intangibles exchanges are different to tangible ones, they need to be treated specific to their characteristics. Tangible exchanges include goods, services, and revenue, in particular physical objects, contracts, invoices, return receipts of orders, requests for proposals, confirmations, and payments. They also include knowledge, products, or services that directly generate revenue, or that are expected (contractual) and paid for as a part of a service or good.

Intangible exchanges comprise knowledge and benefits. Intangible knowledge and information exchanges occur supporting the core product and service value chain, but are not contractual. Intangibles are extras network actors in a certain role provide to others to help keep business operations running. For instance, a service organization asks sales experts to volunteer time and knowledge on organizational development, in exchange for an intangible benefit of prestige by affiliation.

Network actors involved in intangible transactions help in building relationships by exchanging strategic information, planning knowledge, process knowledge, technical know-how, and in this way, sharing collaborative design work, performing joint planning activities, and contributing to policy development. Intangibles, like other assets, are increased and leveraged through deliberate actions. They affect business relationships, human competence, internal structure, and social culture. VNA considers intangibles as assets and negotiables that can actually be delivered by network actors engaged in knowledge exchange. They can be held accountable for the effective execution of that exchange, as they are able to articulate them accordingly when following the VNA's structured procedure.

Albeit various attempts to develop new measures and analytical approaches for calculating knowledge assets and for understanding intangible value creation, traditional scorecards need to move beyond considering people as liabilities, resources, or investments. Responsible network actors need to understand how intangibles create value, and most importantly, how intangibles go to market as negotiables in economic exchanges. As a prerequisite, they need to understand how intangibles act as deliverables in key transactions with respect to a given business model.

Value networks represent organizations or network of organizations as a web of relationships that generates tangible and intangible value through transactions between two or more roles. These roles stem from any public or private organization or sector and stand for individuals, groups, entire organizations, or networks. The network, instead of representing hierarchical positions, structures the dynamics of processing and delivering tangibles and intangibles. Although the roles need to the related to the organization at hand, suppliers, partners, and consumers regardless of their physical location, need to become part of the network once they generate value or receive transactional deliverables.

When modeling an organization as value network several assumptions apply (Allee 2008):

- An exchange of value is supported by some mechanism or medium that enables the transaction to happen. As organizations can also be considered socio-technical systems, typical enablers are information and communication technologies. For instance, a sales briefing is scheduled by utilizing some specific web application, such as doodle.com.
- There is provided value: For instance, the provided value of the briefing is based on a tangible exchange of inputs of customer service, and response to inquiries between organizers and participants. The intangibles are targeted news and offerings as well updates on services and customer status (knowledge), and a sense of community (benefit).
- There is return value: For instance, the value in return is efficiency in terms of short handling time of customer requests as tangible, and informed customer request and feedback on latest developments (knowledge), and customer loyalty (benefits) as intangibles.

Value exchanges are modeled in a special type of concept map (Novak and Canas 2006), termed holomap. The VNA mapping from the observed reality to a holomap is based on the following elements:

- Ovals represent functional roles of network actors, termed Participants of the value network, that is, the nodes of the network.
- Participants send or extend deliverables to other Participants. One-directional arrows represent the direction in which the Deliverables are moving during a specific Transaction. The label on the arrow denotes the Deliverable.

When network actors create holomaps, they think of Participants as persons they know carrying out one or more roles in the organizational system at hand. Holomapping is based on the assumption that only individuals or groups of people have the power to initiate action, engage in interactions, add value, and make decisions. Hence, VNA Participants can be individuals, small groups or teams, business units, whole organizations, collectives such as business networks or industry sectors (networked networks), communities, or even nation-states. VNA does not consider databases, software, or other technology as Participant. It is the decision-making capability about which activities to engage in that qualifies only humans as VNA Participants.

Transactions or activities are represented by an arrow that originates with one Participant and ends with another. The arrow represents movement and denotes the direction of addressing a Participant. In contrast to Participants, which tend to be stable over time, Transactions are temporary and transitory in nature. They have a beginning point, a middle, and an end point.

Deliverables are those entities that move from one Participant to another. A Deliverable can be physical or tangible, like a document or a physical object. A Deliverable can also be non-physical, such as a message or request that may only be delivered verbally. It can also be an intangible Deliverable of knowledge about something, or a favor.

In VNA, an exchange only occurs when a Transaction results in a particular Deliverable coming back. A gap is considered in a case when something is provided without anything being received in return. However, focusing on the exchange as the molecular element of value creation is a generic concept that enables capturing a variety of organizations as value networks. Tangible and intangible exchanges establish patterns typical of business relationships. In many cases, tangible exchanges comprise exchanges of matter and energy (goods and money), while the intangible exchanges capture cognitive and emotive exchanges such as favors and benefits.

In the following, we exemplify a VNA case in Sales and Presales from different organizational units of a networked service company providing innovative instruments (methods and technologies) for knowledge acquisition and sharing. Due to a merger with another company, Presales should complement the service chain of the company providing all other services, including Sales. In order to understand the overall patterns of exchange and determining the impact of tangible and intangible inputs for each Participant, the merging companies decided to perform a VNA. It should not only help in analyzing the state of affairs, but also leverage potential changes for each Participant. Sales and Presales aim to improve their ability to utilize operation and customer feedback in further developing their services, although stemming from different organizations.

The first step Participants need to consider in the modeling process are all the roles, organizational units or work groups, both internally and externally, that are considered of relevance in the activities of the Sales and Presales group. In this case, three network actors (Participants) inside the organization, namely Sales, Product Development, and Customer Service, and two network actors, Presales and free-lanced Interviewers of different organizations are identified. They represent the nodes in the holomap in Fig. 3.12.

For modeling, first, network actors need to think about tangible exchanges that take place between the Participants. What are the Transactions adding value? What are the tangible Deliverables in the work system? Figure 3.12 shows tangible Deliverables such as product information, feedback from market, requests, and updates. For these cases, the transaction and communication channel is considered a tangible Deliverable because it either comprises core data relevant for operat-

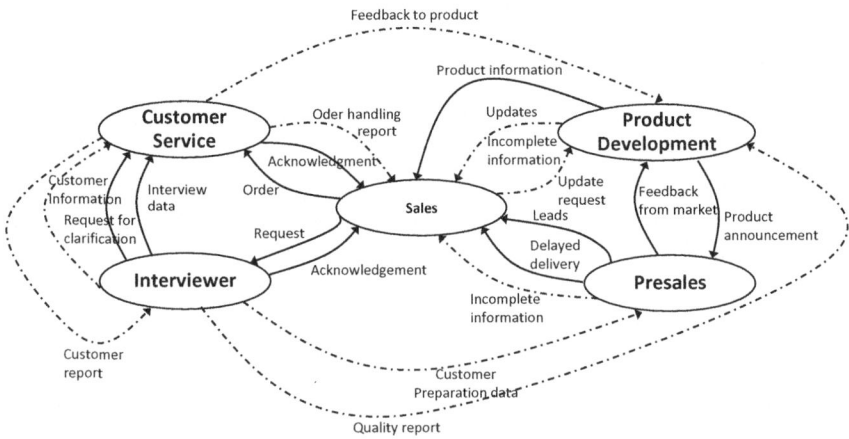

Fig. 3.12 Sample holomap for developing Sales and Presales relations

ing the business, or affects essential relations to organizational units for product and (customer) knowledge management.

Intangible transactions or exchanges are modeled the same way. In order to distinguish the intangible Deliverables from the tangible Deliverables, modelers use a different line style (dotted line in Fig. 3.12). For the original service provider, intangibles are incomplete information, order handling report, customer report, customer preparation data and so on, which various network actors make available through reporting and active sharing of knowledge (see Fig. 3.12). They are considered intangible because there is no direct monetary income related to them. They are neither contracted by the provider nor expected by the recipients. They are extra offerings to Participants to keep the operation running, and product development informed, mainly based on informal learning, and experiential knowledge.

As shown in Fig. 3.12, several tangible exchanges occur, for example, Product Development provides product announcements to Presales in exchange for feedback from the market, Sales provides requests to Interviewers who acknowledge them. In addition, several intangible exchanges occur, such as Interviewers provide customer information to Customer Service in exchange to customer reports. The latter complements the formal, role-specific exchange specified through the pair 'request for clarification—interview data'. It documents the intention to provide a comprehensive picture of customers in order to build trustful relationships to customers (representing the benefits of the exchange).

However, several one-sided transactions with respect to tangibles and intangibles become evident, as also shown in the holomap in Fig. 3.12: For instance, concerning intangibles, the Interviewers provide customer preparation data and quality reports to Presales and Product Development, respectively, without any intangible return. Concerning tangibles, for example, Product Development provides both, product information and updates to Sales without any return.

Once all exchanges and Deliverables are captured in the holomap, a diagram of how the business is perceived from a network actor perspective is established. The value network view of an inter-organizational network helps understand the role of knowledge and intangibles in value creation. The modeling process allows capturing strategically critical intangible exchanges from a network actor perspective, thus, enabling further targeting opportunities for value creation. This issue is addressed through analyzing the value network as represented by the holomap using three different types of analyses and will be discussed in Chap. 5.

3.5 Cross-Cutting Issues

Table 3.1 gives a structured overview on the reviewed techniques of this chapter. It structures each approach according to its

- focus revealing its objective
- understanding of organization as a cognitive construct
- means of representation, in order to document work knowledge
- procedure to follow for articulating work knowledge

Value-based articulation can thereby range from natural language-based documentation to highly structure-determined approaches. They require role-specific behavior recognition and various levels of detail in specifying individual and collective behavior.

Considering the requirements and subsumed procedural cornerstones from Chap. 2, we can reflect on the results of this section in a structured way. The reflection takes into account individual engagement of actors, as well as the activities on the collective level with respect to organizing work. In Table 3.2, we revisit the list of requirements as given in Chap. 2 and elaborate on them according to relevant properties for each presented articulation technique.

From a procedural perspective, subject-oriented articulation can be assigned to the following phases:

1. The preparations require (i) determining the scope of articulation, for example, a specific business case, an organizational structure of work, (iii) identifying the actors as role carriers, since their behavior needs to be specified in terms of subjects, (iii) explaining the subject-oriented notation and tools for articulation, for example, how to structure natural language sentences, paper, pencil, and a diagrammatic editor for

Table 3.1 Value-oriented articulation approaches

	Subject-oriented articulation (as presented in Sect. 3.1)	Card-based elaboration (as presented in Sects. 3.2–3.4)	Value network specification (as presented in Sect. 3.5)
Focus	Integration of functional and interactional activities in a work process	Recognizing interactional requirements for successful handling of work process completion while detailing individual workflows	Integrated representation of formally acknowledged and informally valued transactions exchanged between stakeholders when acting in a specific role
Understanding organization as	Interacting behavior encapsulations	Interaction of individual workflows in specific organizational situations	A Complex Adaptive System
Means of representation	Natural language Subject Interaction Diagram (SID) Subject Behavior Diagram (SBD)	Individual workflow diagram Interaction relations Context diagrams (Concept Maps)	Holomap nodes Holomap tangible relations Holomap intangible relations
Articulation procedure	1. Natural language expressions 2. SID-Specification by Construction by Restriction 3. SBD Specification by refining SID	1. Role identification 1a. Context Specification (optional) 2. Role refinement 3. Individual Workflow Specification (in-situ or ex-post) 4. Interaction Specification	1. Role definitions (node identification) 2. Relation specification Tangible transactions Intangible transactions

Table 3.2 Elicitation requirements and subject-oriented articulation

Elicitation requirement	Subject-oriented articulation
Awareness of role(s) and their management	Roles are constitutive elements of subject-oriented articulation. Articulation of work knowledge requires thinking of a set of communicating actors and their roles. Thereby, each actor can have various roles in work processes. Managing roles is done implicitly initially, as the articulation and arrangement of subjects in the course of articulation lays ground to manage roles. Roles become visible through representing subjects as behavior encapsulations which can be structured according to their modeled refinement and patterns of interaction.
Situation awareness	Since the selection of roles depends on the situation to be modeled, each actor determines subjects as he/she perceives the situation. There is no explicit construct in subject-oriented articulation featuring situations. However, starting with natural language description, each role is refined according to role-specific task activities, thus providing for each action situation-specific context of work.
Conceptual understanding of complex systems	As the ultimate concept is the subject as the entity encapsulating behavior, a system can be composed of an arbitrary set of entities. This can either be achieved by construction adding up to a complex system, that is, adding subject specifications successively, or by reducing complexity, that is removing interactions not relevant for the modeled work context. Since the only connection between subjects are message-passing activities, complex systems can be managed by handling a single type or relationship.
Creating a reflective practice for situations-to-be	Subject-oriented articulation can start either with a situation as-it-is or with a situation to-be, or even with a mixture of existing and envisioned patterns of work. It depends on the person articulating knowledge and his/her mental model-building process. Natural language specification facilitates reflection processes.
Focusing while utilizing multiple perspectives	The target of eliciting work knowledge to becomes more focused when looking through different glasses on work is achieved not only by structuring natural language sentences, but also by the bipartite specification of functional and interactional aspects of role-specific behavior. In case several actors are involved in an articulation session, each actor could represent a specific role to capture different mental models and possible mismatches through incompatible message exchange.

(continued)

Table 3.2 (continued)

Elicitation requirement	Subject-oriented articulation
Articulating intangible assets	Subject-oriented elicitation mainly tackles explicit knowledge, namely how work tasks are accomplished when collaborating with other roles or actors. Actually, many stakeholders are not aware when eliciting knowledge on work processes how often and how much information they exchange with others throughout collaboration. When making these interactions explicit, this knowledge is shifted to the same level of awareness like functional role behavior.
Engage in alignment for collective intelligence	Since subject-oriented articulation assumes a collaborative work setting, involving only a single stakeholder in articulation leads to specifying only a personal perspective on a work process, even when this person plays the role of other subjects. Involving the actual taker of each role provides a more balanced articulation, and meets the requirement of engaging the relevant stakeholders. Since in subject-oriented modeling only five symbols need to be used, the intelligibility of the models by the notation (once the modelers are familiar with it).

documentation, and (iv) providing a facilitator to effectuate the articulation procedure.

2. Situation-sensitive articulation features are subject constellations (represented in Subject-Interaction Diagrams), enabling stakeholders to externalize their knowledge on roles and work tasks as they experience it, from a functional and interactional perspective.
3. *Looking beyond what-is/addressing situations-to-be*: The various points in time to articulate work knowledge allow flexible application of the approach. Facilitation should encourage stakeholders to revisit existing patterns, to rethink the role assignments to subjects, and to generate novel patterns of work interaction capturing situations of relevance. The facilitator can develop proposals to trigger modification of existing models.
4. *Representational alignment*: The subject-oriented notation and specification scheme enables consolidating individual perceptions and specifying interaction patterns. Both enable stakeholders to change patterns

of behavior, either internally for each subject, or externally by redefining interaction patterns.
5. Organizational alignment can be achieved through consensus-finding among the concerned subject carriers or stakeholders, once elicited knowledge is represented and discussed how to embody the generated knowledge in the workspace of the organization at hand.

Table 3.3 discusses the requirements with respect to card-based elaboration.

From a procedural perspective, card-based elaboration comprises several phases:

1. A preparation of the setting, actors, and instruments. In case of card-based articulation this step includes (i) determining the scope of elicitation, which is mainly a business case involving several stakeholders, (ii) a physical surface and/or digital media support as articulation environment, (iii) the cards, paper, markers, and/or building blocks as tangible material, and (iii) actors willing to articulate role-specific behavior for the selected business case, and engage in sharing and reflecting on the underlying mental models, (iv) a facilitator to guide the articulation procedure and introduce the corresponding material and environment(s).
2. Situation-sensitive articulation features comprise the (physical) surface as articulation environment, the notational elements (cards, paper, markers, building blocks, relations) in order to describe and document work knowledge in the course of articulation.
3. Facilitation is required (i) to set the stage involving stakeholders as role carriers, (ii) to ensure the correct use of notational elements, and (iii) to identify situation correspondence (as-it-is, to-be), and (iv) tutor the use of (digital) media.
4. Representational alignment might need to be facilitated when the participants aim to consolidate their findings into a shared representation.
5. Organizational alignment needs to be documented when the participants envision how elicited work knowledge should become part of future organizational designs.

Table 3.3 Elicitation requirements and card-based elaboration

Elicitation requirement	Card-based elaboration
Awareness of role(s) and their management	Roles constitute the lines of articulation and representation along the elicitation process. Although articulation of work knowledge targets towards aligning the interaction among role carriers, it primarily helps externalizing the functional flow of operation per role (i.e., line of articulation). For each role in a work process, an actor taking this role specifies functional behavior before detailing the interaction with other roles. Role management is started with determining the various lines denoting role behavior over time, and might lead to re-arrangements of roles or their behavior. Using digital tools, re-arrangements can be facilitated.
Situation awareness	The role- or task-specific activities are framed by information of the situation an actor is part of, as the selection of roles depends on the situation to be modeled. There is no explicit modeling construct for a situation in card-based articulation—it remains implicit by the selected work behavior specifications: Each role is refined to specific task activities. Context can be provided through an additional modeling step, in which the relevant situational influence factors are identified and represented in a concept map. Alternatively, modeling can happen in-situ with appropriate tool support, implicitly contextualizing the articulation process.
Conceptual understanding of complex systems	It is a linear (for the roles) and networked (by interaction between roles) articulation procedure. Hence, each line representing a role is continuously developing in the course of articulation. The model might easily exceed the physical limits of the modeling surface and thus, need to be re-arranged by re-sorting the cards, if not mapped to digital media and being encapsulated. When modeling interactions between roles and crossing lines of other roles, numbering interactions helps in identifying the correct entry points for information exchange on the physical surface. Overall, a system could be composed of an arbitrary set of roles and information exchanges. Sequences of task activities can be encapsulated with the help of digital support.

(continued)

Table 3.3 (continued)

Elicitation requirement	Card-based elaboration
Creating a reflective practice for situations-to-be	When continuously placing cards, actors become aware of how they accomplish tasks in certain roles, both, from a functional and interactional perspective. Their explicit work knowledge can address a situation as-it-is, or a situation to-be, or transitions from existing to an envisioned organization of work. When engaging in collaborative articulation settings, they develop a shared understanding of their individual perspectives through collaborative reflection.
Focusing while utilizing multiple perspectives	In case the articulation involves several actors, each representing a specific role, eliciting work knowledge becomes focused when looking through the role perspective for each actor while recognizing the other roles both, from a functional and interactional perspective. Mismatches become evident on the boundaries of behavior specifications.
Articulating intangible assets	Card-based elaboration has its focus on task-aware behavior in a certain situation, thus explicit knowledge on work. In the course of articulating this knowledge, actors could become aware of implicit knowledge, for example, when interaction patterns with other actors are elicited. In those cases, implicit knowledge becomes explicit.
Engage in alignment for collective intelligence	Since articulating role carriers are considered an integral part of a work organization, they model their task behavior by describing task activities in the interaction context with other actors. Their engagement is bound to the modeling task and the willingness to describe their activities in an intelligible way while documenting them on the cards. The set of cards and adjacent color scheme should facilitate sharing and communicating the documented knowledge.

Finally, we discuss how the requirements are addressed in value network specification in Table 3.4.

From a procedural perspective, the elicitation phases are instantiated as follows:

1. The preparation comprises the setting, actors, and instruments. The scope is a subset of a specific business operation, usually a core business case, and some physical space or digital tool as articulation envi-

Table 3.4 Elicitation requirements and value network-based articulation

Elicitation requirement	Value network specification
Awareness on role(s) and their management	Since value networks are constituted by role definitions, the participants need to be aware of the roles to be represented in a holomap. Once the set of roles is specified, roles can only managed by manipulating their (tangible or intangible) exchange patterns with other roles.
Situation awareness	Roles are framed by interaction patterns with other roles, thus constituting the situation to be at the center of articulation. The situation of interest determines the selection of roles. The notation does not contain an explicit element for denoting a situation—it is given by the selected set of interacting roles.
Conceptual understanding of complex systems	The complexity of a system of interest is given by the selected situation, thus by a set of roles (establishing the value network) and their interaction patterns constituting the network. The concepts addressed in the course of reflection and changes are the interactions which may change their quality (intangible or tangible) and appearance from existing to envisioned transactions between roles.
Creating a reflective practice for situations-to-be	The mental modeling is focused on reflecting an existing network of interacting roles. They lay ground for articulating future interactions. When placing transactions between roles, participants consider formal and informal relations observed in the course of operating a business as it is.
Focusing while utilizing multiple perspectives	After determining the roles acting in a selected situation, the participants focus on the type of interactions. Since these can be either tangible or intangible, two perspectives are taken in the course of articulating work knowledge: a formal and an informal.
Articulating intangible assets	Setting up a value network of interacting roles allows taking into account intangible transaction between roles. They correspond to informal relations between persons, usually in order to keep the business operation running smoothly aside formal regulations.
Engage in alignment for collective intelligence	Value networks can be set by individuals or groups of participants. In the latter case, they need to agree on a common set of network nodes (i.e., roles) and check the intelligibility of labels for them and their transactions.

ronment. The participants need to be willing to articulate formal and informal work knowledge, and engage in common reflection, eventually guided by a facilitator explaining the topology, node, and relation types.
2. Situation-sensitive articulation features are roles and their formal and informal relations, termed tangible and intangible transactions. Value network can be designed individually or in a shared environment involving different people externalizing their knowledge on roles and their interactions.
3. Facilitation includes encouraging stakeholders to look beyond well-known connections between role carriers, besides explaining the network topology, nodes (i.e., actors), and relation types representing deliverables.
4. Representational alignment is only required in case a consolidated network representation needs to be achieved for further development.
5. Organizational alignment is not concerned as the network representation is constructed for a situation as-it-is.

Cross-checking the presented articulation technique, each of the presented techniques has its focus on role-specific behavior and allows representing interaction patterns between roles. For complex networked systems, subject-oriented elicitation provides a comprehensive while structured way to approach elicitation by human-centered means, as it starts with natural language before focusing on a dual representation of work knowledge. Card-based elaboration follows the same line, but might be limited to physical constraints when being performed without digital support—a dilemma it shares with subject orientation in case the granularity and choice of media is not appropriated to both. Value networking follows a declarative perspective throughout articulation, in contrast to subject-oriented and card-based elicitation. As a result, the exchange patterns refer to work deliverables likely subsuming the data-driven exchange of subject-oriented or card-based elicitation. In addition, intangible assets can be represented in addition to tangible ones, whereas subject-oriented or card-based elicitation mainly target explicitly encoded ones. In terms of articulation procedure, subject-orientation and value networking start from an interactional perspective and (in the case of subject-orientation) detail on behavioral implications of interactions in a subsequent step. Card-based modeling initially focuses on individual behaviors contributing to an overall work process and derives interactions in a follow-up step when matching the mutual dependencies encoded in individual behavior models.

References

Allee, Verna. 1997. *The Knowledge Evolution: Expanding Organizational Intelligence*. Butterworth-Heinemann.

———. 2002. *The Future of Knowledge: Increasing Prosperity Through Value Networks*. Butterworth-Heinemann.

———. 2008. Value Network Analysis and Value Conversion of Tangible and Intangible Assets. *Journal of Intellectual Capital* 9 (1): 5–24 (Emerald Group Publishing Limited).

Allee, Verna, Oliver Schwabe, and Marilyn Krause Babb. 2015. *Value Networks and the True Nature of Collaboration*. Meghan-Kiffer Press.

Arias, E.G., and G. Fischer. 2000. Boundary Objects: Their Role in Articulating the Task at Hand and Making Information Relevant to It. In *Proceedings of the International ICSC Symposium on Interactive and Collaborative Computing (ICC'2000)*.

Bach, Volker. 2000. Business Knowledge Management: Wertschöpfung Durch Wissensportale. In *Business Knowledge Management in Der Praxis*, 51–119. Springer.

Britton, Carol, and Sara Jones. 1999. The Untrained Eye: How Languages for Software Specification Support Understanding in Untrained Users. *Human-Computer Interaction* 14 (1–2): 191–244. https://doi.org/10.1080/07370024.1999.9667269.

Brocke, vom Jan, S. Zelt, and T. Schmiedel. 2015. Considering Context in Business Process Management: The BPM Context Framework. *BPM Trends* 1: 272.

Brocke, vom Jan, Sarah Zelt, and Theresa Schmiedel. 2016. On the Role of Context in Business Process Management. *International Journal of Information Management* 36 (3): 486–495 (Elsevier).

Davies, Islay, Peter Green, Michael Rosemann, Marta Indulska, and Stan Gallo. 2006. How Do Practitioners Use Conceptual Modeling in Practice? *Data & Knowledge Engineering* 58 (3): 358–380 (Elsevier).

Fahland, Dirk, and Matthias Weidlich. 2010. Scenario-Based Process Modeling with Greta. *BPM Demos*. Citeseer, 52–57.

Fjuk, A, and L Dirckinck-Holmfeld. 1997. Articulation of Actions in Distributed Collaborative Learning. *Scandinavian Journal of Information Systems* 9 (2): 3–24 (Unknown).

Fleischmann, Albert, and C. Stary. 2012. Whom to Talk to? A Stakeholder Perspective on Business Process Development. *Universal Access in the Information Society* 11: 125–150 (Springer Berlin/Heidelberg). https://doi.org/10.1007/s10209-011-0236-x.

Fleischmann, Albert, Werner Schmidt, C. Stary, Stefan Obermeier, and Egon Börger. 2012. *Subject-Oriented Business Process Management*. Springer.

Frederiks, P.J.M., and Th.P. van der Weide. 2006. Information Modeling: The Process and the Required Competencies of Its Participants. *Data & Knowledge Engineering* 58 (1): 4–20. https://doi.org/10.1016/j.datak.2005.05.007.

Genon, Nicolas, Patrick Heymans, and Daniel Amyot. 2011. Analysing the Cognitive Effectiveness of the BPMN 2.0 Visual Notation. In *Software Language Engineering*, Lecture Notes in Computer Science, vol. 6563, 377–396. Berlin and Heidelberg: Springer Berlin Heidelberg. https://doi.org/10.1007/978-3-642-19440-5_25.

Groeben, Norbert, and Brigitte Scheele. 2000. Dialogue-Hermeneutic Method and the 'Research Program Subjective Theories'. *Forum: Qualitative Social Research* 1 (2).

Han, Byung-Chul. 2015. *Im Schwarm: Ansichten Des Digitalen*. Matthes & Seitz Berlin Verlag.

Herrmann, Thomas, and Alexander Nolte. 2014. Combining Collaborative Modeling with Collaborative Creativity for Process Design. In *COOP 2014—Proceedings of the 11th International Conference on the Design of Cooperative Systems, 27–30 May 2014, Nice (France)*, 377–392. Cham: Springer International Publishing. https://doi.org/10.1007/978-3-319-06498-7_23.

Hjalmarsson, Anders, Jan C. Recker, Michael Rosemann, and Mikael Lind. 2015. Understanding the Behavior of Workshop Facilitators in Systems Analysis and Design Projects: Developing Theory from Process Modeling Projects. *Communications of the AIS* 36 (22): 421–447.

Hoppenbrouwers, S., I. Wilmont, D. van Loon, T. van der Geest, and S. Oppl. 2018. Measuring Process Experience: A Collaborative Modelling Instrument for Determining the Impact of a New Law on Public Service Experience. In *Proceedings of the 10th International Conference on Subject-Oriented Business Process Management*, 13. ACM.

Jamshidi, Mo. 2008. *System of Systems Engineering*. Hoboken, NJ: John Wiley & Sons, Inc.

Kabicher, Sonja, and Stefanie Rinderle-Ma. 2011. Human-Centered Process Engineering Based on Content Analysis and Process View Aggregation. In *Advanced Information Systems Engineering. CAiSE 2011*, ed. H. Mouratidis and C. Rolland, 467–481 (Chap. 35). Lecture Notes in Computer Science, vol. 6741. Berlin and Heidelberg: Springer Berlin Heidelberg. https://doi.org/10.1007/978-3-642-21640-4_35.

Krenn, Florian, and Christian Stary. 2016. Exploring the Potential of Dynamic Perspective Taking on Business Processes. *Complex Systems Informatics and Modeling Quarterly*, no. 8: 15–27. https://csimq-journals.rtu.lv/issue/view/87.

Krogstie, J., Guttorm Sindre, and H.D. Jørgensen. 2006. Process Models Representing Knowledge for Action: A Revised Quality Framework. *European Journal of Information Systems* 15 (1): 91–102. https://doi.org/10.1057/palgrave.ejis.3000598.

Lai, Han, Rong Peng, and Yuze Ni. 2014. A Collaborative Method for Business Process Oriented Requirements Acquisition and Refining. In *Proceedings of ICSSP 2014*, 84–93. New York: ACM Press. https://doi.org/10.1145/2600821.2600831.

Lerchner, H., and C. Stary. 2016. Model While You Work: Towards Effective and Playful Acquisition of Stakeholder Processes. In *Proceedings of the 8th International Conference on Subject-Oriented Business Process Management*, 1. ACM.

Muehlen, zur Michael, and J.C. Recker. 2008. How Much Language Is Enough? Theoretical and Practical Use of the Business Process Modeling Notation. In *Advanced Information Systems Engineering. CAiSE 2008*, Lecture Notes in Computer Science, vol. 5074, ed. Z. Bellahsène and M. Léonard, 465–479. Berlin and Heidelberg: Springer Berlin Heidelberg. https://doi.org/10.1007/978-3-540-69534-9_35.

Neubauer, Matthias, and Christian Stary. 2017. *S-BPM in the Production Industry*. Springer.

Novak, Joseph, D., and A.J. Canas. 2006. *The Theory Underlying Concept Maps and How to Construct Them*. Florida Institute for Human and Machine Cognition.

Nüttgens, M., and F.J. Rump. 2002. Syntax Und Semantik Ereignisgesteuerter Prozessketten (EPK). *Promise*, 64–77.

Oppl, Stefan. 2015. Articulation of Subject-Oriented Business Process Models. In *Proceedings of S-BPM ONE 2015*, 1–11. New York: ACM Press. https://doi.org/10.1145/2723839.2723841.

———. 2016. Towards Scaffolding Collaborative Articulation and Alignment of Mental Models. *Procedia Computer Science* 99: 124–145. https://doi.org/10.1016/j.procs.2016.09.106.

———. 2017. Business Process Elaboration Through Virtual Enactment. In *Proceedings of S-BPM ONE 2017*. New York: ACM Press. https://doi.org/10.1145/3040565.3040568.

———. 2018. Which Concepts Do Inexperienced Modelers Use to Model Work?—An Exploratory Study. In *Proceedings of MKWI 2018*.

Oppl, Stefan, and Stijn Hoppenbrouwers. 2016. Scaffolding Stakeholder-Centric Enterprise Model Articulation. In *The Practice of Enterprise Modeling*, Lecture Notes in Business Information Processing, vol. 267, 133–147. Springer International Publishing. https://doi.org/10.1007/978-3-319-48393-1_10.

Oppl, Stefan, and Thomas Rothschädl. 2014. Separation of Concerns in Model Elicitation—Role-Based Actor-Driven Business Process Modeling. In *S-BPM ONE—Setting the Stage for Subject-Oriented Business Process Management*, Communications in Computer and Information Science, vol. 422, ed. Hagen Buchwald, Albert Fleischmann, Detlef Seese, and Christian Stary, 3–20. Springer International Publishing. https://doi.org/10.1007/978-3-319-06191-7_1.

Oppl, Stefan, Christian Stary, and S. Vogl. 2017. Recognition of Paper-Based Conceptual Models Captured Under Uncontrolled Conditions. *IEEE Transactions on Human-Machine-Systems* 47 (2): 206–220. https://doi.org/10.1109/THMS.2016.2611943.

Recker, J.C., and A. Dreiling. 2007. Does It Matter Which Process Modelling Language We Teach or Use? An Experimental Study on Understanding Process Modelling Languages Without Formal Education. In *18th Australasian Conference on Information Systems*.

Rittgen, Peter. 2009. Collaborative Modeling of Business Processes: A Comparative Case Study. In *Proceedings of the 2009 ACM Symposium on Applied Computing*, 225–230. New York: ACM Press. https://doi.org/10.1145/1529282.1529333.

Santoro, Flávia Maria, Marcos R.S. Borges, and José A. Pino. 2010. Acquiring Knowledge on Business Processes From Stakeholders' Stories. *Advanced Engineering Informatics* 24 (2): 138–148. https://doi.org/10.1016/j.aei.2009.07.002.

Scholz, M., and A. Holl. 1999. Objektorientierung Und Poppers Drei-Welten-Modell Als Theoriekerne Der Wirtschaftsinformatik. In *Wirtschaftsinformatik*

Und Wissenschaftstheorie—Grundpositionen Und Theoriekerne, ed. R. Schütte. Universität Essen.

Stamper, Ronald. 1996. Signs, Information, Norms and Systems. In *Signs of Work: Semiosis and Information Processing in Organisations*, 349–397. Berlin and New York: De Gruyter.

Stary, Christian. 2017. System-of-Systems Design Thinking on Behavior. *Systems* 5 (1): 3 (Multidisciplinary Digital Publishing Institute).

Stary, Chris, and Dominik Wachholder. 2015. System-of-Systems Support—A Bigraph Approach to Interoperability and Emergent Behavior. *Data & Knowledge Engineering* 105: 155–172.

Van Someren, M.W., Y.F. Barnard, and J.A.C. Sandberg. 1994. *The Think Aloud Method: A Practical Guide to Modelling Cognitive Processes*. Academic Press.

White, S.A., and D. Miers. 2008. *BPMN Modeling and Reference Guide: Understanding and Using BPMN*. Future Strategies Inc (*The Journal of Strategic Information Systems*, vol. 3). https://doi.org/10.1016/0963-8687(94)90004-3.

Open Access This chapter is licensed under the terms of the Creative Commons Attribution 4.0 International License (http://creativecommons.org/licenses/by/4.0/), which permits use, sharing, adaptation, distribution and reproduction in any medium or format, as long as you give appropriate credit to the original author(s) and the source, provide a link to the Creative Commons licence and indicate if changes were made.

The images or other third party material in this chapter are included in the chapter's Creative Commons licence, unless indicated otherwise in a credit line to the material. If material is not included in the chapter's Creative Commons licence and your intended use is not permitted by statutory regulation or exceeds the permitted use, you will need to obtain permission directly from the copyright holder.

4

Alignment of Multiple Perspectives: Establishing Common Ground for Triggering Organizational Change

This chapter introduces methodological support for transitioning from as-is to to-be work processes via direct actor involvement. It suggests direct actor involvement in the alignment and validation of novel work practices, in particular when digital workflows or instruments are involved that fundamentally impact the modes of individual operation and collaboration.

Alignment is required for consolidating various inputs for further processing. In particular, actively involving process participants in process modeling creates a challenge for consolidated digital work design. Process participants are not expected to have modeling skills, and usually, as also stated in Prilla and Nolte (2012), they are not willing to learn a modeling language with a strict syntax and semantics and many different symbols. What they would prefer would be to externalize their knowledge through diagrams that are as simple as possible in terms of both syntax and semantics. As we have already argued for in Chap. 1, this desire calls for supporting 'natural modeling' processes (Zarwin et al. 2014). Such natural modeling processes are usually collaborative and focus on knowledge externalization, sharing, and negotiation of a common understanding about the topic of modeling. In the course of modeling, if appropriately supported and facilitated, alignment processes are carried out. This alignment leads to accommodation of novel perspectives on a work process according to the participants' individual mental models, eventually causing the development of common ground (Convertino et al. 2008). Such common ground is a necessary prerequisite for informed design of to-be work processes and their implementation in organizational practice.

The purpose of this chapter is to introduce alignment from a conceptual and methodological perspective, referring to the gap resulting from enabling natural modeling practices, while at the same time, maintaining a well-defined bridge towards techno-centric (formal) modeling. Through the adoption of natural modeling principles, we present an approach called CoMPArE/WP (Collaborative Model Articulation and Elicitation of Work Processes). It achieves effective involvement of process participants and supports consolidating elicited work knowledge. Effectiveness in this context refers to the extent the participants are facilitated in externalizing their tacit knowledge and reflect on the business process model based on that knowledge. Effectiveness also refers to the acceptance of the approach by the participants. CoMPArE/WP builds upon the card-based articulation method introduced in Chap. 3 and deals with the transition of the model developed by process participants to a techno-centric process model, meaning that it can be processed and enacted using a Business Process Management System (BPMS), paving the way to acting on new work designs, which we elaborate on in Chap. 5.

After introducing fundamental alignment principles, the discrete components of the CoMPArE/WP approach are analytically described. Finally, an illustrative case is presented as proof of concept.

4.1 Alignment Concept and Principles

Alignment was introduced as an issue relevant in management by Kaplan and Norton (1996) based on their book and concepts of the Balanced Scorecard. It was considered novel to strategy management, tackling the adjustment of strategy with organization and management processes which is considered hard to achieve. It provides structured support to meet the need already recognized in 1982 for the alignment of corporate strategy with structure, systems, staff, style (culture), skills, and shared values (Peters and Waterman 1982).

Although Kaplan and Norton targeted strategic management, their plan of action and process for alignment is quite comprehensive. They suggest a thorough diffusion of adjustment activities into an organization, when suggesting the development of strategy maps and balanced scorecards ranging from corporate office to customers and suppliers, and addressing the variety of intermediate organizational units. Hence, alignment can be considered to be omnipresent, going beyond linking financial, customer, internal and learning, and growth objectives.

As an organizational communication device, an alignment process is supposed to be implemented in a variety of management activities, such as handling project meetings and multi-faceted development planning—whenever value should be created beyond what individual perspective or technical units could achieve on their own. However, as this process is supposed to be driven by specific interests due to various stakeholders that need to be involved for overall benefit generation, it requires particular management and facilitation skills and techniques. They comprise tackling complexity explicitly when required, in particular when trying to eschew complexity for simplicity. They also need to emphasize organizational adaptation capabilities to change through interventions around finance, customers, and people—how they organize their work and accomplish business tasks.

Several approaches have been made to provide technological support for that alignment process. In Business Process Management, the semantic heterogeneity between business processes has been addressed. Alignment has been focused on business ontologies for integration (Jung 2009; Fan et al. 2016). Two types of alignment processes are researched, namely, manual alignment for building a comprehensive business process ontology in a business process management (BPM) system, and automated alignment between business processes stemming from different BPM systems. Automation support is based on detecting the optimal integration of a business process into another has to be discovered, in order to maximize the summation of a set of partial similarities between semantic components consisting of the business processes.

An ontology (i.e., specification of a concept) captures knowledge with terms, definitions, and axioms related to a specific domain while representing real-world phenomena. Besides choosing a proper notation for representing the domain, an ontology aims towards improving the understanding of phenomena represented through the notation and clarifying (ambiguous) semantics. Process models could make use of ontologies for checking whether a process model covers completely the constructs in business process ontology, and thus, measuring whether a process model clearly represents the real-world phenomena.

Semantic ambiguities result from domain knowledge and its development processes. They could lead to cognitive overload and finally, inaccurate models. Domain ontologies could help ease semantic ambiguity, reducing cognitive load. When modelers use ontologies to represent domain knowledge for business process management, they define the semantics of existing business process models for process model verification and automation (Jung 2009). They also could use it to ground business process models on domain knowledge (Fan et al. 2016).

Design ambiguities in process modeling are structural or semantic:

- Structural ambiguity refers to the notation or language used for modeling, when lacking a formal definition of modeling constructs.
- Semantic ambiguity occurs when a model does not represent the business logic correctly, as it is supposed to be.

The latter is caused by the lack of accurate mapping between two items or concepts, one stemming from observing reality and the other from a formal or visual representation scheme. Consider a holiday approval process of an organization. A holiday application could be handled in several steps:

1. the responsible verifies the validity of application
2. the human resource department verifies the vacancy contingent of the applicant

In case the modeler is not clear about the granularity of work process to be represented (domain knowledge), the developed model could represent a holiday approval as single activity including both verifications. In such cases, a modeler needs to align the representation with domain experts to ensure correct models.

In order to automate alignment processes, semantic components could be extracted from annotations of business process representations (Jung 2009; Lin and Krogstie 2012). Figure 4.1 shows a potential architecture of such an approach. An ontology-based BPM system is composed of a resource repository. It contains resources, such as documents, videos, and so on, and business processes or service APIs. The latter process the resources, while ontologies as additional resource serve as a point of reference or baseline for clarification. Ontology-based BPM systems semantically describe their resources and business processes to support execution of processes.

When using ontologies through semantic annotations of business process, meaning can be shared with other participants; this alignment can be supported constructively. Thereby, concepts and items from other ontologies or additional work knowledge can be brought in when considered relevant for the participating stakeholders. Since the relations of concepts in ontologies can be quite heterogeneous, several adjustments could occur (Jung 2009):

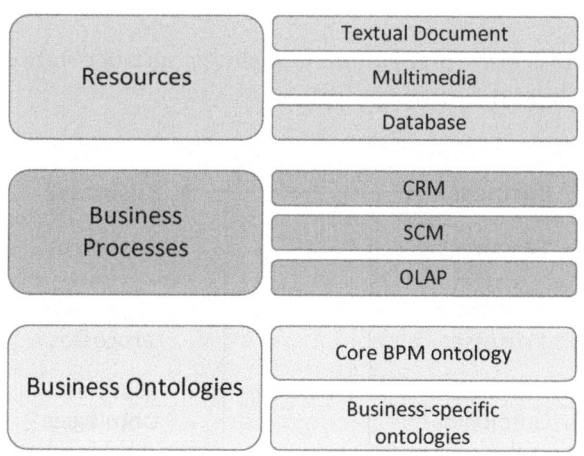

Fig. 4.1 Architecture of ontology-based BPM systems (adapted from Jung 2009)

- Lexical heterogeneity may occur due the labeling, for example, when concepts are named in different ways while being consistent with respect to semantics, for example, Human Resource Department is also termed HR.
- Structural heterogeneity occurs once relations between two concepts differ.
- Conceptual items are missing.

One way to automatically couple even heterogeneous BPM systems is to align business process representations for the sake of semantic interoperability. Figure 4.2 shows the layered approach when developing an ontology matching algorithm. It aims at discovering semantic correspondences between model elements, such as activities.

Figure 4.3 shows exemplary results of ontology-based alignment. When merging parts of ontologies, manual alignments between fragments (ref dotted line) and annotations (arrows) for a work process need to be set.

When the objective of ontology development is to come up with developing a process ontology allowing the resolution of semantic ambiguities along business process modeling, approaches such as the Process Ontology Based Approach (Fan et al. 2016) are helpful. It has been designed to help reducing semantic ambiguity by avoiding cognitive overload. It tackles

- construct overload, that is, one modeling construct stands for two or more ontological constructs, and

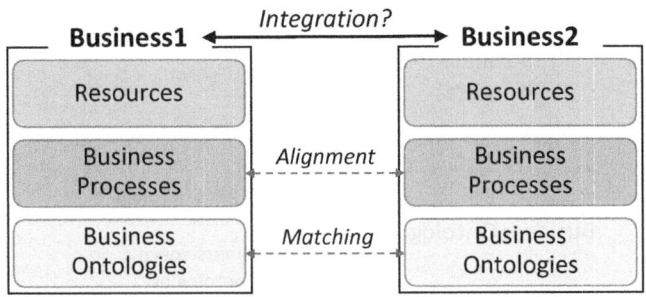

Fig. 4.2 Ontology-based alignment (adapted from Jung 2009)

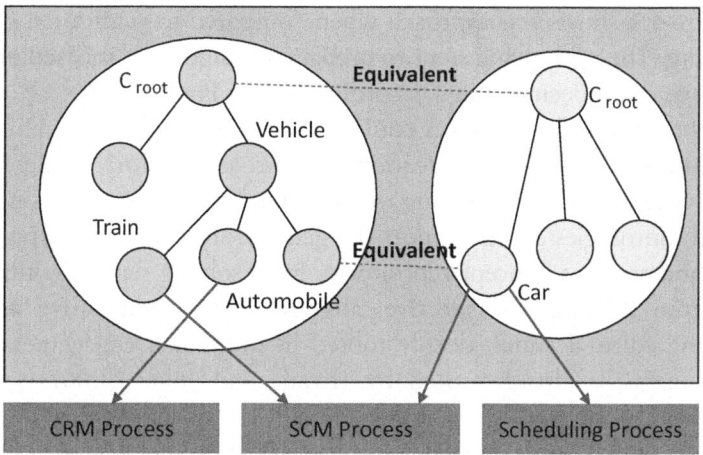

Fig. 4.3 Alignment through merging ontology fragments (adapted from Jung 2009)

- construct redundancy—more than one modeling construct is used to represent a single ontological construct

When creating a systematic approach to ontology-based modeling in business process management that reduces semantic ambiguity, classes (i.e., various types of terms and relationships) for domain process ontology need to be formally defined to guide business process modeling. Relationships can be differentiated according their validity across different domains or not, referring to domain-specific relations. For business process modeling, different types of domain relationships have been identified (Fan et al. 2016):

- Activity-performing relations which connect two roles involved in an activity performed by one of the roles, for example, a customer sending a request to customer service.
- Temporal relations denoting sequencing activities performed by a role, for example, an employee goes on vacation after the superior's approval.
- Conditional relations as they specify conditions when performing specific activities for a role, for example, vacations require superior approval.

Figure 4.4 shows the approach when compared to traditional process modeling. The underlying steps to prepare for alignment is based on validations and has been detailed accordingly—see Fig. 4.5.

A first empirical evaluation could demonstrate that the domain process ontology, although its creation requires some effort, could reduce cognitive effort while enhancing the perceived quality of process models. Overall, ontologies could support the generation of accurate representations, and serve as concept repositories for resolving semantic ambiguities. From a human perspective, alignment is a cooperative activity, involving cultural issues, deeply rooted in individual engagements and mindfulness, as already noted by (Evans and Jukes 2000) (see also Fig. 4.6).

Although existing approaches aim to consolidate elicited work knowledge, they require explicit ontology building. In the following, we introduce a support technique allowing direct consolidation (in line with direct manipulation), thus, reducing the semantic distance between actors (modelers) and the content to be consolidated.

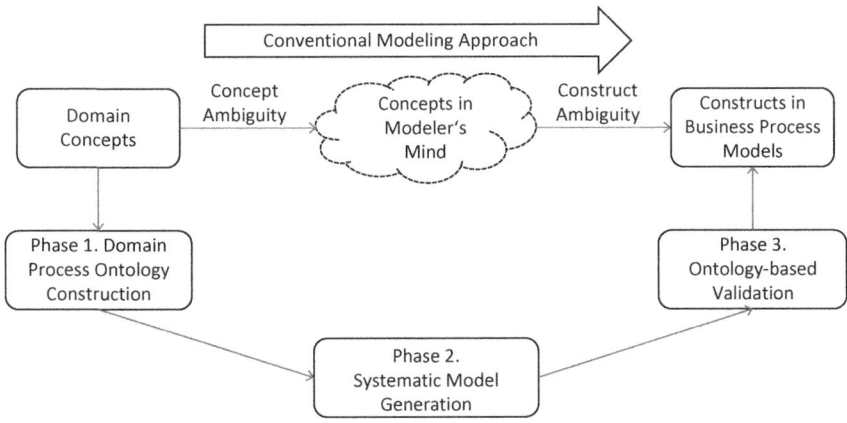

Fig. 4.4 Facilitating resolving semantic ambiguities in process modeling based on ontologies according to Fan et al. (2016)

Alignment of Multiple Perspectives: Establishing Common... 141

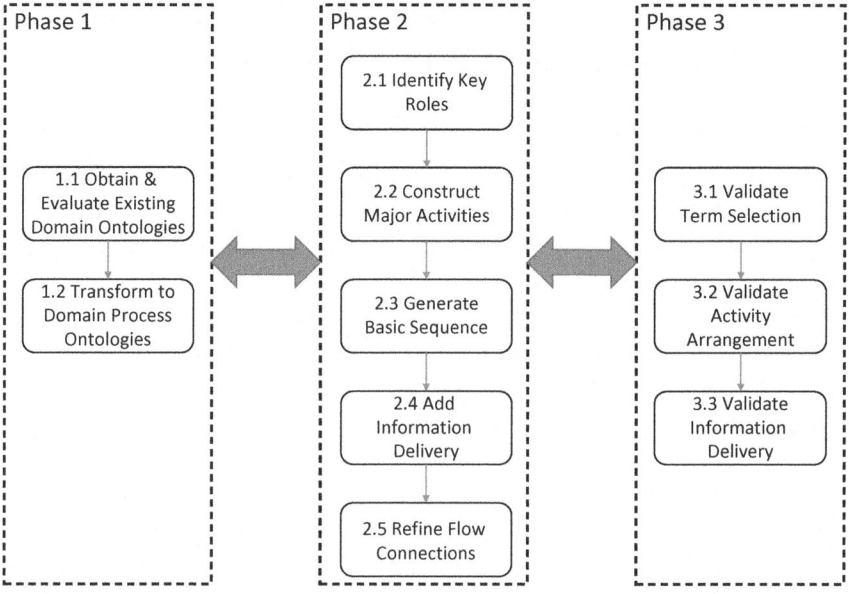

Fig. 4.5 Developing a domain process ontology instance (according to Fan et al. 2016)

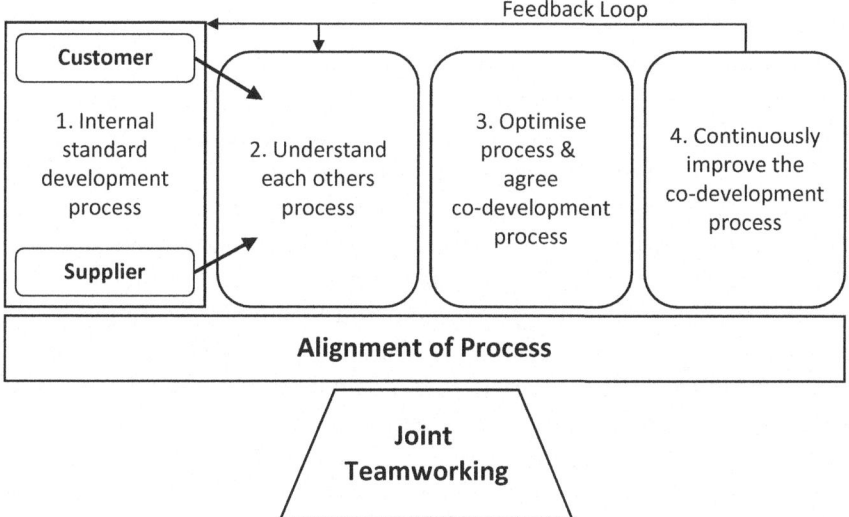

Fig. 4.6 Alignment of business processes as part of co-developing organizations

4.2 Towards Direct Stakeholder Support—Minimizing Semantic Distance

In this section, we first review participatory elicitation approaches and detail the instrument, CoMPArE/WP, which aims to minimize structure inputs for modeling, while guiding actors to express their mental work models and process understanding collaboratively using scaffolds. As in participatory design (Kensing and Blomberg 1998), in the proposed instrument, actors are actively involved in process design. In this procedure, actors are guided by the process analyst who acts mainly as a facilitator. Modeling is not performed sitting in front of a PC screen and using some kind of software for process modeling. Instead, participants hold cards with different colors which are assigned specific semantics during the design procedure. Like in card sorting (J. R. Wood and Wood 2008), participants create structures using the cards. Employing tangible means to conduct process modeling has already been proposed in the literature (Weske and Luebbe 2011). Using tangible means like cards instead of sophisticated software also allows technologically illiterate actors or, in general, actors who do not feel comfortable with technology to take part in modeling and, overall, makes modeling more enjoyable and appealing to modeling participants.

Participatory Design is also the foundation of the work of Türetken and Demirörs (2011), who propose a decentralized process elicitation approach ('Plural') in which individuals describe their own work. Plural is based on a multi-perspective modeling paradigm (Mullery 1979), which focuses on representation of individual work contributions in models and subsequently merges them into a common model by agreeing on the interfaces among the individual models. It uses Extended Event-driven Process Chain (eEPC) (Nüttgens and Rump 2002) as a modeling language and assumes that actors are familiar with this (techno-centric) language. Plural uses tool support built upon a commercial modeling environment, which identifies inconsistencies between individual models. The authors mention tool support for resolution of inconsistencies between models, but do not elaborate further on how scaffolding for inexperienced modelers could be implemented.

Multi-perspective modeling is also proposed by Front et al. (2017) in their ISEA (Identification, Simulation, Evaluation, Amelioration) approach to involve process participants in business process elicitation. Perspectives here are with respect to different constructs used to describe organizational reality (which is different to PLURAL and CoMPArE/WP, where multiple users conceptually describe their perspective on organizational reality using the same constructs). Similar to CoMPArE/WP, it emphasizes the needs of process participants for a 'simplified domain-specific language', which, at the same time is kept executable to allow for interactive validation through role-plays. While the intended outcome of the method is similar to that of CoMPArE/WP, the methodological focus of the two methods is different. ISEA focuses on eliciting business process models by reviewing them from different semantic perspectives, while CoMPArE/WP focuses on methodologically supporting the identification and resolution of different viewpoints in terms of construct semantics and collaboration when implementing a business process.

Herrmann et al. (2000) have also adopted the idea of participatory design for process elicitation proposing a methodology ('Socio-technical walkthrough'—STWT) that allows the creation of semi-structured and incomplete models. Workshops following the STWT methodology (Herrmann et al. 2007) target domain experts who do not necessarily need to have modeling experience. The STWT uses SeeMe (Herrmann et al. 2000) as a modeling language, which comprises three core-modeling elements with context sensitive semantics and is designed to represent models of socio-technical systems. It represents vague information, which explicitly captures disputed or unclear parts of a business process and thus is very close to the principles of natural modeling. No explicit scaffolds for model creation or alignment, however, are embedded in the methodology or the modeling language. The resulting models are intended for use in information system design, but are not executable in BPMS.

A similar approach is proposed in CPI modeling (Barjis 2011). Modeling is performed in a workshop setting similar to the STWT and focuses on validation of the process in the course of modeling by revisiting the model concepts in facilitated discourse. The approach claims to use an intuitive modeling language, which appears to be a simplified version of activity diagrams, to let process participants collaboratively create a 'trustworthy and complete' model of an enterprise. Again, the focus is on process elicitation and no bridge towards execution of the created models is discussed. In an attempt to make BPMN (Business Process Modeling Notation)—as a techno-centric language—more accessible for participatory design by process participants, T-BPM (Luebbe and Weske 2011) uses tangible modeling elements in a collaborative workshop setting. The modeling methodology focuses on articulation using BPMN notation elements, which, the authors claim, are intuitively understandable by participants after a brief introduction using examples. The result of modeling can be manually transcribed to a digital representation for further processing.

CoMPArE/WP in its final component provides tool support for guiding collaborative model creation among participants. This approach is also promoted by Collaborative Modeling Architecture (COMA) (Rittgen 2009) and Cooperative Editor for Processes Elicitation (CEPE) (Santoro et al. 2000). COMA focuses on providing support for articulating and consolidating models during collaborative modeling with a language-agnostic negotiation approach. The COMA tool provides support for UML (Unified Modeling Language) and enables actors to communicate via the software in a structured way specified by the COMA methodology. It provides scaffolds for model consolidation (i.e., the negotiation process), but presupposes that the involved participants are technology-proficient. As a result, participants, who have an important input to a process but do not feel comfortable with such software tools, might express unwillingness to be involved in a software-based collaborative elicitation-modeling procedure.

CEPE also supports collaboration during modeling with a particular focus on BPM. The modeling language proposed uses a limited set of elements to describe tasks, responsibilities, and decisions in a process. Further technical processing of the resulting models, however, is not addressed. The associated tool provides awareness features that support collaborative modeling. Aside of these features, no dedicated methodological or conceptual support for collaboration of process participants is provided. In a more recent research, Santoro et al. (2010) propose to use storytelling techniques in the early phases of process elicitation and further develop these stories to BPMN models of the described process.

They describe a method to support the abstraction process necessary to derive models from stories, and to finally create formal representations in BPMN. As such, it takes a complementary approach to CoMPArE/WP, where the need for explicitly creating formal representations is avoided by refinement via virtual enactment.

CoMPArE/WP is not the first approach to tackle collaborative modeling by process participants for eliciting business process knowledge. Existing approaches supporting collaborative articulation and modeling, however, either target inexperienced modelers, or aim at producing a model that can be directly executed. This is a reasonable approach given the conflicting requirements in those areas (Zarwin et al. 2014). From a BPM perspective, however, it remains desirable to satisfy requirements in both areas with a single methodological approach. The present work goes beyond the state-of-the-art by proposing a methodology that involves transitioning from natural modeling towards refinement of technically interpretable models. To enable this transition, the representation used for articulation and alignment support is syntactically and semantically compatible with techno-centric modeling languages like BPMN.

4.3 Alignment Scheme

In the following, we introduce CoMPArE (Collaborative Model Articulation and Elicitation) (Oppl 2017b) as a generic approach for collaborative articulation and alignment of individual understandings about collaborative work—independent of the actual focus of modeling, which in the case of CoMPArE/WP is work processes. CoMPArE facilitates collaborative articulation of different aspects of work using conceptual modeling techniques. As identified in related work, collaborative conceptual modeling is a recognized means to facilitate the development of a common understanding between people about a subject of discourse. The conceptual models serve as externalized artifacts, representing the participants' mental models, and so act as mediators for the development of a shared understanding (Groeben and Scheele 2001).

CoMPArE offers structural and procedural guidance in a multi-step modeling approach (cf. Fig. 4.7). The first step makes sure that every

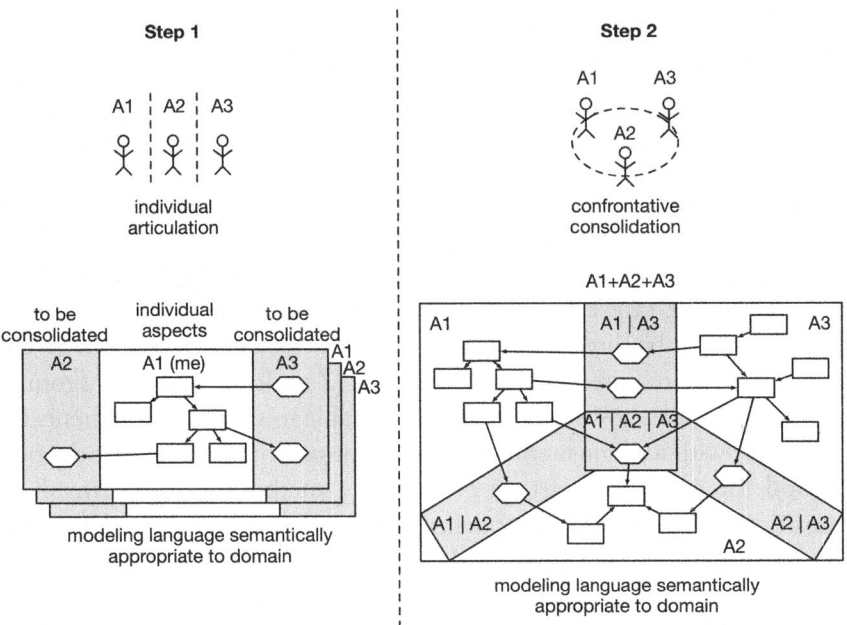

Fig. 4.7 CoMPArE articulation scheme

involved participant is able to contribute his or her individual view on the work process. The second step aims at avoiding the unreflected acceptance of inconsistent or conflicting views by explicitly confronting the participants with these issues. Figure 4.7 shows a generic scheme for this process. The steps are described in the following in more detail.

The guidance measures aiming at facilitating alignment activities need to be integrated in the modeling approach. This, however, cannot be done generically for all potential modeling languages. Work processes in organizations can be described with different foci (Curtis et al. 1992) that require conceptual modeling languages to provide different language constructs to describe appropriately the respective aspect (referred to as "semantic appropriateness to the modeling domain" by Krogstie et al. 1995). As an example, creating a model of the interaction in a collaborative work process requires different constructs than describing the flow of materials through a production chain. The used modeling language thus needs to be tailored to the targeted aspect of articulation. It needs to provide constructs that allow a description of the relevant aspects of the work process.

Independently of the aspects to be represented, the language needs to adhere to certain structural requirements in order to facilitate alignment activities. The aim of step 1 is to allow individual articulation of the view of every participant and have it represented in an individual conceptual model. These individual models are used again in step 2, in case conflicting representations need to be consolidated, in order to create ultimately an agreed-upon conceptual model representing a view on the work process shared by all participants. The modeling language can support the consolidation process by providing structural guidance. In line with the work of Türetken and Demirörs (2011), guidance measures are incorporated in the modeling notation in order to make visible the parts of the individual models that are subject to negotiation during the consolidation process, and which parts should remain the genuine responsibility of the contributing individual (cf. modeling areas and elements for modeling individual aspects and aspects to be consolidated in Fig. 4.7).

As an example, the individual ways of working in a collaborative work process might not be subject to negotiation as long as the collaboration interfaces are agreed upon. Consequently, the modeling language comprises elements to describe individual work and elements to describe the collaboration aspects. The former are specified to remain the responsibility of the contributing individual, whereas the latter are used to describe the relevant collaboration aspects from an individual perspective in step 1, and are subject to negotiation in step 2.

4.4 Alignment Approaches

The Subject-oriented Business Process Management (S-BPM) modeling language introduced in Chap. 3 provides a good starting point for designing a work modeling approach following the CoMPArE scheme. Section 3.2 describes how individual articulation can be supported with a representation scheme that can be transformed to S-BPM models for further processing. Its use for alignment activities, however, has not been discussed so far. We present a detailed procedural model designed for this purpose in the next subsection. Here, we first discuss the conceptual considerations that need to be elaborated for communication-oriented work process models in the light of the CoMPArE scheme.

Separating a process along the involved roles has implications for modeling support. Modelers need support for interlinking and aligning different contributions to a business process and ultimately deriving a commonly agreed upon model of the business process (Oppl 2013).

Each role's contribution to work is created as a separate part of the model. As noted above, one role can be taken by several actors in an organization. Different actors introduce different viewpoints about how one role's contribution can be implemented (Herrmann et al. 2002). These different viewpoints require alignment in order to derive a unified, commonly agreed upon view on a business process. Consequently, collaboration support for modeling role behavior has to be provided. All participating actors in this case share the same part of the model.

The role-based process parts are interconnected by communication acts, which are represented by flows of discrete messages. Communication among roles occurs whenever results of work (information and/or physical goods) have to be passed on from one role to another. The following modeling activities can occur in this context (using the concept 'message' to represent transmitted results of work): (a) send a message to another role; (b) get notified that a message has been sent to one's own role; (c) request a message from another role to be able to proceed with one's own part of the process; and (d) get notified that another role requests a message to be able to proceed with its part of the process.

The first two communication acts (a and b) occur regularly in the course of modeling. They are sufficient to describe all communication situations if the business process is modeled in fully sequential manner across all involved roles. This, however, requires actors to wait for another role to send a message, before they can proceed with modeling their own process part. Communication acts c and d are introduced to avoid these delays in modeling and to explicitly allow to express expectations on modeling that might require further discussion. Actors can specify messages they expect to arrive from another role and continue modeling as if this message already would have arrived. Elicitation support has to address

Alignment of Multiple Perspectives: Establishing Common... 149

the specification of these different types of messages as well as the resolution of inconsistent communication acts across roles.

Modeling of role behavior is realized using a generic activity modeling element that is used for representing activities as well as acts for sending and receiving messages. The actual semantics of the element (i.e., do something, send, receive) is determined during modeling time by whether it has incoming or outgoing message elements attached to it.

Modeling of communication acts implement all four modes of communication modeling described in the previous section. Message elements are used to either send a message (outgoing message element) to another role or request a message from another role (incoming message element). Their respective incoming or outgoing message counterparts are added to the communication partner's modeling surface to enable linking the models. Incoming messages or message requests, however, do not necessarily need to be processed by the communication partner immediately. For that reason, they are pooled in tray areas that visualize all unprocessed messages for each communication partner (cf. Fig. 4.8).

The uses of the three modeling elements are visualized in Fig. 4.8, which shows an elicitation process in an intermediate stage for illustration purposes. The depicted scenario consists of two interacting roles. The behavior of role 1 is modeled by three actors; two actors provide input for role 2. The modeling surfaces include trays for coupling to the respective other role on one of their borders.

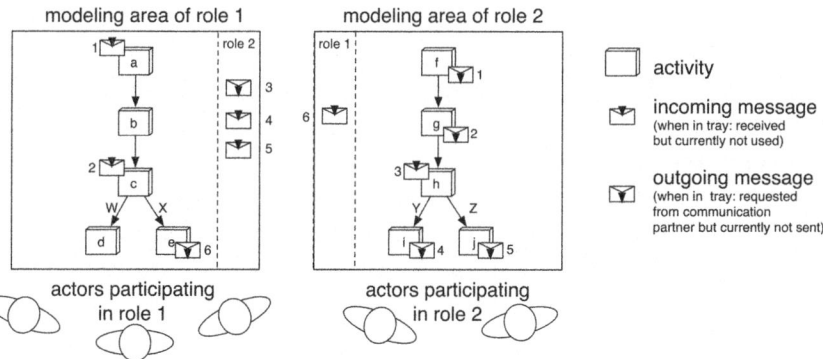

Fig. 4.8 Example setting of role-distributed models in an intermediate stage during modeling

The model of a role's behavior is created in the main area of each surface. Activities (labeled with lower-case letters in Fig. 4.8) are placed on the surface and are associated following their sequential order. Optional paths are represented by decision parameters placed next to the according association link (labeled with upper-case letters in Fig. 4.8).

The two model parts are interlinked using message elements (labeled with numbers). Following the coupling concept, messages always exist in pairs of two, with an outgoing message element on one surface and an incoming message on a second surface. The semantics of a message element changes depending on whether it is incorporated into the actual model (i.e., attached to an activity element) or kept in the tray area (i.e., so far not being used in the model). There are four different cases: (a) the combination of an incoming message attached to an activity (e.g., activities a, c, or h in Fig. 4.8) represents the act of processing a received message; (b) the combination of an outgoing attached to an activity (e.g., activities e, i, or j in Fig. 4.8) represents the act of sending a message to a communication partner; (c) an incoming message placed in an tray area represents a message that is offered by a communication partner, but has not yet been used in any way in one's own role behavior model; and (d) an outgoing message placed in a tray area represents a message that is expected by a communication partner, but has not yet been created and sent in one's own role behavior model.

As noted above, message elements are always created pairwise. If a role's behavior includes sending a message, the outgoing message element is directly attached to the sending activity (e.g., activity e with message 6 in Fig. 4.8). The corresponding incoming message is placed in the tray area of the receiving role's surface (e.g., message 6 on the surface of role 2 in Fig. 4.8). From there, it can be incorporated in the receiving role's model by attaching it to a receiving activity (e.g., activity c with message 2 in Fig. 4.8, which was sent from activity g earlier). Requesting a message is performed in a similar way. When an incoming message is required to proceed with a role's model and it has not yet been provided by the communication partner, the incoming message can be preliminarily used in the model by attaching it to a receiving activity (e.g., activity h with message 3 in Fig. 4.8). The corresponding outgoing message is considered a request to the communication partner to provide the necessary message. The outgoing message element is therefore placed in the tray area of the communication partner (e.g., message 3 on the surface of role 1 in Fig. 4.8). From there, it again can be incorporated in the designated sender's model by attaching it to a sending activity (e.g., activity f with message 1 in Fig. 4.8, which was requested from activity a earlier).

The messages kept in the tray areas make mutual expectations and potential communication flaws explicitly visible. Requested messages or unused incoming messages that remain in one of the trays always point at a mismatch between the expectations and the current behavior of the communication partners. During elicitation, this visualization of communication problems triggers negotiation and alignment activities that allow for the specification of a sound overall model.

Three different procedural approaches for distributed model elicitation can be identified following the concept of behavior and communication specification described above. They differ in the point in time when message specification happens. In *ex-ante communication negotiation*, all messages are specified collaboratively by the involved actors before the roles' behaviors are described. The messages are initially placed in the tray areas for each role and a then used during behavior modeling. In *ex-post communication negotiation*, each role's behavior including all outgoing and required incoming messages is modeled separately. In a consolidation step, the communication among the roles is then aligned by mutually matching requested and sent messages. In *ongoing communication negotiation*, messages are put into the trays of communication partners immediately when they are specified during behavior modeling. Inconsistencies or different understandings are discussed immediately.

Each of the three approaches stresses different aspects of the modeling process and appears to be suitable for different modeling purposes.

- *Ex-ante communication alignment* creates an initial common overall picture of the work process and leaves identification of non-suitable communication to the subsequent distributed modeling phase. Uncovered communication problems might then require an additional iteration of alignment of the communication acts among the involved roles.

- *Ex-post communication alignment* by contrast does not create an overview of the entire work process upfront and forces modelers to only focus on their own contribution to the work process. The identification of inconsistent communication acts is most likely here, as communication partners need to describe their communication completely independently of each other. The alignment of communications acts could lead to the need for a subsequent revision of roles' behavior models, in case fundamental inconsistencies, for example, conflicting communication sequences, are identified.
- *Ongoing communication alignment* avoids the need for fundamental revisions of either behavior models or communication acts, as both are specified simultaneously. Different viewpoints are immediately visible and can be discussed ad-hoc. This immediacy, at the same time, can be challenging for modelers, as they are continuously confronted with incoming messages or message requests while at the same time describing their own behavior.

4.4.1 Example: Ex-ante Communication Alignment

In S-BPM, modeling of interaction is based upon identification of the relevant subjects and the messages they exchange in the course of performing their collaborative work process. In scenarios where representatives for all involved subjects are available on-site, the elicitation of interactions in a certain work process can be performed using a methodology similar to storytelling (Swap et al. 2001). The involved actors assemble around the modeling surface (cf. Fig. 4.9), each one representing one role. A part of the surface is assigned to each role.

The involved actors agree upon a scenario that serves as an example for the work process to be modeled. Then they start to collaboratively describe their roles and activities in the work process and their mutual interactions.

For each interaction, a message element is placed on the surface (cf. Fig. 4.9). These message elements are named and additional information can be assigned. Assignment is performed using the elements as containers and putting inside physical representations of digital information

Alignment of Multiple Perspectives: Establishing Common... 153

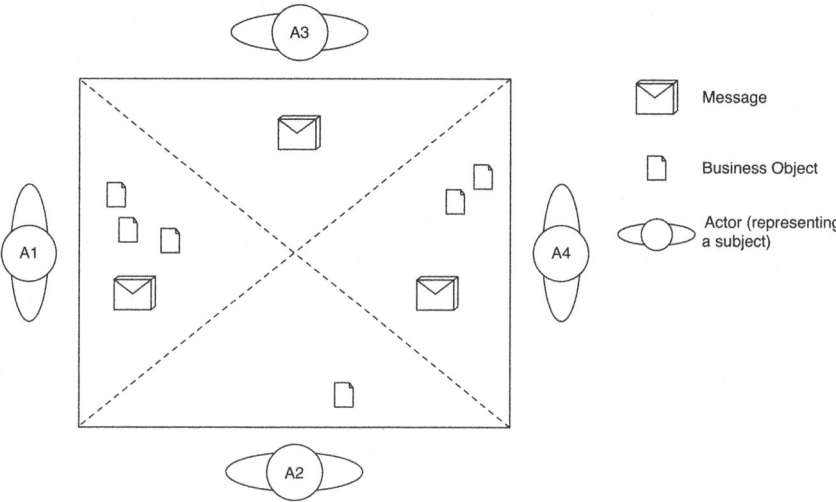

Fig. 4.9 Co-located creation of interaction models on a shared surface

(business objects). The message elements are then passed to the representative of the receiving subject. The receiver continues to act according to the received information. In cases where different messages can be passed from one subject to another (e.g., depending on a decision of the sending subject), these cases are acted out one after another. As incoming messages stay on the surface in the area of the receiving subject as long as they have not been handled, messages cannot get lost or be overlooked. For each outgoing or incoming message, the representatives can take (digital) notes of what activities triggered the message or are triggered by the message. This information is used to provide context for modeling internal behavior later on.

After modeling their collective view on interaction, the representatives of the subjects have to model their internal behaviors to react upon the incoming messages.

The involved individuals use the interactive support system to model their behavior one after another, handling one or several incoming messages at a time. The main building blocks for modeling internal behavior are states. States are visualized using physical building blocks and can represent functions (i.e., activities which create some work result) or message handling (receiving and sending states). While state elements are generic before they are placed on the surface, they take one specific role (function, sending, receiving) as soon as they are used. The modeling surface shows messaging ports to all other subjects at its borders when modeling internal behavior (cf. Fig. 4.10). The ports display all incoming and outgoing messages for the respective subject, visually marking those that have not been handled so far. Placing the state element on an incoming message and dragging it to its position creates a receiving state. Temporarily dragging a state element to a messaging port (and putting it back into place again afterwards) creates a sending state.

Placing a state element without any interaction at the borders of the surface creates a function state, which then can be described textually. The control flow of the internal behavior can be established by associating the elements with each another.

Displaying the incoming and outgoing messages provides the global context for a subject, even across several models of internal behavior. Information that was captured during modeling the interaction among subjects (e.g., notes about what happens when a certain message is

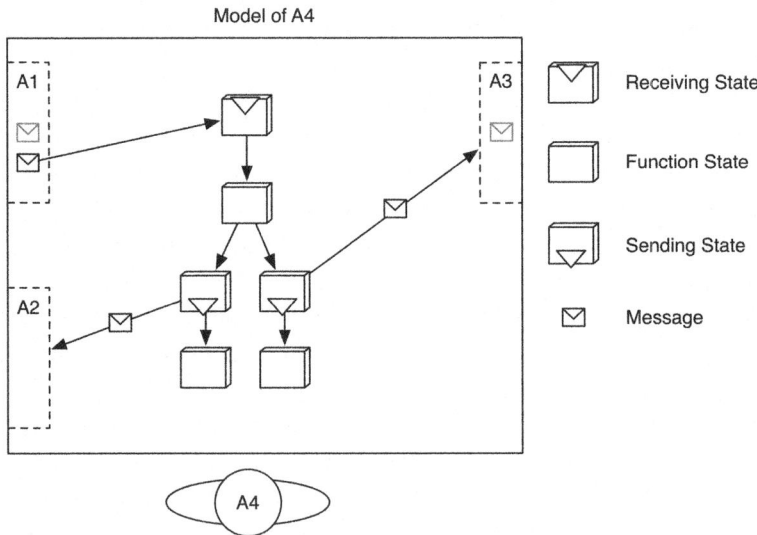

Fig. 4.10 Modeling of internal behavior on an interactive surface

received) is additionally provided during modeling. The representatives of the subjects in this way can focus on internal behavior without losing the big picture provided by the interaction model. The resulting models can be mapped directly onto an S-BPM representation without any further steps of interpretation.

4.4.2 Example: Ongoing Communication Alignment

The former modes of modeling support are tailored to settings, where all representatives of the involved subjects are gathered at the same place at the same time, where modeling of internal behavior can be performed asynchronously.

In scenarios, where modeling should be performed with ongoing communication alignment, several interactive support modeling surfaces can be connected and used to elicit subject-oriented process representations in one single step (the use of several support platforms at the same site would also allow single-step elicitation in a co-located scenarios). The ensemble of surfaces involving four subjects is visualized in Fig. 4.11. It can be supported using the interactive tabletop modeling instruments (Wachholder and Oppl 2014; Oppl and Rothschädl 2014) described in Chap. 3.

Each support platform acts as a modeling environment for the internal behavior and interaction of a single subject in the work process. For the individuals representing the subjects, the modeling experience is similar to modeling individual behavior in the co-located setting. The major difference is that the messaging ports of two subjects (allowing mutual communication) are connected directly and synchronized live. During operation, a sent message from one subject appears as an incoming message at the receiving subject's side without any noticeable delay and is ready to be handled. Using this mechanism, the work process can be performed like in the real world.

Moving a state element to a messaging port generates an outgoing message. Incoming messages are visualized differently depending on whether they have already been handled or not. In this way, users can easily distinguish messages that require additional modeling activities,

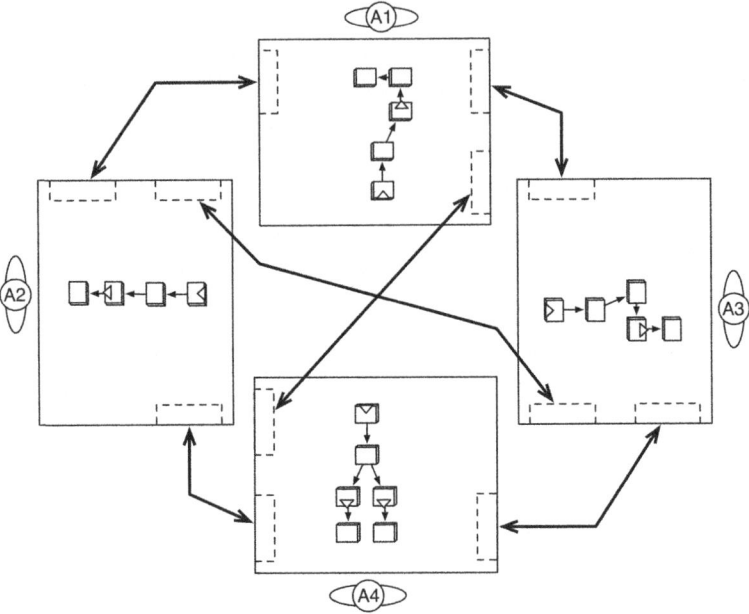

Fig. 4.11 Multi-surface setup for distributed modeling of subject-oriented models (bold arrows indicate linked messaging ports)

from those that have already been used in another model of internal behavior for the same subject.

4.5 Alignment Practice: Ex-post Communication Alignment with CoMPArE/WP

CoMPArE/WP is an instance of the CoMPArE scheme presented above, which is based on natural modeling practices and which at the same time maintains a well-defined bridge towards techno-centric (formal) modeling (Oppl and Alexopoulou 2016). It adopts an ex-post approach for communication alignment. In the following, the description of CoMPArE/WP is structured along these aspects. We start with an overview of the whole method, and subsequently detail each component.

Alignment of Multiple Perspectives: Establishing Common... 157

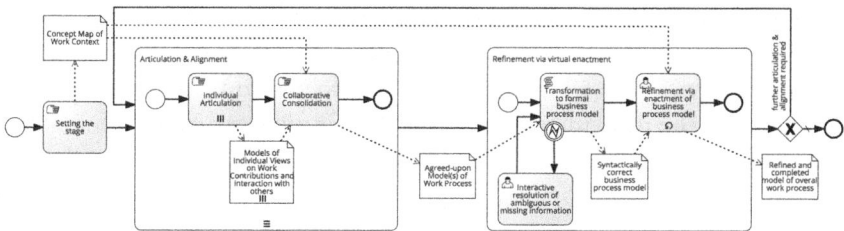

Fig. 4.12 The CoMPArE approach represented as a BPMN process

CoMPArE/WP comprises three components as depicted in Fig. 4.12. The first component ('Setting the Stage') is based on a semantically flexible modeling scheme, with the semantics of the cards being left open, in order to identify and agree upon the concepts that are relevant for the situation at hand. This component is in line with the alignment principles described in Sect. 4.1, where we have argued for semantic alignment among stakeholders. When implementing this component, modeling participants try to find a common understanding about the scope of the business process and the notions to use to refer to the relevant concepts. Scope herein refers to where the business process starts, where it ends, and which aspects are to be addressed when implementing it.

Groups of modeling participants with heterogeneous backgrounds in particular might have an issue with wording when aligning their different views. The notions used to refer to different aspects of the business process are thus explicitly captured. A semantically unconstrained notation similar to concept mapping is used in this component to allow modeling participants to express their concepts without requiring them to initially adapt to a given modeling language. This addresses the first requirement of natural modeling. This stage explicitly meets the third principle of natural modeling (i.e., 'no predefined meaning of symbols').

Component 2 consists of the two steps, which form the core of the CoMPArE concept, namely 'Describing Individual Work Contributions' and 'Collaborative Consolidation', which together lead towards semantically more constrained models eligible for business process representation. During this phase, the participants use the results of phase 1 as a point of reference and implement a multi-perspective articulation approach to process modeling (Mullery 1979). The first step in this phase is dedicated to individual modeling according to the perspective of each modeling participant, while the second step focuses on collaborative consolidation of the individual perspectives. As it will be further elucidated, this separation of individual articulation and collaborative consolidation facilitates knowledge sharing and promotes negotiation and commonly agreed-upon decisions, thus meeting the second requirement of natural modeling (i.e., 'collaborative modeling').

As already described for card-based modeling in Chap. 3, the modeling notation chosen for component 2 is reduced to the very fundamental concepts for the description of a business process, namely, the active entities, the actions performed by these entities and the exchange of tangible or intangible resources between entities by any means. As we adopt a case-based approach to modeling, the notation does not require decision constructs or elements for exception handling.

When actively involving process participants, it seems to be appropriate to limit the number of available modeling elements a priori to those appropriate for the intended modeling perspective and targeted outcome, that is, case-based models of business processes, as in scenario-based elicitation techniques. In this way, models are kept simple and comprise the most fundamental constructs used for the description of work and therefore the first requirement of natural modeling is met ('i.e., intuitive constructs'). This reduction of complexity, however, interferes with the requirement of creating semantically complete formal models of business processes. Component 3 conceptually addresses this shortcoming by elaborating the model in an interactive way towards a comprehensive representation of the business process. This is achieved through refinement during virtual enactment, that is, engaging modeling participants in identifying problems and gaps of their initially agreed upon model by playing through it and elaborating it concurrently.

The whole modeling framework is iterative, enabling the flexible combination of design components as the shared understanding about the business process evolves over time and potentially uncovers additional aspects to be addressed. Flexibly combining the three components enables the adaptation of the design procedure to the business process at hand (higher complexity requires more overall iterations), to the amount of divergent views that is present in the group of modeling participants (more divergence requires more iterations of component 2), and to their skills in abstraction and modeling (higher skills enable more complex changes to be made during virtual enactment). Selecting the appropriate steps in an ongoing design process is the task of a modeling facilitator. The selection is made based on the observed situation in the group of the modeling participants and the desired outcome in terms of elaborateness of the resulting model.

All components are carried out in a workshop setting, where the modeling participants work on creating a shared artifact. However, component 2 comprises an initial step of individual activity without any interaction to capture the different participants' views on the business process, before collaboratively consolidating those views to an agreed upon model. The methodology enables process participants to gradually develop a comprehensive model of their business process in a cooperative way without requiring them to be familiar with techno-centric modeling languages.

4.5.1 Component 1—Setting the Stage

Process participants do not necessarily share a common understanding of the organizational setting of the business process and which concepts to use for describing it (Sarini and Simone 2002). Component 1 aims at 'setting the stage' to enable co-operatively creating a business process model in the later components. It establishes a common understanding of the scope of the business process and of the concepts used for referring to its relevant aspects.

The modeling method used for setting the stage is based upon research on collaborative concept mapping as a means to create common ground (van Boxtel et al. 2002; Gao et al. 2007). Concept mapping is a method for externalizing and reflecting knowledge about real world phenomena, which reflects cognitive structures of the creator (van Boxtel et al. 2002).

Concept maps allow arbitrary model element types. This ensures avoiding misrepresentation or loss of information of individual work perceptions due to lack of support of what people want to express (Sarini and Simone 2002). Creating concept maps without any semantic restrictions supports actors not used to thinking in distinct concepts and helps to verbalize their work perception. It guides them towards conceptual thinking and sets a common frame of reference for all members of the group. This frame of reference facilitates consolidating the different individual views on collaboration later on.

The modeling participants perform the following steps as a group to build collaboratively a concept map:

1. They collect a set of elements (depicted on the cards) they consider relevant in the context of the business process under design. The types of the elements remain unconstrained. All modeling participants assign names to each of their elements individually. Then they group together elements that are of the same type (e.g., persons, tools, and documents), making the first step towards conceptual abstraction.
2. Each modeling participant presents each of his/her elements separately, one after the other. The element is added to a shared modeling surface accessible to all actors. The other modeling participants are asked to check, if they have also created an element representing the same real-world concept (independently whether they used the same name or not). In case an element is added to the shared modeling surface with the assumption that it is equivalent to the element by another participant, the equivalence of both elements need to be discussed by comparing the (verbal) descriptions provided by the actors. In case different names have been used, all of them remain in the model for future reference. In case the same name has been used for different concepts, a clarification is added. This step is repeated until all concepts have been added.
3. Concepts are correlated by the modeling participants by (a) spatial clustering of elements and (b) explicit associations depicted by connecting two elements and naming the connection. If the card-based models are developed on top of a shared paper surface, markers are used to draw the arrows between the cards. If the spatial arrangement of cards is done directly on top of a table (i.e., without a paper intervening), the incoming/outgoing arrows can be drawn on the cards themselves. Initial clustering and association specification can be per-

formed while adding concepts in step 2. A final round of collaborative clustering and association specification after all elements have been added completes the setting-the-stage design step.

As the semantics of the modeling language is not predetermined but evolves during the design procedure, semantic compatibility to the subsequent semantically more constrained phases cannot be taken for granted. One might even argue that leaving semantics unconstrained in phase 1 makes it incompatible with the following steps and superficial for the overall modeling result. A more efficient approach might be to provide the participants with the structure of the notation used in phase 2 to have a well-defined gateway between unstructured and structured modeling.

This approach, however, does not consider the cognitive requirements of process participants who are not skilled in structured business process modeling (Genon et al. 2011), and moreover it a priori directs the participants' mind which might result in constraining externalization of their tacit knowledge. Furthermore, a shared set of language constructs used by all involved participants to describe their mental models is a prerequisite for alignment on content level (Sarini and Simone 2002; Roschelle 1996). The existence of a common ground (Clark and Brennan 1991) in this respect, however, cannot be taken for granted—particularly when people with a diverse professional background are involved (Sarini and Simone 2002). Semantically open modeling has been shown to be an appropriate approach to address this issue (Faily et al. 2012; Engelmann and Hesse 2010; Trochim et al. 1994).

4.5.2 Component 2—Articulation and Alignment

The presented arguments for semantically open modeling in an initial phase of business process elicitation, however, leave open the question of how the results of component 1 can be used in component 2 and 3 beyond the indirect effects caused by the upfront alignment of the participants' mental models. Although the modeling constructs are semantically not constrained in component 1, clusters of concepts that are instances of the same semantic construct generally emerge during modeling and can be identified and named (Trochim et al. 1994). Following the assumption that a business process can be described by naming the active entities, the actions performed by these entities and the exchange of tangible or intangible resources between these entities (ibid.), it is likely that concepts using these semantic constructs will naturally emerge already in component 1. A dedicated step of asking the participants to identify the concepts, which are instances of the constructs used in component 2 has two potential effects: (a) it triggers another iteration of reflection on the outcome of component 1 and prepares the transition to the semantically more constrained modeling approach in component 2, and (b) it allows the identification of the concepts that can be reused in component 2 and therefore provides a means for reflecting on the completeness of the model in the course of collaborative consolidation. This is done by matching the elements of component 1 with those having emerged from collaborative consolidation in component 2.

Still, there might be clusters of concepts that bear semantics, which is not used in component 2. These cases cannot be directly incorporated in the models resulting from collaborative consolidation. They are, however, still available for another iteration of reflection in component 3, where semantically more comprehensive modeling approaches, such as BPMN, are used. This might allow matching further constructs having emerged in component 1 to the resulting model (e.g., data used within an activity of a single participant, which are not part of the modeling language used in component 2, but can be represented in BPMN). If concepts remain that still cannot be matched to semantic constructs of the formal languages after component 3, they have to be considered to describe the process context, that is, provide further information about how the model has to be interpreted and/or can be put to practice. This additional information is also considered of value for model understanding of process participants (Herrmann and Nolte 2014; Santoro et al. 2010).

4.5.2.1 Step 1: Individual Articulation

The first step of component 2 focuses on individual articulation of the participants' own perceived work contributions. The participants are provided with cards of different colors for modeling, with each color representing different semantics. The spatial arrangement of the cards based on their colors acts as a structural scaffold, which enables guiding the consolidation process in a structured manner via dedicated areas for describing different aspects of the process (cf. Fig. 4.13). Scaffolding is a concept widely used in education to describe structures or methodologies that support learners in self-directed efforts to understand something new (Van de Pol et al. 2010). Using the structural scaffold, modeling participants can independently of each other describe their own activities, the actors or organizational entities they are interacting with, and how this interaction manifests itself in terms of information or artifact exchange.

The detailed semantics of the modeling elements in the stage of individual modeling is hard to be determined upfront, as the people involved in modeling are not necessarily accommodated to explicitly follow specific semantics when describing work. As long as people use the fundamental process element classes (WHO, WHAT, EXCHANGE), a common level of conceptual abstraction can be achieved in the next, collaborative phase. The modeling elements in individual articulation should consistently be used as follows to provide for easier consolidation in the collaborative phase:

- WHO-items (represented by blue cards) indicating the role represented by the modeler herself/himself and those roles the modeler perceives to directly interact with.
- WHAT-items (represented by red cards) describing individual activities and their sequence indicating a causal and/or temporal relationship.

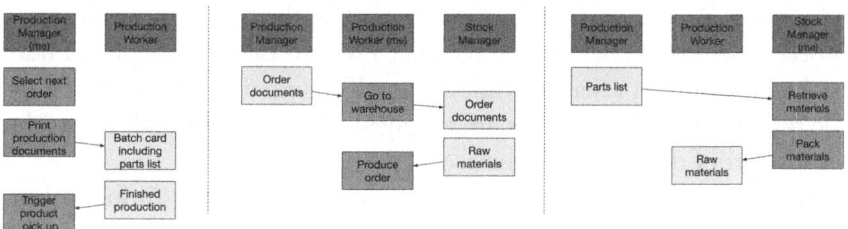

Fig. 4.13 Result of individual articulation

- EXCHANGE-items (represented by yellow cards) incoming to the participant's stream of activities indicating tangible or intangible resources expected from others.
- EXCHANGE-items (represented by yellow cards) outgoing from the participant's stream of activities indicating tangible or intangible resources offered to others.

Figure 4.13 shows three sample models created individually in step 1 of component 2, which together form a foundation for later consolidation. The labels in the models refer to a (exemplary) production process, in which a production manager, a production worker, and a stock manager are involved. The models indicate several fundamentally different understandings of how the production process should be implemented. While those differences might not occur in such a drastic way in reality, the scenario has been chosen to illustrate different aspects of consolidation below.

Modeling starts with a blue card bearing a name for one's role, which is used by the individual modeling participants to refer to themselves. The card is placed at the top border of the modeling surface. Modeling participants then describe what they are doing in order to complete their contribution to the business process. They describe their work by means of a sequence of distinct activities. Each activity is represented by a red card, named by the participant to indicate what the activity is about (referred to as WHAT-item in the following). The cards are placed vertically below the blue card representing the participant's own role. Their vertical ordering indicates their sequence, the top-most card consequently representing the first activity of the participant.

Subsequently, modeling participants determine people or roles they have to collaborate with to finish their work in the course of the business process. For each collaboration partner, a named blue card is placed next to the blue card representing him or herself (referred to as WHO-item in the following). All blue cards are arranged along a horizontal line at the top border of the modeling surface.

Finally, modeling participants determine what artifacts (information, material, etc.) they exchange with others in order to complete their work.

Alignment of Multiple Perspectives: Establishing Common... 165

In particular, they distinguish what they require from others in order to carry out certain activities, and what they can provide to others as a result of their activities. For each exchange, a yellow card is placed vertically below the blue card representing the respective collaboration partner (referred to as EXCHANGE-item in the following). The cards are vertically arranged to match the activities, for which the exchange is required or by which it is provided to others. Yellow cards indicating required exchanges are connected to the red cards representing the dependent activity using an arrow from the yellow to the red card. Provided exchanges consequently are indicated by an arrow from the respective red card to the yellow card.

4.5.2.2 Step 2: Collaborative Consolidation

The resulting models of step 1 are consolidated into a common model in step 2. The individual models are merged and aligned according to the following scheme (Fig. 4.14 shows the merging process for two of the sample models depicted in Fig. 4.13).

The modeling participants agree upon people or roles, who are or should be involved in the business process. Each process participant is represented by a named blue card. The name is mutually agreed upon. All blue cards are arranged along a horizontal line at the top border of the

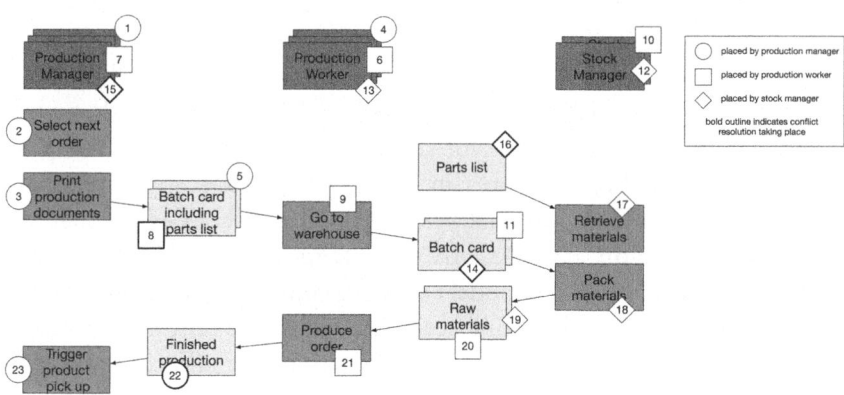

Fig. 4.14 Result of component 2.2: Collaborative Consolidation

modeling surface. Additionally, modeling participants articulate how each of them implements their contribution to the overall business process. All activities are represented by named red cards. The name is determined by the modeling participant responsible for the activity, but has to be understandable by the other modeling participants as well. The cards are placed vertically below the blue card representing the person or role responsible for enacting it. Their vertical ordering indicates the sequence in which they are enacted by the person or role. The top-most card consequently represents the first activity.

Finally, modeling participants agree upon how to collaborate in the course of the business process and which information, material and so on is exchanged in the course of this collaboration. All exchanged information, materials, and so on are represented by named yellow cards. The name is agreed upon by the modeling participants involved in the exchange but has to be understood by the other modeling participants as well. Each card is placed between the source lane (i.e., the sequence of red cards headed by the blue card representing the providing person/actor) and receiving lane. If the lanes are not adjacent, the card is placed next to the lane the exchange originates from. The cards are vertically arranged to match the activities, for which the exchange is required and by which it is provided. Arrows are used to connect the red cards representing the providing and requiring activities to the yellow card.

Consolidation is performed according to the following scheme (modeling steps described in brackets refer to the example depicted below):

1. One of the modeling participants starts by placing the WHO-item representing him/herself on the shared modeling surface. If known a priori, the actor responsible for starting the real-world business process starts modeling (cf. step 1 in Fig. 4.14). The process start is indicated by an individual model, which contains WHAT-items that are not dependent on any EXCHANGE-items to be received. If more than one such individual model exists, this indicates a business process with multiple parallel starting activities, which are only synchronized at a later point in time. In such cases, any of the affected modeling participants can start modeling.

Alignment of Multiple Perspectives: Establishing Common...   167

2. The same participant describes his/her own contribution to the business process by placing WHAT-items below his/her own WHO-item. Others do not intervene during this stage (cf. steps 2–3 in Fig. 4.14).
3. As soon as the participant places the first EXCHANGE-item (step 5 in Fig. 4.14), the targeted communication partner steps in and matches his/her own perception of the business process (steps 6–8). Matching can take the following forms:

   - The communication partner has a matching EXCHANGE-item (i.e., an EXCHANGE-item that matches the already placed item). In this case, the matching elements are merged (cf. steps 19–20 in Fig. 4.14).
   - The communication partner has no matching WHO-item (i.e., he/she has not perceived any collaboration with the original modeling participant at all). This is a fundamental difference in the perception of the business process. Participants need to agree how to resolve this issue (cf. steps 15–16 in Fig. 4.14, where the stock manager expected to receive a part list of parts from the production manager directly, whereas the production manager passed it on via the production worker).
   - The communication partner has no matching EXCHANGE-item (i.e., he/she did not share the perception of collaboration or did not consider it relevant). Such a difference again needs to be resolved by the affected participants (cf. step 22 in Fig. 4.14, where the production worker considered it to be finished after the order was produced, whereas the production manager expected an explicit notification that the production process had finished).
   - The communication partner considers one of his/her own EXCHANGE-items to match. The involved participants, however, have a different understanding of the content or form of the exchanged information or artifact. Such differences need to be addressed by the participants (cf. steps 5 and 8 as well as steps 11 and 14 in Fig. 4.14, where in the first case the production manager provided a more detailed description of the EXCHANGE than the production worker, and in the second case the EXCHANGE between stock manager and production worker was modified due to upfront communication of the parts list).

4. Consolidation continues in this way until all points of collaboration are agreed upon. Once one actor has completed his or her contribution, others with remaining elements not yet incorporated in the common model take over and provide further input to the consolidation process (cf. step 22–23 in Fig. 4.14).

The limited set of modeling elements used in component 3 prevents the occurrence of co-operation and externalization problems due to lack of participants' experience in modeling (Genon et al. 2011; Britton and Jones 1999). When actively involving process participants, it seems to be appropriate to limit the number of available modeling elements a priori to those appropriate for the intended modeling perspective and targeted outcome (Muehlen and Recker 2008), that is, case-based models of business processes, as in scenario-based elicitation techniques. In this way, models are kept simple and comprise the most fundamental constructs used for the description of work and therefore the first requirement of natural modeling is met ('i.e., intuitive constructs').

Figure 4.14 shows the merging process for the sample models depicted in Fig. 4.13. The numbering indicates the sequence of consolidation steps, the outlines of the numbers indicate the different modelers, and the stroke of the outline indicates whether conflicting viewpoints needed to be resolved.

4.5.3 Component 3—Refinement via Virtual Enactment

Completing the modeling components described above leads to models that are semantically incomplete representations of business processes. Most notably, these models do not account for different variants of a business process. Refinement through virtual enactment is a means to complete a process description without the need to create comprehensive process models as in the case of traditional conceptual modeling. This is enabled by transforming the results of component 2 to an executable process model (as described in Chap. 3) to play through complex decision processes via workflow enactment (Oppl 2017a). By incrementally adding process variants, the model evolves as virtual enactment continues. Complex models of business processes are documented in this way without the need to ever translate one's perceptions of a business process to abstract process descriptions in a single step. The model permanently maintains a syntactically valid state during refinement, which allows for further processing, such as live validation of dead- or live-locks or mathematical simulation of capacities. The conceptual details on and instruments used for virtual enactment are presented in Chap. 5.

For refinement during virtual enactment, an instance of the process is started. As stated earlier, this model initially only reflects one single variant of the process, omitting more sophisticated control flow constructs such as decisions or loops. It also does not contain the content and format of the exchanged information or resources. The aim of refinement through virtual enactment is to create a semantically correct and complete representation of the business process in all its variations as perceived by the involved actors. During the process of virtual enactment, the modeling participants enact the process step by step. For each step, the responsible modeling participant assesses the semantic correctness and completeness of the represented information above.

If any of these assessments leads to the need for changes in the process, these changes are made directly during execution. It should be stressed at this point that participants during the virtual enactment do not perform modeling. The system rather presents web-based dialogue forms to the participants, allowing them to describe the deviations from the currently enacted process. Potential changes include adding, altering, or removing activities of a process participant, shifting activities between participants, adding or removing messages required from or provided to another participant, and so on. The forms support the description of the new or altered process steps by providing the current process context (i.e., what was done, before the deviation was started), as well as information about potential interaction partners.

Modeling participants identify any steps in the business process that are described in a way they consider erroneous or cannot agree upon content-wise. Such steps are modified in a way that all affected participants can agree to. For each task, the participants assess whether there are any alternative ways of acting, and, if so, under which conditions these alternatives are to be executed. Both, the additional activities and the conditions need to be specified by the affected participant and have to be understandable to all other participants, as such changes might trigger cascaded changes that need to be addressed by them. As a result of these modifications, but also due to incomplete representation in component 2, gaps might by identified in the business process. These gaps need to be addressed by agreeing on and adding further activities, exchanges, or even new roles. Fundamental changes might trigger the need to go back to component 2 and explicitly address the newly identified part of the business process.

4.5.4 Transition from Modeling to Enactment

Components 1 and 2 from a representational aspect are implemented using physical cards. In order to enable execution of the models in component 3, the card-based models need to be converted into digital model representations. To this end, the card-based model initially is captured as a pixel-based image via taking a picture, for example using a mobile phone. The modeling cards bear visual markers that can be recognized and uniquely identified in the picture. The optical marker recognition engine (Oppl et al. 2017) used for this purpose is based upon the ReacTIVision system (Kaltenbrunner and Bencina 2007). Based upon the coordinates of each marker, the cards contained in the image can be identified and extracted. The extracted information is also used for identification of potential connections that are drawn between cards. The model layout is subsequently analyzed in the next step regarding its adherence to the CoMPArE/WP notation. If modeling rules are violated, missing, or ambiguous, then the information needed for the transformation can be added interactively. IT-based guidance through the interactive parts of the transformation process is described in (Oppl 2015). Once the transformation process is finished, the resulting model can be used for refinement through virtual enactment.

4.6 Conclusion

Considering the requirements and subsumed procedural cornerstones developed in Chap. 2, we can reflect on the results of this section. The reflection takes into account individual engagement of actors, as well as the activities of organizations. In Table 4.1, we follow the provided list of requirements and elaborate on each according to relevant achievements for each presented articulation technique.

Table 4.1 Elicitation requirements and CoMPArE/WP

Elicitation requirement	CoMPArE/WP
Awareness on role(s) and their management	Along with interacting with other participants, consolidation, alignment and consolidation role-specific argumentation is at the center. This is a condition-sine-qua-non for getting and keeping stakeholders involved in work knowledge elicitation and further processing.
Situation Awareness	Since the core of modeling are role- or task-specific activities including communication with other actors, the participants are aware of the business case or situation that forms the frame for those activities. In addition, all refinements occur within that frame of reference. Additional information is kept separately for further processing.
Conceptual understanding of complex systems	The networked development of socio-technical settings increases the complexity of systems which requires concepts to handle it for reflection and change.
Creating a reflective practice for situations-to-be	Alignment may start with considering various individual mental model representations referring to a situations-as-it-is. Consequently, any refinement of models can refer to reflecting on existing work practices. However, at some point in the course of sharing knowledge on work settings, integrating individual existing work practices, or developing a collective novel structure of work refers to situations-to-be. This shift may require additional articulation steps to develop a common understanding of work knowledge.

(continued)

Table 4.1 (continued)

Elicitation requirement	CoMPArE/WP
Focusing while utilizing multiple perspectives	Bringing together different stakeholders to reflect on and discuss models of task accomplishment allows focusing on a work process while utilizing the individual perspectives given by the individual mental models of the stakeholders. Another set of perspectives is given by the various types of procedural interventions, such as ex-ante refinements that could facilitate co-creating models through resolving conflicts before further consolidation. The equal handling of situations- to-be and as-it-is provides a robust baseline for switching perspectives. Finally, the probing of processes through proper technical execution adds another perspective on accomplishing tasks, as technical execution enables life experience of envisioned processes. However, execution can also be utilized in the course of reflecting on existing work practices.
Articulating intangible assets	The approach mainly tackles explicit knowledge on work. Implicit aspects are elicited and explicitly encoded when addressed in the course of verbal reflection, laddering, and model refinements.
Engage in alignment for collective intelligence	The setting of the methodological support scenarios foster participatory design of work processes. Each participant is an active part in a work system or situation that is referred to in the course of externalizing knowledge. They play a dual role, as providers of knowledge about individual work processes and observers needing to reconstruct work knowledge from other members of the addressed work system. Actively taking the latter role ensures intelligible and purposeful representations for individuals and the collective they are part of.

From a procedural perspective, the presented alignment procedure and its variants are addressing the various steps as follows:

1. Along with preparation, the setting including participating actors, existing models, and alignment instruments is configured and provided for use. The scope is given by the previously elicited knowledge, mainly focusing on individual mental models of a certain

business case. The environment is motivating in case graspable material is used for reflecting individual perspectives. It also facilitates negotiating when sharing the represented knowledge while coming up with a common model that participants could agree upon. Co-creation instruments for executing process models when probing them in the course of generating work knowledge also need to be prepared.
2. Situation-sensitive articulation features are provided, in particular when additional knowledge on role behavior or work tasks is externalized. For ex-post alignment, in-depth articulation turns out to be of value. Its results can be aligned with existing representations in a structured way. Hence, the procedure remains traceable and transparent for stakeholders.
3. Facilitation needs to be provided to structure the alignment and consolidation procedure, in particular when several models need to be aligned or different strategies of negotiation support are applied to specific cases. Intervention may be helpful when interpreting cross-boundary topics or work patterns, together with suggesting executing models for collecting implementation or practical experience in case of complex work situations.
4. Representational alignment as a consolidated representation serves as the baseline for documentation and further exploration. CoMPArE offers incremental, structured alignment support due to its focus on role-specific work process designs.
5. Organizational alignment has to follow representational alignment, for example, through playing roles or executing the consolidated work knowledge and process models in the operational context of the business. This is the point in time, when elicited knowledge becomes part of workspaces of an organization.

Overall, the presented approach can be advised for all development settings where elicited work knowledge needs to be aligned taking into account different mental models and requiring strategic intervention for consolidating stakeholder knowledge in an accountable way.

References

Barjis, J. 2011. CPI Modeling: Collaborative, Participative, Interactive Modeling. In *Proceedings of the Winter Simulation Conference*, 3099–3108. Winter Simulation Conference.

Britton, Carol, and Sara Jones. 1999. The Untrained Eye: How Languages for Software Specification Support Understanding in Untrained Users. *Human-Computer Interaction* 14 (1–2): 191–244. https://doi.org/10.1080/07370024.1999.9667269.

Clark, Herbert H., and Susan E. Brennan. 1991. Grounding in Communication. *Perspectives on Socially Shared Cognition* 13 (1991): 127–149.

Convertino, Gregorio, Helena M. Mentis, Mary Beth Rosson, John M. Carroll, Aleksandra Slavkovic, and Craig H. Ganoe. 2008. Articulating Common Ground in Cooperative Work: Content and Process. In *CHI '08: Proceeding of the Twenty-Sixth Annual SIGCHI Conference on Human Factors in Computing Systems*, 1637–1646. New York: ACM. https://doi.org/10.1145/1357054.1357310.

Curtis, B., Marc I. Kellner, and Jim Over. 1992. Process Modeling. *Communications of the ACM* 35 (9): 75–90 (New York: ACM Press).

Engelmann, T., and F.W. Hesse. 2010. How Digital Concept Maps About the Collaborators' Knowledge and Information Influence Computer-Supported Collaborative Problem Solving. *International Journal of Computer-Supported Collaborative Learning* 5 (3): 299–319.

Evans, Stephen, and Sarah Jukes. 2000. Improving Co-Development Through Process Alignment. *International Journal of Operations & Production Management* 20 (8): 979–988 (MCB UP Ltd).

Faily, S., J. Lyle, A. Paul, A. Atzeni, D. Blomme, H. Desruelle, and K. Bangalore. 2012. Requirements Sensemaking Using Concept Maps. In *Proceedings of the 4th International Conference on Human-Centered Software Engineering*. Springer.

Fan, Shaokun, Zhimin Hua, Veda C. Storey, and J. Leon Zhao. 2016. A Process Ontology Based Approach to Easing Semantic Ambiguity in Business Process Modeling. *Data & Knowledge Engineering* 102: 57–77 (Elsevier).

Front, A., D. Rieu, M. Santorum, and F. Movahedian. 2017. A Participative End-User Method for Multi-Perspective Business Process Elicitation and Improvement. *Software & Systems Modeling* 16 (3): 691–714.

Gao, H., E. Shen, S. Losh, and J. Turner. 2007. A Review of Studies on Collaborative Concept Mapping: What Have We Learned About the Technique and What Is Next? *Journal of Interactive Learning Research* 18 (4): 479–492. https://doi.org/10.1207/s15430421tip4101_7.

Genon, Nicolas, Patrick Heymans, and Daniel Amyot. 2011. Analysing the Cognitive Effectiveness of the BPMN 2.0 Visual Notation. In *Software Language Engineering*, Lecture Notes in Computer Science, vol. 6563, 377–396. Berlin and Heidelberg: Springer Berlin Heidelberg. https://doi.org/10.1007/978-3-642-19440-5_25.

Groeben, Norbert, and Brigitte Scheele. 2001. Dialogue-Hermeneutic Method and the "Research Program Subjective Theories". *Forum Qualitative Sozialforschung/Forum: Qualitative Social Research* 1 (2): 1–10. http://nbn-resolving.de/urn:nbn:de:0114-fqs0002105.

Herrmann, Thomas, and Alexander Nolte. 2014. Combining Collaborative Modeling with Collaborative Creativity for Process Design. In *COOP 2014—Proceedings of the 11th International Conference on the Design of Cooperative Systems, 27–30 May 2014, Nice (France)*, 377–392. Cham: Springer International Publishing. https://doi.org/10.1007/978-3-319-06498-7_23.

Herrmann, Thomas, K.U. Loser, and I. Jahnke. 2007. Sociotechnical Walkthrough: A Means for Knowledge Integration. *The Learning Organization* 14 (5): 450–464.

Herrmann, Thomas, M. Hoffmann, G. Kunau, and K.U. Loser. 2002. Modelling Cooperative Work: Chances and Risks of Structuring. In *Cooperative Systems Design, a Challenge of the Mobility Age. Proceedings of COOP 2002*, 53–70. IOS Press.

Herrmann, Thomas, M. Hoffmann, K.U. Loser, and K. Moysich. 2000. Semistructured Models Are Surprisingly Useful for User-Centered Design. In *Designing Cooperative Systems. Proceedings of COOP 2000*, ed. R. Dieng, A. Giboin, L. Karsenty, and G. De Michelis, 159–174. Amsterdam: IOS Press.

Jung, Jason J. 2009. Semantic Business Process Integration Based on Ontology Alignment. *Expert Systems with Applications* 36 (8): 11013–11020 (Elsevier).

Kaltenbrunner, Martin, and Ross Bencina. 2007. reacTIVision: A Computer-Vision Framework for Table-Based Tangible Interaction. In *TEI '07: Proceedings of the 1st International Conference on Tangible and Embedded Interaction*, 69–74. New York: ACM Press.

Kaplan, Robert S., and David P. Norton. 1996. *The Balanced Scorecard-Translating Strategy into Action*. Harvard Business School Press.

Kensing, Finn, and Jeanette Blomberg. 1998. Participatory Design: Issues and Concerns. *Computer Supported Cooperative Work (CSCW)* 7 (3–4): 167–185 (Springer).

Krogstie, J, Odd Ivar Lindland, and Guttorm Sindre. 1995. Defining Quality Aspects for Conceptual Models. *Isco*, 216–231.

Lin, Yun, and J. Krogstie. 2012. Semantic Annotation of Process Models for Facilitating Process Knowledge Management. In *Organizational Learning and Knowledge: Concepts, Methodologies, Tools and Applications*, 733–754. IGI Global.

Luebbe, Alexander, and Mathias Weske. 2011. Bringing Design Thinking to Business Process Modeling. In *Design Thinking*, 181–195. Berlin and Heidelberg: Springer Berlin Heidelberg. https://doi.org/10.1007/978-3-642-13757-0_11.

Muehlen, zur Michael, and J.C. Recker. 2008. How Much Language Is Enough? Theoretical and Practical Use of the Business Process Modeling Notation. In *Advanced Information Systems Engineering. CAiSE 2008*, Lecture Notes in Computer Science, vol. 5074, ed. Z. Bellahsène and M. Léonard, 465–479. Berlin and Heidelberg: Springer Berlin Heidelberg. https://doi.org/10.1007/978-3-540-69534-9_35.

Mullery, G. P. 1979. CORE-a Method for Controlled Requirement Specification. In *ICSE '79—Proceedings of the 4th International Conference on Software Engineering*, pp. 126–135.

Nüttgens, M., and F. J. Rump. 2002. Syntax Und Semantik Ereignisgesteuerter Prozessketten (EPK). *Promise*, 64–77.

Oppl, Stefan. 2013. Towards Role-Distributed Collaborative Business Process Elicitation. In *Proceedings of the Workshop on Models and Their Role in Collaboration*, ed. Alexander Nolte, M. Prilla, Peter Rittgen, and Stefan Oppl, 33–40. CEUR-WS.

———. 2015. Articulation of Subject-Oriented Business Process Models. In *Proceedings of S-BPM ONE 2015*, 1–11. New York: ACM Press. https://doi.org/10.1145/2723839.2723841.

———. 2017a. Business Process Elaboration Through Virtual Enactment. In *Proceedings of S-BPM ONE 2017*. https://doi.org/10.1145/3040565.3040568.

———. 2017b. Supporting the Collaborative Construction of a Shared Understanding about Work with a Guided Conceptual Modeling Technique. *Group Decision and Negotiation* 26 (2): 247–283. https://doi.org/10.1007/s10726-016-9485-7.

Oppl, Stefan, and Nancy Alexopoulou. 2016. Linking Natural Modeling to Techno-Centric Modeling for the Active Involvement of Process Participants in Business Process Design. *International Journal of Information System Modeling and Design* 7 (2): 1–30. https://doi.org/10.4018/IJISMD.2016040101.

Oppl, Stefan, and Thomas Rothschädl. 2014. Separation of Concerns in Model Elicitation—Role-Based Actor-Driven Business Process Modeling. In *S-BPM ONE—Setting the Stage for Subject-Oriented Business Process Management*,

Communications in Computer and Information Science, vol. 422, ed. Hagen Buchwald, Albert Fleischmann, Detlef Seese, and Christian Stary, 3–20. Springer International Publishing. https://doi.org/10.1007/978-3-319-06191-7_1.

Oppl, Stefan, Christian Stary, and S. Vogl. 2017. Recognition of Paper-Based Conceptual Models Captured Under Uncontrolled Conditions. *IEEE Transactions on Human-Machine-Systems* 47 (2): 206–220. https://doi.org/10.1109/THMS.2016.2611943.

Peters, Thomas J., and Robert H. Waterman. 1982. *In Search of Excellence: Lessons From America's Best-Run Companies*. New York: Warner.

Prilla, M., and Alexander Nolte. 2012. Integrating Ordinary Users into Process Management: Towards Implementing Bottom-Up, People-Centric BPM. In *Enterprise, Business-Process and Information Systems Modeling*, 182–194. Springer.

Rittgen, Peter. 2009. Collaborative Modeling—A Design Science Approach. In *2009 42nd Hawaii International Conference on System Sciences*, 1–10. IEEE. https://doi.org/10.1109/HICSS.2009.112.

Roschelle, J. 1996. Designing for Cognitive Communication: Epistemic Fidelity or Mediating Collaborative Inquiry? In *Computer, Communication and Mental Models*, ed. D. Day and D.K. Kovacs, 15–27. Taylor & Francis.

Santoro, Flávia Maria, Marcos R.S. Borges, and José A. Pino. 2000. CEPE: Cooperative Editor for Processes Elicitation. *Hicss 2000*, vol. 1. IEEE Computer Society: 10. https://doi.org/10.1109/HICSS.2000.926587.

———. 2010. Acquiring Knowledge on Business Processes from Stakeholders' Stories. *Advanced Engineering Informatics* 24 (2): 138–148. https://doi.org/10.1016/j.aei.2009.07.002.

Sarini, Marcello, and C. Simone. 2002. The Reconciler: Supporting Actors in Meaning Negotiation. In *Proceedings of the Workshop on Meaning Negotiation (MeaN-02) at AAAI-02*, Edmonton, AB, Canada.

Swap, W., D. Leonard, M. Shields, and L. Abrahams. 2001. Using Mentoring and Storytelling to Transfer Knowledge in the Workplace. *Journal of Management Information Systems* 18 (1): 95–114 (ME Sharpe, Inc.).

Trochim, William M.K., Judith A. Cook, and Rose J. Setze. 1994. Using Concept Mapping to Develop a Conceptual Framework of Staff's Views of a Supported Employment Program for Individuals with Severe Mental Illness. *Journal of Consulting and Clinical Psychology* 62 (4): 766 (American Psychological Association).

Türetken, Oktay, and Onur Demirörs. 2011. Plural: A Decentralized Business Process Modeling Method. *Information & Management* 48 (6): 235–247. https://doi.org/10.1016/j.im.2011.06.001.

van Boxtel, Carla, Jos van der Linden, Erik Roelofs, and Gijsbert Erkens. 2002. Collaborative Concept Mapping: Provoking and Supporting Meaningful Discourse. *Theory into Practice* 41 (1): 40–46. https://doi.org/10.1207/s15430421tip4101_7.

Van de Pol, Janneke, Monique Volman, and Jos Beishuizen. 2010. Scaffolding in Teacher—Student Interaction: A Decade of Research. *Educational Psychology Review* 22 (3): 271–296 (Springer).

Wachholder, Dominik, and Stefan Oppl. 2014. Interactive Coupling of Process Models: A Distributed Tabletop Approach to Collaborative Modeling. In *ECCE '14: Proceedings of the 2014 European Conference on Cognitive Ergonomics*, September, 1–8. New York: ACM Request Permissions. https://doi.org/10.1145/2637248.2637262.

Weske, Mathias, and Alexander Luebbe. 2011. Tangible Media in Process Modeling—A Controlled Experiment. In *Advanced Information Systems Engineering*, Lecture Notes in Computer Science, vol. 6741, ed. Haralambos Mouratidis and Colette Rolland, 283–298. Berlin and Heidelberg: Springer Berlin Heidelberg. https://doi.org/10.1007/978-3-642-21640-4_22.

Wood, Jed R., and Larry E. Wood. 2008. Card Sorting: Current Practices and Beyond. *Journal of Usability Studies* 4 (1): 1–6 (Usability Professionals' Association).

Zarwin, Z., M. Bjekovic, J.M. Favre, J.S. Sottet, and Erik Proper. 2014. Natural Modelling. *Journal of Object Technology* 13 (3): 4:1–36. https://doi.org/10.5381/jot.2014.13.3.a4.

Open Access This chapter is licensed under the terms of the Creative Commons Attribution 4.0 International License (http://creativecommons.org/licenses/by/4.0/), which permits use, sharing, adaptation, distribution and reproduction in any medium or format, as long as you give appropriate credit to the original author(s) and the source, provide a link to the Creative Commons licence and indicate if changes were made.

The images or other third party material in this chapter are included in the chapter's Creative Commons licence, unless indicated otherwise in a credit line to the material. If material is not included in the chapter's Creative Commons licence and your intended use is not permitted by statutory regulation or exceeds the permitted use, you will need to obtain permission directly from the copyright holder.

5

Acting on Work Designs: Providing Support for Validation and Implementation of Envisioned Changes

This chapter introduces methods to enable to act on the results of articulation and alignment processes. As such, it outlines paths towards sharing and anchoring new work practices in organizations. The proposed methods allow to qualitatively validate novel work practices and to agree on strategies for their wider roll-out. Drawing from current practice-oriented research in organizational development, Computer Supported Collaborative Work (CSCW), and knowledge management, this chapter also includes an account on supporting technology that could be used to aid method deployment.

5.1 Creating Executable Models Through Scaffolding Articulation and Alignment

Articulation and alignment of knowledge that guides collaboration in work processes has been addressed in nearly every knowledge management approach proposed over the last decades. As already described in earlier chapters, "articulation" refers to the process of encoding individual mental models (Pirnay-Dummer and Lachner 2008) about a particular work process in a tangible form ("artifacts") that enables reflection (Roberts 2009; Prilla 2015) and sharing (Arias and Fischer 2000). Processes of "alignment" consequently build upon these artifacts and are used in collaborative settings to alter and extend the individuals' mental models. This enables them to operatively work together (Convertino et al. 2008; Oppl 2016) and modify the artifacts accordingly to represent the aligned understanding.

The representation of mental models in externalized artifacts is often approached via creating diagrammatic conceptual models. The capability to use conceptual modeling for the purpose of articulation and alignment must, however, not be assumed to be present for all participating actors (Arias et al. 2000; Hjalmarsson et al. 2015). Inexperienced participants require more support and guidance than experienced ones. However, the latter might be hampered by an enforcement of modeling structures of procedures, leading to less articulation and alignment activities (Franco and Rouwette 2011).

The topic of how to facilitate the development of skills in conceptual modeling in organizational research has been addressed as early as in the 1960s, when Morris (1967) stated that "if one grants that modeling is and, for greatest effectiveness, probably ought to be, an intuitive process for the experienced, then the interesting question becomes the pedagogical problem of how to develop this intuition". This question has also been picked up in enterprise and work modeling as the discipline continued to mature (Sandkuhl et al. 2014), and has moved away from being considered an "art" that requires "intuition," to a more scientifically grounded discipline (Willemain 1995).

In recent years, literature recognizes a trend towards a strong and active involvement of stakeholders in the modeling process (Tavella and Papadopoulos 2014) (Recker et al. 2012), who are usually not formally educated in the modeling skills (Powell and Willemain 2007). Models are considered to act as boundary objects (Franco 2013) that enable people who collaborate within organizational structures to articulate and align the understanding of their work systems (Briggs et al. 2013). Research in this domain has focused on how to facilitate modeling activities under involvement of such "novice modelers" (Tavella and Papadopoulos 2014) towards the ends of generating models appropriate for the respective aims of modeling (Hjalmarsson et al. 2015). In contrast, the question of how to support the development of skills to work with, and on the basis of, conceptual models for this group of people, who usually do not have the opportunity to dedicate effort and time to formal modeling education, has hardly been a subject of research. The potential added value of such skills for this group, however, has been recognized repeatedly over the last decades in terms of pursuing a deeper understanding of the domain and phenomenon being subject of modeling (Mayer 1989; Davies et al. 2006; Hjalmarsson et al. 2015).

5.1.1 Scaffolding

Scaffolding is a concept introduced in the field of educational tutoring by Wood et al. (1976). It originally refers to having an experienced person help an inexperienced learner to acquire knowledge about a particular topic. Scaffolding is a metaphor adopted from construction industry and refers to a temporary means of support that is present until the entity supported by scaffolds (here: an actor participating in conceptual modeling) can accomplish a given task independently (Van de Pol et al. 2010). It is usually motivated by keeping actors in their "zone of proximal development" (ZPD) during a learning process (Vygotsky 1978), that is, putting them in a situation, which is challenging, yet attainable, to them. In order for scaffolds to be acceptable for actors and provide added value to them, they need to be appropriated to their current skill level (Dennen 2004).

Scaffolding can take different forms. Based on a meta-study of scaffolding research Jumaat and Tasir (2014) distinguish conceptual scaffolds, procedural scaffolds, metacognitive scaffolds, and strategic scaffolds. *Conceptual scaffolds* help learn to decide what to consider to be worth learning. In particular, they can help to prioritize fundamental concepts. *Procedural scaffolds* assist students in using available tools and methods and point them at potentially useful resources. *Strategic scaffolds* suggest alternative ways to tackle problems in learning. Finally, *metacognitive scaffolds* guide students in how to approach a learning problem and what to think about when elaborating on a problem. Orthogonally to these categories, Bulu and Pedersen (2010) identify differences in the sources of scaffolding. *Scaffolds provided by teachers* are considered the original form of scaffolding. *Scaffolds provided in interactions among learning peers* refer to the phenomenon that scaffolding can arise from the collective knowledge of a learning group. Scaffolds can also be provided as *textual or graphical representations*, similar to a manual. *Technology-driven scaffolding* uses (information) technology to provide scaffolds. This includes interactive systems that try to intervene appropriately in the learning process based on observing learners' behaviors or static intervention rules.

Independently of which form of scaffolding is pursued, it is always characterized via the presence of three principles that have been identified by Van de Pol et al. (2010): The first common principle is *contingency*, which is often referred to as responsiveness or calibrated support. Scaffolds need to be adapted dynamically to the learners' current level of performance. The second principle is *fading*, which refers to the gradual withdrawal of the scaffolding. As learners develop their skills, support becomes less necessary and is decreased over time. This is closely connected to the third principle *transfer of responsibility*. Via fading, responsibility for the performance of a task is gradually transferred to the learner. The responsibility for learning is transferred when a student takes increasing control of the learning process. The implementation of these principles is based on *diagnosis* of a learner's need for support, which is usually done by a teacher (Stender and Kaiser 2015), but also can be implemented in interactive systems (Su 2015).

On an operative level, scaffolding is implemented via different means. Van de Pol et al. (2010) list a (non-exhaustive) set of measures such as giving feedback, providing hints, instructing, explaining, modeling (i.e., demonstrating the skill to be acquired), and questioning. They differ in their depth of intervention and the reduction of freedom in students' learning processes. How to appropriately select and implement scaffolding as interventions in the learning process is disputed (ibid.). The described categories and means thus should be considered a framework for observing and designing learning settings, rather than attributing them any normative value.

5.1.2 Scaffolds for Stakeholder-Centric Work Modeling

Our review of related work has shown that, while scaffolding has already implicitly and, to some extent, explicitly been deployed in the field of conceptual modeling, a structured approach to describe and design scaffolding in modeling activities is not available. As argued earlier, augmenting the design of stakeholder-centric modeling activities with a scaffolding perspective could help improve the understanding and creation of enterprise models in the target group. In the following, we therefore review scaffolding approaches proposed in other disciplines, which require skills similar to stakeholder-centric conceptual modeling, in particular with respect to articulation of abstract concepts describing real-world phenomena (Frederiks and van der Weide 2006) and the support of developing a common understanding about these phenomena (Prusak et al. 2012). Based on these approaches, we develop a framework for scaffolding in enterprise modeling, which constitutes our nascent design theory.

5.1.2.1 Scaffolding the Articulation of Models

The process of articulating abstract concepts, being a main activity in conceptual modeling, has been widely examined regarding potential scaffolding support in the field of mathematical and science education.

Ozmantar and Roper (2004) consider teacher interventions as the major means of scaffolding (in the context of mathematical abstraction problems). Their study focuses on examining the activities of the person providing scaffolds. They identify three major facets that they could observe. First, they observed that scaffolding strategies were chosen in an ad-hoc manner based on continuous monitoring and analyzing the actors' performance. Which scaffold is appropriate in which situation cannot be determined ex-ante. It requires continuous monitoring of the learning process to check whether a provided scaffold achieves the intended effect and allow for adapting one's scaffolding strategy. Second, they identify a major means of scaffolding to provide metacognitive scaffolds by organizing the main goal of the learning activity into hierarchical sub-goals. Third, they could observe fading and transfer of responsibility to take place when models went beyond their initial construction and had begun to stabilize via consolidation activities.

Land and Zembal-Saul (2003) examine how reflection and articulation processes on scientific explanations can be supported by scaffolding. This focus is conceptually close to what other authors refer to as conceptual modeling. They examine means to scaffold reflection and articulation on a longer-term time scale and focus on means of scaffolding via peers. Their scaffolds are deployed via a software platform and are mainly metacognitive, based on task-specific prompting. They could show that their design was useful to learners and led to sophisticated explanations, indicating the construction of elaborate abstractions. They, however, found that the utility of "static" scaffolds as provided by their platform was dependent on the background knowledge of the learners. They thus suggest combining their approach with human instruction that provides more explicit scaffolding, especially for novices.

Stender and Kaiser (2015) discuss the importance of scaffolding the process of developing mathematical models related to real-world problems and validate their appropriateness for problem-solving. Rather than describing concrete scaffolds, they focus on the diagnosis of student needs and fading support measures to facilitate independent problem-solving by students. Based on existing research on adaptive teacher interventions, they identify the invasiveness of different types of scaffolding in terms of restricting student's freedom of choice on action. Their empirical results allow them to suggest scaffolding interventions in the model articulation process to facilitate problem solving. Their results indicate, among others, the usefulness of decomposition of the modeling task, availability of model simulation tools, and referring to existing knowledge via metacognitive scaffolds.

5.1.2.2 Scaffolding Argumentative Collaboration for Alignment

Several authors have also examined how to scaffold collaborative articulation, in particular with focus on peer facilitation of argumentative processes.

Abdu et al. (2015) examine how the process of peer facilitation can be supported through scaffolding with whole-group interventions in classroom settings. Their focus consequently is on argumentative design that should prevent unguided model creation. They propose to provide strategic scaffolds by demonstrating solution paths upfront, before peer interaction starts. Furthermore, they establish explicit prompting practices for peer collaboration to establish collective responsibility for the learning process.

Chin and Osborne (2010) show how discursive interaction (i.e., argumentation) can be supported by scaffolding in science education. They propose to provide question prompts for enabling peers to ask questions that allow exploring a problem or a proposed solution in depth (King and Rosenshine 1993). They also suggest providing conceptual scaffolds in the form of additional resources on the topic of interaction and procedural scaffolds in the form of guiding structures (such as writing stems or diagram templates). Finally, the authors propose to work with heterogeneous groups, where participants have different views, to facilitate confronting argumentation (Prusak et al. 2012).

Ge and Land (2004) focus on the development of domain-specific question prompts to scaffold problem-solving in peer interaction settings. They establish guidelines for designing such scaffolds that are based on a combination of generic discursive prompting (King and Rosenshine 1993) and domain-specific prompts that they claim should be developed in cooperation with domain experts. They also suggest structuring a discourse via explicit assignment of roles to learners, which they should take in their interaction.

Vogel et al. (2017) present a meta-study on how collaboration scripts can be used for scaffolding in IT-supported learning environments. Collaborations scripts (Kobbe et al. 2007) are strategic scaffolds that specify a sequence of learning activities to be completed to achieve the aim of a particular task. They found that collaboration scripts have positive effects on the acquisition of domain-specific knowledge in relation to the task and collaboration skills in general. They claim that repeated participation and practice in activities supported through scaffolds by collaboration scripts leads to an internalization of the required performative knowledge, and gradually allows withdrawing the script guidance while still maintaining the developed problem-solving strategies, not only in terms of collaboration, but also regarding domain-specific skills.

5.1.3 A Framework for Scaffolding Model Articulation and Alignment

Backed by the results of prior research outlined earlier, we hypothesize that scaffolds can be developed for the purpose of skill development in stakeholder-centric modeling settings. The modeling process in this context is considered a process of articulation. It is indivisibly embedded in its social context that requires viewing conceptual modeling as a process of co-construction, ultimately leading to a shared understanding about the topic of modeling among the participating actors. Consequently, a conceptual model always only can represent the agreed upon abstractions of the perceived real-world phenomena considered relevant by the participants. Its value is further determined by the chosen representational system that needs to be selected according to the intended purpose, that is, goal of modeling.

Following this understanding, modeling approaches from a scaffolding perspective need to address the following meta-requirements (Walls et al. 1992): (A1) provide scaffolds for the level of model representation (i.e., encoding abstractions in an external representation), (A2) provide scaffolds for the level of model articulation (i.e., developing an understanding about the real-world phenomenon that is the topic of modeling), and (A3) provide scaffolds for the level of collaborative model alignment (i.e., the process of mutually supporting the development of a shared understanding about the topic of modeling and the modeling process itself).

Furthermore, in order to allow for contingency, fading, and transfer of responsibility, (A4) scaffolds need to be provided with different degrees of invasiveness to allow to adapt modeling support to the current needs of the modelers.

Based on these requirements, we draw from the results of our literature review in the following and propose a meta-design (Walls et al. 1992) in the form of a scaffolding framework which should support the process of enterprise modeling on all three levels identified earlier. The framework is visualized in Fig. 5.1.

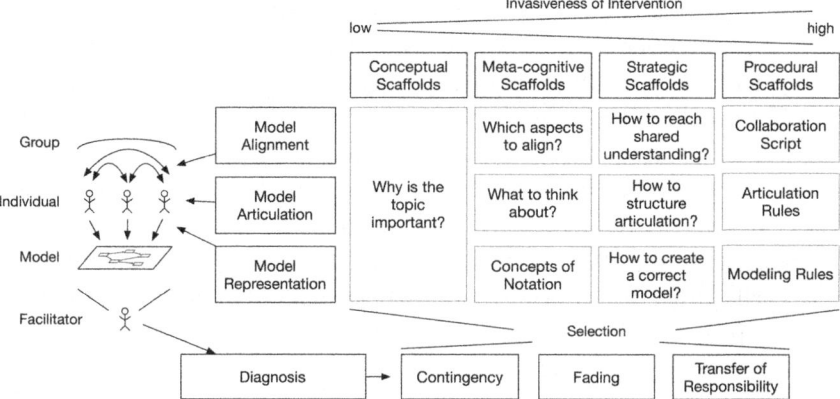

Fig. 5.1 Dimensions of scaffolding during work modeling

The fundamental constituents of the framework are visualized on the top, bottom, and left margin of Fig. 5.1. Its starting point can be found on the left, where our conceptualization of enterprise modeling being activities of a group of actors to create a common conceptual model is shown. As identified earlier, this requires performing model representation, model articulation, and model alignment activities, and is usually supported by a facilitator.

The top margin of Fig. 5.1 structures different types of scaffolds (Van de Pol et al. 2010) according to their invasiveness of intervention (Stender and Kaiser 2015) (cf. A4). Depending on the diagnosis of current needs of the group of stakeholders engaged in modeling, the facilitator is deploying different types of scaffolds following the principles of scaffolding visualized at the bottom margin of Fig. 5.1. The more responsibility is transferred to the modelers, the more support measures are faded out, and scaffolds are deployed (if any) that are less invasive. In case of contingency, the facilitator is free to temporarily fade in stronger support to keep the modelers oriented towards the aim of modeling.

The center of Fig. 5.1 shows the aspects of modeling that should be addressed by different types of scaffolds for model representation (cf. A1), articulation (cf. A2), and alignment (cf. A3). In general, conceptual scaffolds, independently of the addressed level, motivate the topic of modeling, show its relevance, and allow to validate the model with respect to their appropriateness in real-world use. Metacognitive scaffolds support the understanding of the structure of the modeling task and indicate how to conceptualize the real-world phenomenon. On the level of model alignment, metacognitive scaffolds indicate which aspects of a model are subject to alignment (i.e., interfaces between model parts, in contrast to those aspects that remain in the responsibility of individual modelers). Strategic and procedural scaffolds aim at supporting the modeling process itself, either by showing potential behavioral strategies in the first case, or providing stricter guidance in the second case. On the level of model representation, such scaffolds focus on syntactic aspects of modeling, whereas articulation and alignment focus on semantic aspects.

Scaffolds of either form and on either level can be delivered via different channels. They can be provided by procedural guidance or by artifacts (static or interactive media) designed to mediate the modeling process. Procedural guidance can be provided by a human facilitator or an IT-based system, in case the latter is capable of monitoring the modeling progress and analyzing the challenges participants currently face on their individual skill level. Procedural guidance by humans can be provided by expert facilitators or peers, when the latter are provided with further scaffolds that guide the facilitation process itself. Designed artifacts can be provided for domain-specific support and for collaboration support, whereas in both cases their value is determined by an anticipated fit between the skill level of the addressed actors and the support provided by the artifact. As this fit usually cannot be taken for granted, pre-designed scaffolds are usually combined with a form of procedural guidance.

Figure 5.2 gives examples of these different types of scaffolds, distinguishing between different sources of scaffolding. The examples are taken from research cited earlier.

The examples should be considered a non-tentative overview about how the different forms of scaffolding can be provided via different delivery channels. They are deliberately not assigned to the different levels of support indicated in Fig. 5.1 (model representation, model articulation, model alignment), as existing literature does not distinguish these levels.

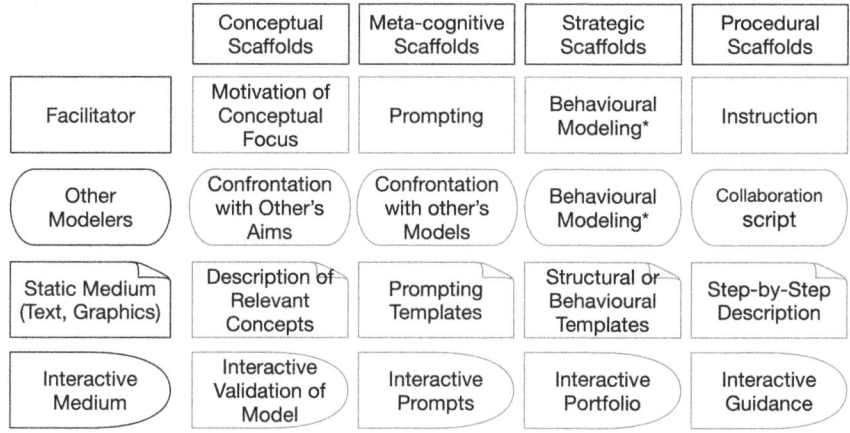

Fig. 5.2 Examples of different forms of scaffolds for work modeling

5.1.4 Scaffolding Articulation and Alignment in CoMPArE/WP

The proposed framework can be used to augment the CoMPArE/WP (Collaborative Model Articulation and Elicitation of Work Processes) method (as described in Chap. 4), which explicitly aims at supporting articulation and alignment of stakeholders' views on their contributions to enterprise processes and the collaboration necessary to implement them. The method follows a multi-step modeling approach, in which participants first collaboratively create a concept map to agree on the notions used to refer to the relevant aspects of their work, then individually model their views on their own contributions and interfaces with others, and finally consolidate these models in a discursive way to create an agreed-upon representation of the overall work process. If modeling rules are adhered to, the resulting models are technically interpretable by a workflow engine, and in this way can be validated through simulated execution (Oppl and Alexopoulou 2016).

CoMPArE/WP offers support measures to enable modeling by stakeholders who do not have any prior knowledge in conceptual modeling. These support measures are briefly described in the following and then classified using the proposed framework (cf. Fig. 5.3). The *global multi-step modeling procedure* is introduced by a facilitator, who is expected to be trained when implementing the method. The participants are also provided with a *one-page written/graphical summary of the global procedure*. The modeling notations for individual articulation and collaborative alignment are pre-specified and are provided via *cardboard model elements* that follow a coloring scheme encoding the semantic elements of the used modeling language. The same coloring scheme is used in *poster-sized printed templates* that indicate the expected model layout that needs to be adhered to in order for the results to be unambiguously interpretable via technical means. Printed *model examples* are provided for reference in case of uncertainty on how to use the model elements or their semantics. The participants have access to *written descriptions of the modeling rules* for each step. In the course of the workshop, the facilitator observes individual model articulation and provides *role-specific prompts* to aid model development. If necessary, the facilitator *demonstrates model development using an example*. During consolidation, the participants are expected to contribute their individual models and support the *identification of model*

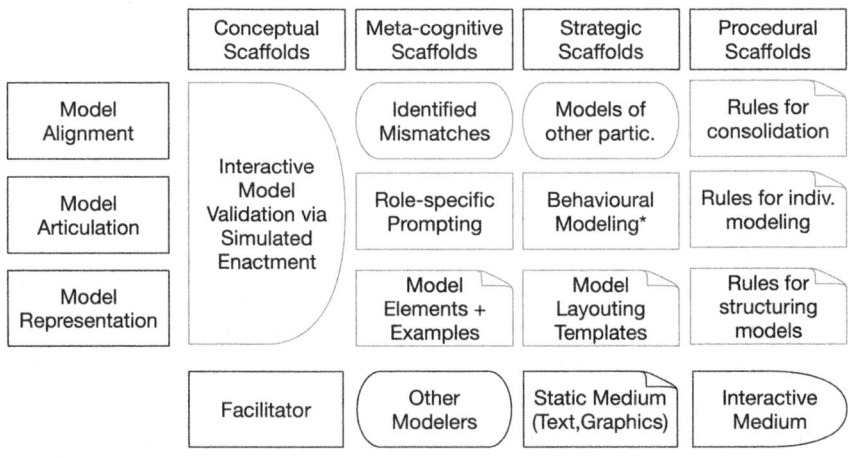

Fig. 5.3 Scaffolds deployed in CoMPArE/WP (references indicate the foundation for design)

Acting on Work Designs: Providing Support for Validation... 191

parts that indicate divergent views on how to collaborate. The identification process is supported via the *contributed models of all participants* that should contain semantically identical model elements in case of agreement on how to collaborate. The resulting model is transformed into an executable version by means of image recognition (Oppl 2015) and *interactively validated via simulated enactment*, enabling participants to identify inadequacies of the developed model.

5.1.5 Example

Applying scaffolding to CoMPArE/WP leads to observable effects caused by how the facilitators guided model representation and alignment. Figure 5.4 shows a *model layout template* used as a strategic scaffold for model representation for consolidation. The three photos in Fig. 5.4

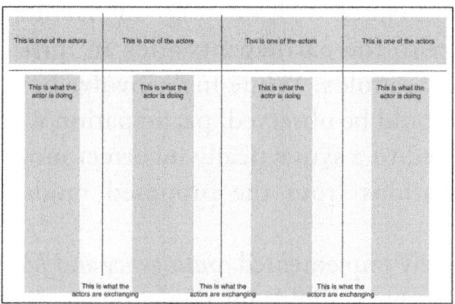

Model Layouting Template

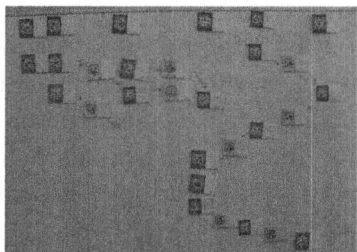

Result Workshop 1

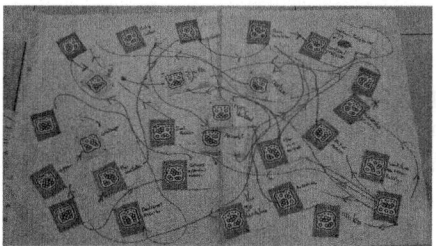

Result Workshop 2

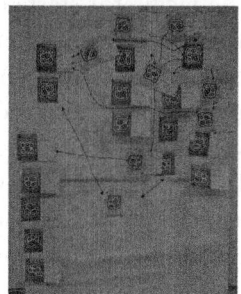

Result Workshop 3

Fig. 5.4 Top left: model layout template; top right and bottom: modeling results of workshops

illustrate the different results of consolidation. On the representation level, the aim is to resemble the layout indicated in the template (blue elements on top, red elements aligned below in lanes, yellow elements placed between lanes). On the alignment level, participants themselves should discover problems in the depicted process (e.g., non-matching communication expectations) and resolve them.

The facilitator in *workshop 1* deployed model representation scaffolds on a strategic and procedural level. Scaffolds for model articulation and alignment were not used. *Fading, transfer of responsibility*, or *contingency* could not be observed. This resulted in a syntactically correct model but led to little involvement of the stakeholders in modeling and articulation and no observable alignment activities. The facilitator in *workshop 2* introduced the global multi-step modeling procedure as a high-level procedural scaffold and provided the participants with metacognitive scaffolds on representation, articulation, and alignment. *Contingency* could not be observed, although participants showed signs of being overwhelmed with the task. The facilitator rather *shifted full responsibility* to the participants after an initial deployment of the metacognitive scaffolds. While high involvement of all participants in articulation could be observed, participation was declining during alignment, and led to a syntactically incorrect modeling result, with semantic deviations from the proposed modeling language.

The facilitator in *workshop 3* actively implemented *contingency* and *fading*. The facilitator started with strategic scaffolds for model representation and articulation, briefly provided procedural scaffolds at the start of the consolidation step, and provided metacognitive scaffolds in case of contingency. The observed modeling process continuously showed high involvement in articulation and alignment, with deviations from the proposed modeling notation in the final modeling result.

This example has shown that scaffolding plays a crucial role in developing models that are both adequate representations of the stakeholder's mental models and syntactically correct descriptions of a work procedure that can be processed further for validation and eventually put to practice.

5.2 Participatory Enactment Support Instrument

Once novel work designs are initially represented as syntactically correct and executable process models, they still require to be validated and potentially need to be elaborated to cover all relevant aspects of the real-world process (Santoro et al. 2010). Validation is usually carried out by chauffeured model walk-throughs (Santoro et al. 2010; Herrmann et al. 2004)—a facilitator presents the model to domain experts and explores the validity of the represented process information—or by simulation games—whereby a process is usually played through in a collaborative setting and potential improvement are identified (Smeds and Alvesalo 2003).

Such simulation games, however, are rarely supported by information systems. They rather take place in a social setting that resembles a simplified version of the actual work environment. Interactive validation through enacting models via a dedicated validation information system, nevertheless, could enable users to play through the model while keeping track of the progress through the model, thus relating work experiences more directly to the properties of an underlying process model. Such approaches are inspired by the idea of user interface prototyping (Floyd 1984; Nielsen 1993) and have been heavily adopted in task-model-based interactive system design (Dittmar et al. 2004; Mori et al. 2002). While proposed for BPM already more than 20 years ago (Berztiss 1996), this idea hardly has been examined scientifically since then, and is only adopted in some commercial tools.

In case of process walk-throughs (Santoro et al. 2010; Herrmann et al. 2004), changes to the work process model are usually performed immediately, allowing users to assess their proposed modifications directly. Immediate changes are usually not possible if validation is based on model execution. Here, design and runtime are kept separate, enabling modifications to the model only in between validation cycles. This is mainly due to conceptual and technical considerations, as changing the model of a process instance during runtime usually has implications on the consistency of the instance information (Floch et al. 2006; Reichert and Dadam 2009).

Validation, in contrast to process execution in production systems, does not lead to actual output. Continuing execution despite missing information or inconsistencies among different process parts thus is an option here, if the validation environment keeps track of such issues and allows manually circumventing or resolving them. Once a process change prevents further execution of an instance, validation can still continue by re-starting an instance.

Direct adaptation of process models in enactment-based validation sessions brings together the advantages of model-based process walk-throughs in terms of immediate reflection of changes, with the advantage of the immediacy of workflow validation via enactment as widely demonstrated in task modeling and UI prototyping. We here introduce a virtual enactment platform for validation and elaboration of process models that addresses this challenge. At the same time, it aims to bring the advantages of facilitated model walk-throughs to the field of validation through model enactment by offering adaptive support in the validation process, offering prompts on what to consider, or pointing at potential further steps to be performed during validation (e.g., as proposed by Herrmann and Loser 2013).

5.2.1 Background: Process Walk-Throughs and Enacted Prototypes

The concept of walk-throughs for exploring the design of a socio-technical system (such as a business process, but also end-user software or enterprise information systems) was first proposed more than 25 years ago. Polson et al. (1992) have proposed to use cognitive walk-throughs to evaluate user interfaces of software. Their approach focuses on individual users and their perception of the socio-technical system. Pinelle and Gutwin (2002) transfer this concept to the area of work support systems and aim at examining groupware usability by means of walk-through techniques. They still focus on the individual users as the main subject in the walk-through and aim at evaluating existing designs rather than actively engaging walk-through participants in design activities. Herrmann et al. (2007) transfer the concept of walk-through to collaborative settings and explicitly integrate design activities in the walk-through process, granting participants an active role in validating and refining a socio-technical system. They draw from earlier work in the area of participatory design (Kensing and Blomberg 1998) and specify the procedures embedded in their approach based on approaches like scenario-based design (Holbrook 1990) or contextual design (Beyer and Holtzblatt 1997). They base their walk-throughs on graphical representations of a model of the socio-technical system, which serve as an anchor for communication and keep the walk-through focused (Herrmann et al. 2002).

Similar concepts have also been proposed for business process elicitation (e.g., Santoro et al. 2010; Front et al. 2017). The proposed approaches, however, focus on knowledge elicitation in models and their transformation to a format along BPM activities that can be processed. They remain vague with respect to the actual procedures in the walk-through and leave guidance to a facilitator, whose role is not discussed in detail. This lack of methodological depth is addressed by Caporale (2016), who proposes an approach in which process model constructs are derived from natural language process descriptions, enabling stakeholders to describe their work using familiar constructs. Hjalmarsson et al. (2015) explicitly examine the role of facilitators in system analysis and design, but focus on identifying generic facilitation strategies rather than describing their activities in detail.

The idea of using prototypes of systems that can be enacted for the purpose of walk-throughs has been adopted early in the area of user interface validation (Nielsen 1993) and is a common practice in this area ever since then. It was eventually picked up for validating work support systems in the area of task-based interactive system design (e.g., Dittmar et al. 2004; Mori et al. 2002). Sousa et al. (2011) propose to combine business-process modeling with task-based user interface design, with involvement of actors to create a better fit between their expectations and the interactive system's properties. A similar approach is proposed by Sukaviriya et al. (2007), who use business process models as the foundation for rapid user interface prototyping.

All these approaches have in common that they distinguish design-time from runtime in model processing and thus do not allow to make changes to a prototype while it is being enacted. More recent research has explored the potential for runtime adaptability of systems (e.g., Hartmann et al. 2008; Floch et al. 2006; Reichert and Dadam 2009; Schiffner et al. 2014), making it possible to modify user interface prototypes (Hartmann et al. 2008) or whole business process architectures (Floch et al. 2006; Reichert and Dadam 2009; Schiffner et al. 2014) while they are being executed.

We adopt the idea of runtime adaptability of processes to extend the ability of performing design activities in the course of walk-throughs, as proposed by Herrmann et al. (2007), to validation and elaboration activities that are based on prototypes of business processes that can be enacted. Such an approach requires facilitation beyond what is required in model-visualization-centric walk-throughs (as not all aspects of the process are visible all of the time), but also what is used in traditional iterative prototyping processes (as changes to the underlying model can occur during running instances). We again adopt the educational concept of scaffolding, as presented earlier, to provide the foundation for developing and deploying appropriate support measures.

5.2.2 Implications of Enacting Dynamically Changeable Prototypes

During participatory enactment, the actors go through the work process step by step. For each step, the responsible actor assesses

- whether the step is correct and described in sufficient detail, and
- whether the next step is the only possible way to progress or if there are alternative ways of continuing with the work process. This can refer to alternative options of progressing, optional activities, or activities that have been omitted in the original model

If any on these assessments lead to the need for changes in the process, these changes would be incorporated in the model underlying enactment immediately. Changes can have different effects that might trigger the need for further changes in the behavior of other actors (here represented in the work models as subjects, as introduced in Chap. 3) or in the overall process. Potential changes in ascending order with respect to their impact on the overall process are:

1. adding, altering, or removing a subject's activities
2. shifting activities from one subject to another
3. altering the sequence of information exchange between subjects
4. adding or removing information required from or provided to another subject
5. involving a new subject in the process

Case 1 refers to situations where only the behavior model of a subject is altered without affecting its interfaces to other subjects. The content, form, and sequence of information exchange remain unchanged. In this case, the changes only affect one actor and do not require changes in the behavior model of other subjects. An example would be extending the behavior model of a subject with an additional optional activity that, in some cases, might be needed to prepare information for an activity already represented in the model.

Case 2 refers to situations where the content, form, and sequence of information exchanges remain unchanged, but responsibilities are shifted from one subject to another. In this case, the affected function state must be incorporated in the behavior model of the target subject. An example would be shifting the responsibility for an already existing activity from one department to another, each of which is represented as a subject in the work model.

Case 3 refers to situations where the sequence of information exchange is altered; however; both content and form remain unchanged. In this case, the subject partnering in the communication needs to adapt its behavior model to fit the new expectations. This might trigger subsequent changes in this subject, which again potentially causes cascaded changes elsewhere in the process. An example without cascaded changes would be altering the sequence of information exchange between the one department and another to optimize for time efficiency in communication. This causes changes in the targeted department, as it has to be prepared to accept and process information already at an earlier point in time in the process.

Case 4 refers to situations where the information exchange is fundamentally altered in a way that adds or removes communication acts to or from the behavior models of the involved subjects. This necessarily causes changes in the targeted subject's behavior model, as it needs to react on new information or provide information that was not expected before the change. This again potentially causes cascaded changes elsewhere in the process. An example would be the decision of one actor to alter communication policies for decisions within an organization. The original way of communication thus is changed and requires the originally involved actors to change the behavior models of their respective subjects.

Case 5 refers to situations when a new subject is added to the process. This requires specifying the communication interface (i.e., the information exchange) with this new subject as well as its behavior if it is known and relevant to the work process. Adding a new subject might have implications on the behavior of the other involved subjects, as additional information exchange might be required. An example would be the addition of an additional actor for decision-making in a work process for specific cases. In this case, the behavior of the other actors needs to change to communicate with the new actor.

In case a change of a behavioral model of a subject triggers the need for changes in the models of other subjects (i.e., in cases 2–5), those changes do not necessarily need to be made immediately. Models of subjects' behavior are only loosely coupled and are basically executed independently. Necessary changes in a subject's behavior, such as addition or removal of activities or messages, need to be kept track of by the enactment instrument and only need to be considered before execution of that respective subject continues. Considering the example used in case 4, this means that the one actor could change his/her process and continue to describe the altered activities from his/her own perspectives. The need for changed communication interfaces and behavior for the other actors would be logged and can be handled when execution of their respective parts of the process continues. The refinement of the process, however, can only be finished, once all open change requests have been resolved by either incorporating them in the targeted subject's behavior model or undoing them in the originating subject's behavior model.

Process validation and elaboration through enactment is a means to complete a process description without the need to create comprehensive formal process models by traditional conceptual modeling. Separation of models and model changes along the different involved subjects reduces complexity and allows focusing on a single subject's behavior at a time. Using the execution engine supports simulating complex decision processes by incrementally adding process variants to the model as the simulation continues. Complex models of collaborative work processes emerge in this way without the need to ever translate one's perceptions of a work process to abstract process descriptions. Still, as the model emerges, it permanently maintains a syntactically valid state that allows for further processing, such as live validation of deadlocks, life-locks, or mathematical simulation of capacities.

5.2.3 Tool Support

We here assume that process models can be validated and elaborated by enacting them in an artificial setting (i.e., not situated in a real-world context impacting actual business cases) and performing changes to model whenever issues are identified. We refer to this process as "virtual enactment" in the following. As is known from facilitated model walk-throughs, domain experts and stakeholders might require support in their model elaboration activities, in particular when they do not have in-depth experiences or expertise in such activities (Hjalmarsson et al. 2015; Herrmann and Loser 2013). The importance of dynamically adapting the level of support depending on the level of experience or expertise of the stakeholders is especially stressed in the study by Hjalmarsson et al. (2015) based on empirical evidence. We therefore augment virtual enactment with the educational concept of scaffolding, which inherently requires supportive interventions to be designed and deployed in an adaptive way (Van de Pol et al. 2010). The overall conceptual approach is thus termed "virtual enactment through scaffolding."

5.2.3.1 Conceptual Considerations

Virtual enactment, as specified here, draws from the concept of facilitated model walk-throughs (Herrmann et al. 2004), which are usually carried out in a co-located group setting, bringing together all involved stakeholders at the same place and at the same time to collaboratively go through the process. A distributed form of model walk-throughs can be imagined (and actually has been considered in our earlier work; Wachholder and Oppl 2012, 2014) but is not subject of the presented approach due to the inevitable loss of communication and negotiation potential that would need to be compensated for by further groupware instruments.

Designing the instrument for co-located synchronous enactment has implications on the necessary support measures. The process fundamentally should be enacted in an actor-centric way, that is, use the involved process actors as the primary dimension for structuring process enactment. At the same time, all involved stakeholders should have the opportunity of observing the progression through the process across all involved process actors, maintaining an overall bird's eye view on what is happening throughout the work process.

Motivated by prototyping research, enactment should allow to go through the work process in an explorative way, as if it were a role-playing game. This implies that more than a single path through the process can be enacted at a time. Hence, there is no need to re-enact the process several times to fully explore it. Enactment following the role-play approach requires that stakeholders can focus on these subjects in the process model, whom they also impersonate in real-world work. This strengthens the point for subject-oriented structuring of process execution. Conceptually separating the behavior of each involved subject (i.e., actor in a specific role) and coupling them by acts of communication and/or exchange of information or physical goods should allow to further strengthen the focus on the individual subjects and support the actors in remaining in their roles, impersonating the subjects. This can be enabled by a process representation that is similar to S-BPM, but adapted to the needs of model elicitation and articulation (Oppl 2016).

Elaboration of the process should be possible during enactment, whenever the need arises, without having to start over the enactment process from the beginning. Such a feature avoids losing the context of the current walk-through and further does not distract stakeholders following their current line of thoughts, when more than one modification should be made. Changes to a process might trigger cascaded need for change, in particular if communication with other process actors is involved (cf. Oppl 2016). Elaboration needs to keep track of unsatisfied dependencies (e.g., information expected by a subject to perform its tasks, which is not provided by any other subject) or other issues introduced by local model changes (e.g., deadlocks or non-terminating loops). Actors need to be pointed to such issues and need to have the opportunity to resolve them.

To deploy scaffolding for supporting the enactment and elaboration process, the provided scaffolds need to be designed in a way that allows for situation-specific support that is adaptable by the stakeholders themselves according to their perceived needs. Scaffolds need to be designed for different areas: the process of enactment and process exploration might require guidance or active intervention, in particular for inexperienced users. Exploration could further be aided by less invasive scaffolds, such as means to display a graphical representation of the model and the current state of enactment on demand. The elaboration process should be provided with scaffolds in a way that does rely on any modeling skills, as these cannot necessarily be expected from stakeholders. Issues and inconsistencies in the model introduced by local model changes through elaboration can also be pointed out via scaffolds that are dynamically generated based on an analysis of the current state of the model. In any case, stakeholders must have the freedom to ignore scaffolds, dismiss them, and ultimately take responsibility to request support when they consider it necessary.

5.2.3.2 Architecture

Based on the concept and requirements described in the previous section, we have developed an online platform for conducting virtual enactment of process models supported by scaffolding measures. In the following, we give an overview about the platform architecture, before we detail the features of the implemented modules.

The virtual enactment platform is implemented in a modular way and accessible to users via a web-based interface. Figure 5.5 gives an overview of the overall architecture of the platform. UI components are shown at the top and bottom of the figure, while functional modules are grouped in the center.

The *VirtualEnactment Core* provides fundamental workflow execution capabilities that are used for enacting a process model. As such, it acts as the anchor for all other components, which enable elaboration of the currently executed model and provide support to users in the selected process.

The *Visualization Engine* renders graphical representations based on the current process and can visualize the execution progress of the current instance. Graphical representations are based on the S-BPM approach, composed of Subject Interaction Diagrams and Subject Behavior Diagrams, as introduced in Chap. 3.

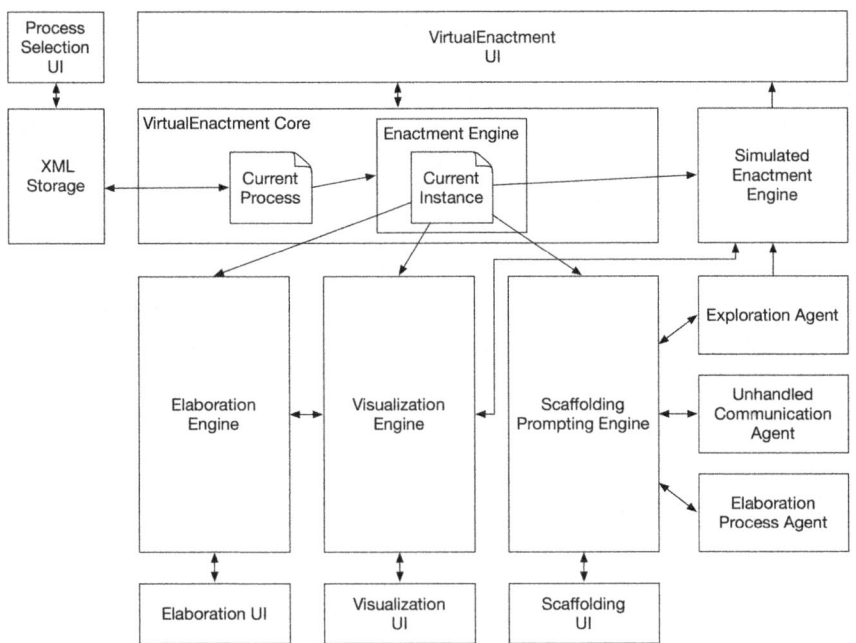

Fig. 5.5 Platform architecture

The *Elaboration Engine* allows for changing a process model while an instance is currently being executed. Retrieving the necessary information is based on prompting. Users can indicate that they consider the currently proposed activity to be inappropriate, and they are then interactively led through the process of providing the information necessary to make the change to the underlying process model.

The *Simulated Enactment Engine* enables to have the system automatically determine a path to a given target state in the model (across the behavior of all subjects) and enact this path in a user-traceable way (using UI-scripting, i.e., state transitions are reflected on the user interface).

The *Scaffolding Prompting Engine* enables to provide scaffolds to users in a flexible way. Users can dynamically change their required level of support, which is reflected in providing more of less concrete scaffolds. Scaffolds are basically based on textual prompts, but can also provide interactive support measures (such as automatically progressing to a particular state in the model using the simulated enactment engine). The scaffolds themselves are generated by *Scaffolding Agents*, which can focus on different aspects of the modeling and elaboration process. Three different agents are currently provided, which are described in more detail later.

The *XML Storage* component provides functionality to upload a new process model and download altered process models in a proprietary XML format. Users can furthermore select from different sample models or start a new process specification from scratch (as the elaboration engine provides bootstrapping features that allow defining new subjects, their behavior, and their interaction using the same prompting mechanism as deployed for elaboration).

5.2.3.3 VirtualEnactment Core

The VirtualEnactment Core is the component used for executing a process during virtual enactment. The execution engine consequently is tailored towards this use case. Virtual enactment does not require distributed user interfaces (i.e., only requires one common interface for all participants, visualizing the current states of all subjects at the same time).

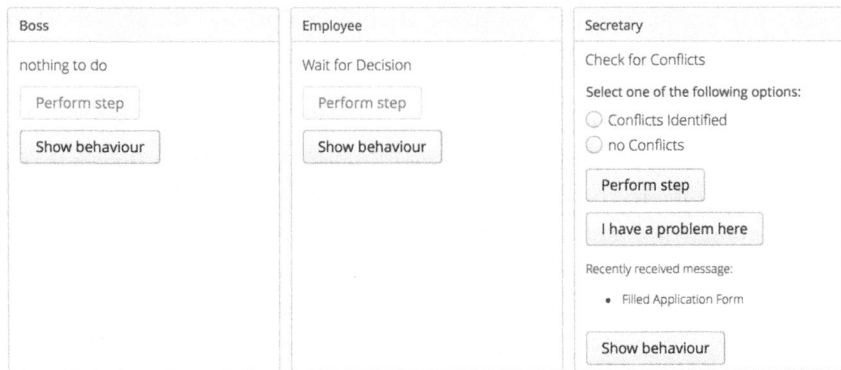

Fig. 5.6 Enactment UI (released under a Creative Commons Attribution 4.0 International License (CC BY 4.0))

Incoming messages are stored in an input pool, from which they are removed as soon as the respective receive state is triggered.

Figure 5.6 shows an example of the enactment UI. As can be seen, subject *Secretary* currently is in action state with two potential outcomes (cf. Fig. 5.8 for a visualization of the respective subject behavior). Subject *Employee* is on hold in a blocking receive state, and the behavior of subject *Boss* has not yet started. The button labeled *Perform step* triggers the shown state to be processed by the workflow engine, the button labeled *I have a problem here* triggers the elaboration engine (see later), while the button labeled *Show behavior* triggers the visualization engine to display the respective Subject Behavior Diagram.

Elaboration can lead to overall processes that contain inconsistent subject behaviors. One subject's behavior might rely on the availability of a message, which is not currently provided by the envisaged sender. Such messages are added to the envisaged sender's pool of expected messages and can be triggered via the UI as shown in Fig. 5.7.

In turn, subjects might also provide messages to other subjects, who do not react at these messages in their current behavior. Such messages are added to the envisaged recipient's pool of provided messages. The pool of expected and provided messages is used as a source of information by the elaboration engine. In this way, expected and provided messages can be

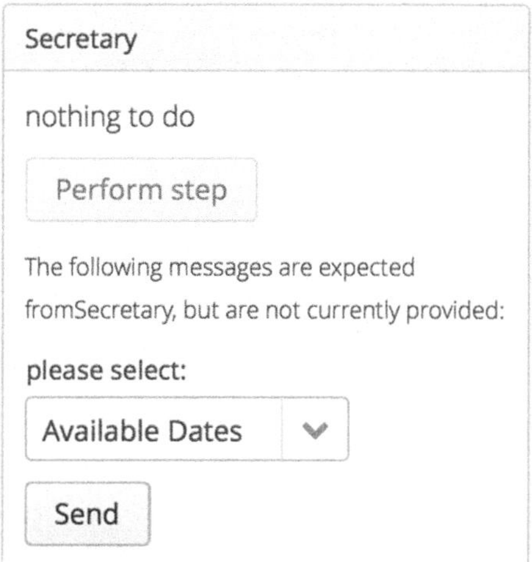

Fig. 5.7 Expected messages in subject UI (released under a Creative Commons Attribution 4.0 International License (CC BY 4.0))

incorporated in a subject's behavior by adding send and receive states, respectively.

Once an instance of the process is finished, it can be restarted to enact it another time. The new instance takes into account all changes to the process model that might have been made via elaboration during prior enactments. In this way, the process model can gradually be explored in all its variants and be elaborated, where necessary.

5.2.3.4 Visualization Engine

To enable users to create a link between the current state of the simulation and the underlying model, visualizations of the model can be displayed at any time during exploration. The visualizations are available in different levels of complexity and from different perspectives on the process (view per actor, overall actor-centric view, overall flow-oriented view), and are augmented with information about the current instance, such as the currently available activities and the path through the process. The visualizations are created dynamically using the GraphViz software suite (Ellson et al. 2001).

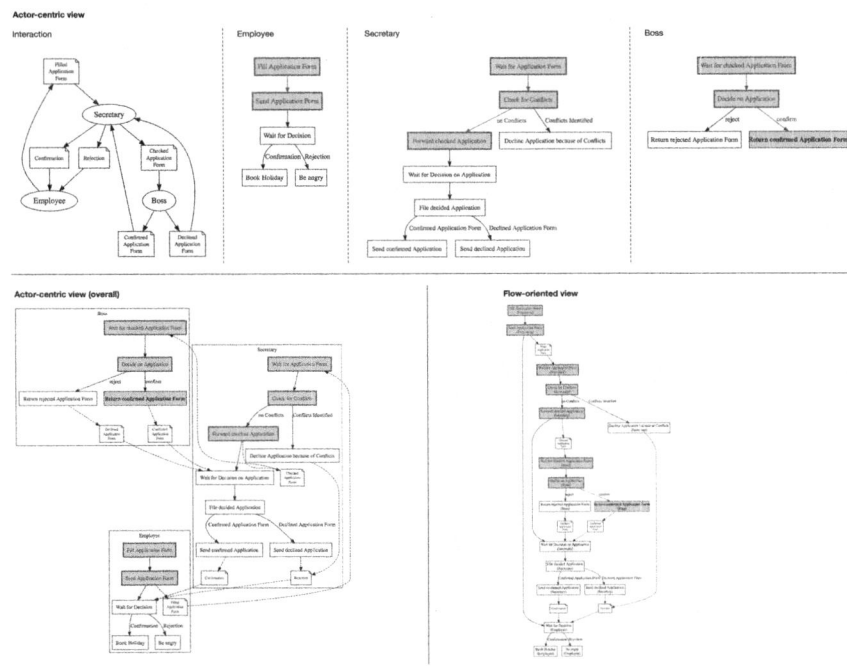

Fig. 5.8 Process visualizations (released under a Creative Commons Attribution 4.0 International License (CC BY 4.0))

Figure 5.8 shows visualizations for an instance that has been simulated halfway through a sample process. The four models at the top of the figure together form the least complex visualization, where the behavior of each actor is shown as a separate model. The model at the extreme left shows the interaction among the actors. The gray boxes indicate already executed activities, whereas green boxes represent currently available activities. The lower left model in Fig. 5.8 compiles the separate actor models in a single visualization and enriches them with connections representing the exchanged messages. The lower right model removes the actors as the primary structuring dimension for the overall model and, in this way, provides a flow-oriented view on the process. Users can switch between the actor-specific behavior models, the interaction overview and the two overall views at any point in time and, in this way, can focus on different aspects of the model in the course of simulation or reflection.

5.2.3.5 Elaboration Engine

The elaboration engine allows modifying a process model while an instance of the process is currently being executed. It is anchored in the enactment UI and is always triggered in the context of a particular state. Whenever users consider a state inappropriate to be executed (for whatever reason), the elaboration engine allows to alter a process in the context of that particular state.

Elaboration is guided via interactive prompting. Users are not confronted with business process modeling concepts or nomenclature but can describe what they want to change in the currently enacted work context. Figure 5.9 shows the sequences of prompts in the interactive elaboration process. Boxes indicate user interaction prompts, while vertical brackets indicate changes made to the process model in the background.

An example for an interactive prompt is shown in Fig. 5.10. It shows the user interaction for the element labeled *specify new step or select existing one* in the topmost branch in Fig. 5.9. Clicking the button labeled *Let me choose from existing steps* would trigger the visualization engine and display the behavior diagram for the respective subject in interactive mode. Alternatively, a new activity can be specified by entering its name in the text field. If the checkbox labeled *This step leads to results I can provide to others* is ticked, the optional path towards element *specify message & recipient* is triggered additionally.

Wherever necessary, the prompts dynamically adapt to the current state of the process. Figure 5.11 shows an example for this feature. It shows the user interaction for *specify message & recipient* as referred to earlier.

In this particular case, the pool of expected messages for the respective subject already contained a message named *Available Dates*. In case the users want to incorporate this message in the subject's behavior, no further input is necessary, as the envisaged recipient of the message has already been specified in the elaboration process, during which the message was defined (via the user interaction element *specify input & sender or source* in the lowermost branch in Fig. 5.9). If users choose to specify a new message to be provided to others (as shown in Fig. 5.11), the list of

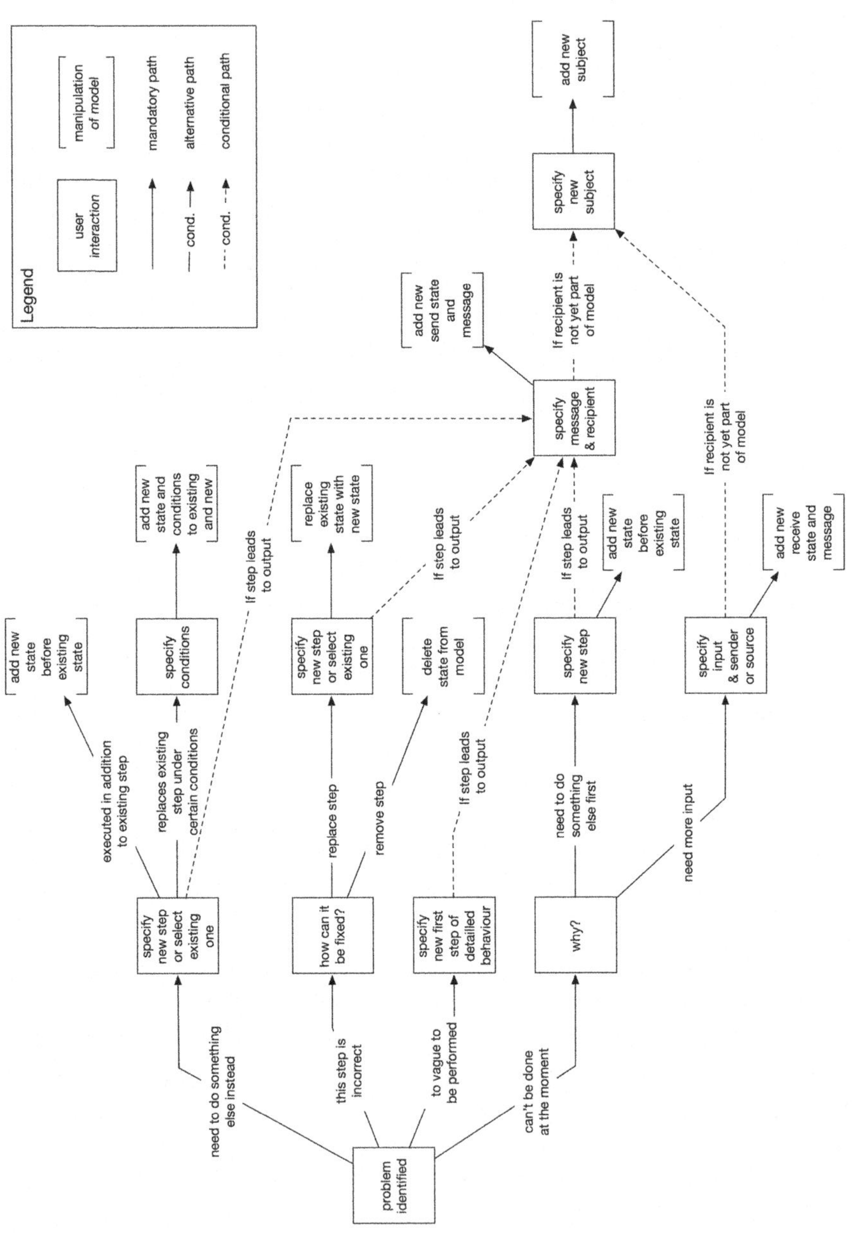

Fig 5.9 Prompting sequence for elaboration

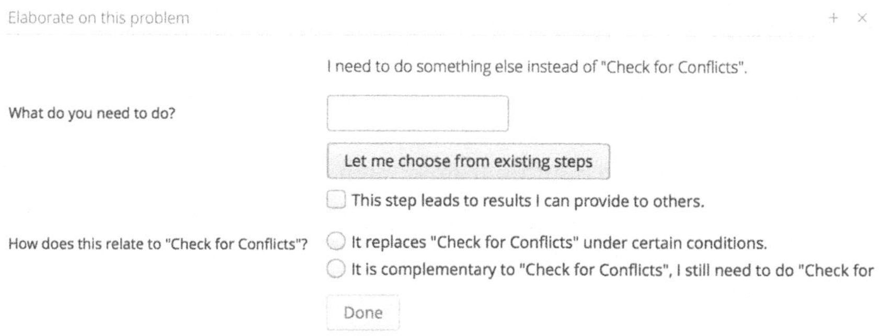

Fig. 5.10 Example for interactive elaboration prompt (released under a Creative Commons Attribution 4.0 International License (CC BY 4.0))

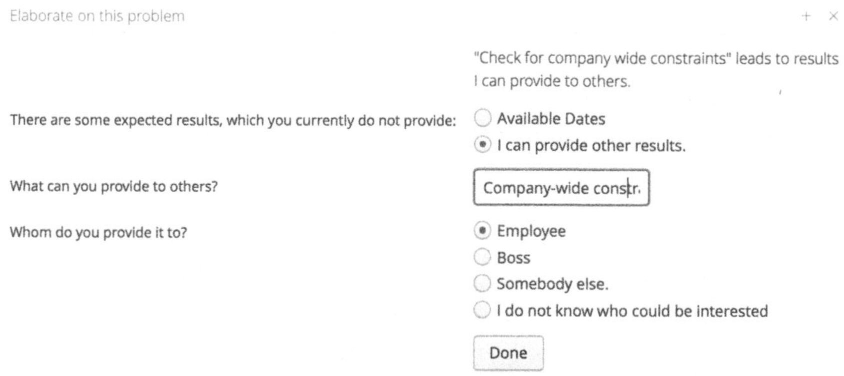

Fig. 5.11 Specification of messages during elaboration (released under a Creative Commons Attribution 4.0 International License (CC BY 4.0))

potential recipients is dynamically created from the set of subjects currently contained in the process model. The option *Somebody else* consequently would trigger the option path leading to the addition of a new subject. The option *I do not know who could be interested* creates an anonymous subject, which is shown on the enactment UI. While the behavior of anonymous subjects cannot be elaborated, they still can be used to trigger sending of expected messages (in case the envisaged sender was unknown during elaboration of required input).

The elaboration engine also has process-model bootstrapping capabilities, that is, it can be used to elaborate an initially empty process model.

In such cases, the enactment UI offers to add an initial subject. The behavior of this subject can then be elaborated by adding an initial state. Whenever the behavior of a subject is finished in a particular instance, the elaboration engine offers to add an additional state. In this way, process descriptions can be built up from scratch, elaborating the behavior of the initial subject and its interaction requirements in a first-process instance, and then gradually refining the model in follow-up instances.

5.2.3.6 Simulated Enactment Engine

Playing through instances of complex processes several times might become tiresome, especially when the initial parts of the process are already agreed upon and elaboration is going on in later parts, which need to be manually navigated to in each new instance. The simulated enactment engine provides functionality to automate this navigation process and start manual enactment only in the still interesting or questioned part of a process.

The simulated enactment engine searches for a path to the specified target state in the respective subjects' behavior. It then recursively traverses the messages of all encountered receive states, searching for paths to the respective send states in the sending subject's behavior. In this way, a subject-spanning path to the requested target state is compiled, considering both, the states to be executed and the decisions to be made.

The sequence of steps constituting the path from the current state of the running process instance to the requested target state is then used for UI-scripting. The steps are executed with short delays in between, making it possible for users to follow the simulated enactment process on the UI. Simulated enactment stops at the requested end state and hands back control to the users for further manual enactment of the currently running instance.

5.2.3.7 Scaffolding Prompting Engine

The scaffolding prompting engine provides dynamic and user-adaptable scaffolds for different aspects of the exploration and elaboration process. The engine offers an extensible architecture, relying on scaffolding agents to provide the actual scaffolds for a given enactment situation. Scaffolding agents are dynamically registered with the engine and can draw from any source of information (in particular, process and instance data). They can choose whether they want to be triggered for providing new scaffolds after each execution step or whether a new instance is going to be started. Concrete examples of different scaffolding agents are described later.

The basic form of a scaffold is a text-based prompt that is displayed in a dedicated area of the UI below the subjects of the current process. Figure 5.12 gives an example of the selection of scaffolds displayed in a table-like form.

The slider bar on the left border of the area can be used to adapt the concreteness of scaffolds to be displayed. Depending on the requested level of concreteness, the engine displays either procedural (most concrete), strategic, metacognitive, or conceptual scaffolds (least concrete). Placing the slider at the bottom turns off the display. A scaffolding agent consequently provides groups of scaffolds of different types on a particular issue. For instance, such a group can contain a procedural scaffold and a metacognitive scaffold, omitting strategic and conceptual scaffolds. The engine then displays the most concrete scaffold for the level requested by the users. If users requested strategic scaffolds, the metacognitive scaffold would be displayed.

A further level of user control with respect to displayed scaffolds is the ability to dismiss scaffolds. The engine keeps track of dismissed scaffolds and does not display them as well as less concrete scaffolds of the same group anymore, although they still might be provided by the scaffolding agents. This avoids annoying users with scaffolds that they deem unhelpful or unnecessary.

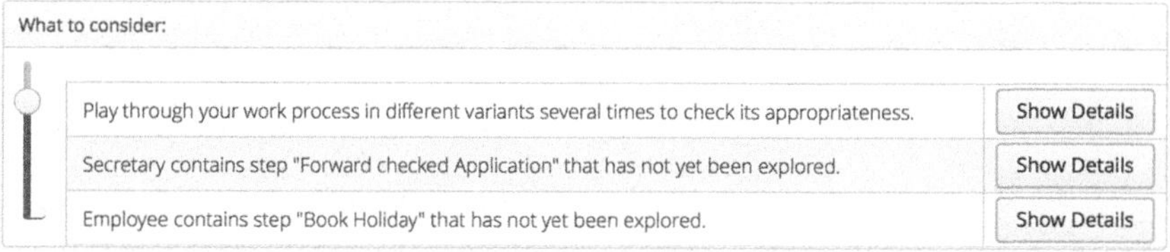

Fig. 5.12 Scaffolding prompts (released under a Creative Commons Attribution 4.0 International License (CC BY 4.0))

The text-based prompts displayed in the table can be detailed in arbitrary ways. The current implementation—besides the basic single-line prompt-only scaffold—offers a type of scaffolds that can display further information in a pop-up window upon request, and a type of scaffolds that can trigger the simulated enactment engine to take users to that part of the process which the scaffold suggests exploring further (cf. Fig. 5.13).

The scaffolding prompting engine is triggered by the process execution engine after each change in the process instance or the underlying process model. The engine informs its registered agents according to their requested update-frequency (per instance or per executed step) and provides them with information about both, the currently used process model and the current instance.

The *elaboration process agent* is a simple agent implementation, which does not provide any dynamically created scaffolds at all. It aims at supporting novice users to handle the platform. Consequently, it provides scaffolds, introducing the features of the platform as being distributed in the first few executed instances. While initially users are only asked to explore the process using the execution UI, the agent gradually offers scaffolds introducing the visualization UI and the elaboration UI. In this way, users are introduced to the platform features step by step.

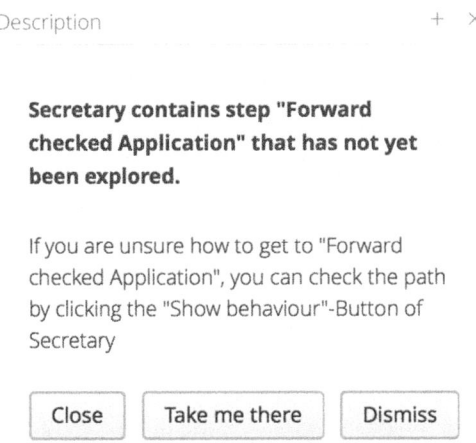

Fig. 5.13 Example for exploration scaffold (released under a Creative Commons Attribution 4.0 International License (CC BY 4.0))

The *exploration agent* keeps track about the already executed states contained in the process model across all executed instances. In this way, it can provide scaffolds on whether a subject behavior still has unexplored parts, which could be visited by the users in the current instance. The agent always only provides one scaffold per subject, only pointing at one per instance, not indicating all unexplored model parts of a subject behavior at once.

Figure 5.13 shows a strategic scaffold pointing at an unexplored part of the behavior of subject *Secretary* (behavior shown in Fig. 5.8). The button labeled *Take me there* triggers the simulated enactment engine, which automatically progresses the current instance to the suggested state, independently of the current state of the instance (assuming the suggested state still can be reached, otherwise a prompt to try again with the next instance is displayed).

The *unhandled communication agent* offers an example which tries to support the actual elaboration process. It keeps track of the pools of expected and provided messages for each subject and provides scaffolds that point users at, or directs them towards, resolving such modeling issues.

Figure 5.14 shows a metacognitive scaffold for the fact that the subject *Secretary* has expected messages, which is currently not provided by its

Fig. 5.14 Example for unhandled communication scaffold (released under a Creative Commons Attribution 4.0 International License (CC BY 4.0))

behavior specification. The respective strategic or procedural scaffolds would contain more concrete directions on how to potentially resolve this issue.

As the availability of provided and expected messages might change at any time due to user elaboration activities, this agent registers not to be updated per completed instance, but per each completed execution or elaboration step.

5.2.4 Conclusive Summary

The goal of exploring a business process is to understand the depicted information fully and to compare this information to the actual perceived context of the work process, that is, whether the portrayed information matches informal specification as established during the information elicitation phase. Usually, there is a rather strict distinction between the tasks of a system analyst and domain experts in a modeling process (Frederiks and van der Weide 2006), which leads to model alterations not being part of the validation process itself, but rather between validation iterations, as a task that system analysts take over. However, as validation through virtual enactment aims at reducing the need for facilitators and system analysts during the validation phase, this step is considered part of validation. Finally, consolidation activities are required in case different domain experts or stakeholders have opposing views on the depicted information or possess different points of views about the reality the model is an abstraction of.

Reflecting these validation steps against the presented instrument's capabilities and functionalities, the following observations can be made: there are multiple possibilities to explore a business process, utilizing the platform. First, the main intended usage of exploring the process is by enacting it akin to a role-playing game. As business processes in most cases are not comprised of only a single, linear path but exhibit multiple decision points resulting in forking paths which eventually are joined together again, it is usually not possible to explore the complete business process during one instantiation. In addition to process enactment, it is also possible to visualize the process in various ways, such as depicting a Subject Behavior Diagram for each actor, a Subject Interaction Diagram, or the whole process as a control flow diagram. The main target audience of the virtual enactment instrument include domain experts and stakeholders who are participating in the represented work process and are assumed not to be experienced in modeling notations or visualization methodologies. Accordingly, visualization methodologies should be seen as an enhancement of process enactment and, with regards to the target audience, not as the main means of process exploration.

Although making changes to the business process model is not an explicit part of the validation from the domain experts' point of view, due to factors already mentioned, it is an integral part of the business process validation in conjunction with the instrument. Changes to the underlying model can only be made while the process is currently enacted. Although constraining the freedom to change the model, it enables users to alter the model directly where the need for changes arises, thereby reducing the cognitive load (Forster et al. 2013) of having to remember those activities that are not reflecting reality in the desired way. Additionally, the instrument does not require its users to possess prior knowledge of modeling concepts, modeling notations, or to completely understand the visualizations of the model. Model alterations are conducted by following a series of elaboration prompts, using natural language, ultimately leading to the anticipated changes to the model.

The instrument does not explicitly implement means to support consolidation activities. However, this does not necessarily mean that it is unable to support the validation of business processes for that reason, since the platform is currently implemented in a way that requires process participants, that is, actors, to be in the same room, or rather in front of the same screen. Therefore, it can be argued that different point of views or varying perspectives which require consolidation activities can be resolved via means of face-to-face discussions.

5.3 S-BPM-Driven Execution of Actor-Centric Work Processes

Subject-oriented Business Process Management (S-BPM) (Fleischmann et al. 2012) explicitly considers the role of actors during process design and execution. The primary elements of structuring are subjects that separate a process along who performs work in the process. How to deal with knowledge, however, is not explicitly taken into account in this approach.

Knowledge-intensive business environments (Dalmaris et al. 2007) claim for support measures that allow for actor-aware and agile process management (Bruno et al. 2011). Agility allows for overcoming contingencies during work on an ad-hoc basis (Minor et al. 2008), and actor awareness enables individualized workflow execution (Prilla and Nolte 2012). Individualized workflow execution does not restrict actors in how they perform their work while still maintaining reliable interfaces to others and pursuing the overall work goals. It furthermore enables situation-specific IT support that is tailored to both, the current work aim and the person executing the work (Monsalve et al. 2010).

The operative aspects of S-BPM are basically specified by a set of actors carrying out different activities in a set of activity bundles. In contrast to other BPM approaches (for a comprehensive overview cf. Weske 2010), the BPM activity bundles are not specified to be carried out in any particular order (whereas most other approaches follow a cyclic approach that is implemented by iteratively stepping though the different phases of business process management).

S-BPM specifies a set of four roles that are relevant when implementing the activity bundles (Fleischmann et al. 2012). They can be linked to the concepts developed by Firestone and McElroy (2003) in the Knowledge Lifecycle (KLC) Framework as described in Chap. 1, thus intertwining work modeling activities with knowledge management processes:

- *Actors* are people that are actively involved in the work process. In S-BPM, they are the source of process knowledge and the ones who put this knowledge to practice again. In relation to the KLC, actors are the primary entities in the business processing environment and thus are the main triggers of learning processes.
- *Governors* are responsible for development, implementation, and monitoring of processes from an organizational perspective. They define the general condition under which a process is implemented and can be altered. While their role is limited in the immediate activities of the KLC, they determine the fundamentals under which processes can be improved in an actor-centric way.

- *Experts* are providers of knowledge that none of the involved actors have in a specific situation. They support activities in all activity bundles of S-BPM with their expertise. Their role in the KLC is also spread across support during business processing and knowledge processing.
- *Facilitators* aid organizational development and provide support during activities, leading to process change. They are not directly involved in operative work activities. In the context of the KLC, their role is mainly important in the knowledge processing environment.

All four roles work together to implement process improvement and adaptation to the current organizational situation. The activities leading to this change are structured into seven activity bundles (cf. Fig. 5.15). As mentioned earlier, these bundles can be executed in arbitrary sequence depending on the current situation in which process change is triggered. Not only sequence but also the actual activities carried out during these bundles are not fully pre-specified and again depend on the situation in which an activity bundle is triggered (Fleischmann et al. 2012). We can implement an instance of the S-BPM activity bundle that is specifically

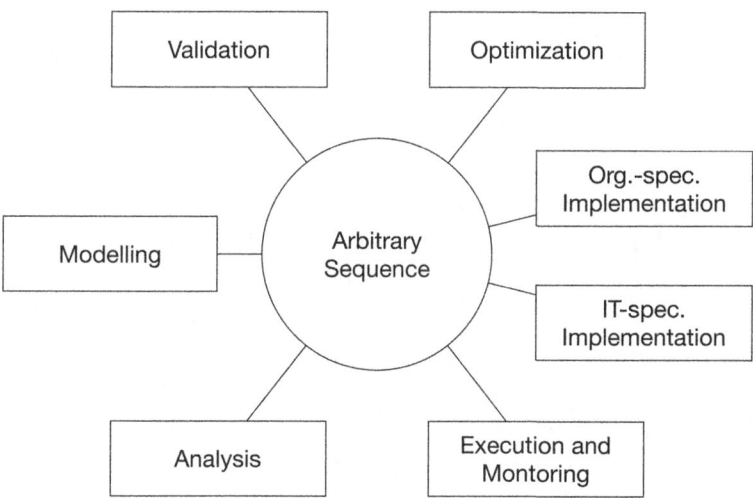

Fig. 5.15 The S-BPM activity bundle (adapted from Fleischmann et al. 2012)

tailored to support the KLC in both, possible sequences and methodological support for activity bundle implementation.

- *Analysis* covers activities concerned with examining a process and the conditions under which it is executed. It collects and structures information about a process and thus provides input of a number of other bundles that are concerned with process change. This bundle might be triggered externally to set up a new process, but also from monitoring during process execution, especially when a mismatch between expected and actual outcome occurs.
- *Modeling* refers to all activities that deal with conceptualizing a process by abstracting it from its real-world execution environment and describing it as a generic phenomenon that can be reproduced given certain conditions. Process models cover different aspects of work (depending on the chosen modeling approach) and conceptualize them in form of a graph. In general, at least information about the activities to be carried out (what?), their sequence (when?), the actors carrying out the activities (who?), and which resources are required (with what?) is represented. Specifically, the who-dimension is used in S-BPM as the primary dimension for structuring the models, which makes it especially suitable for representing processes from an actor's perspective. Actors can focus on their own activities and their interaction with others without the need to have an overall view on the whole process.
- *Validation* is the activity bundle concerned with testing the effectiveness of a process, that is, if its implementation leads to the desired outcomes. Validation is not necessarily carried out in the real work situation but can also be performed in artificial, simulated work situations that allow for testing different relevant aspects of the process (depending on which parameters are considered in setting up the artificial situation).
- *Optimization* refers to the activities that improve the efficiency of processes. Originally described to focus on an economic approach to efficiency, optimization in this context refers to activities that improve the implementation of an already existing process without questioning it. This activity bundle is mainly triggered if parameters of the organizational situation in which the process is carried out in change and the expected outcome cannot be reached anymore.

Acting on Work Designs: Providing Support for Validation... 219

- *Organization-specific implementation* deals with all activities that are necessary to implement a new or changed process within the organization, that is, rolling out information about the changes or changing organizational structures and responsibilities, when necessary.
- *IT-specific implementation* covers activities that are concerned with implementing a new or changed process in an IT-supported environment, for example, using a workflow management engine or groupware systems.
- *Execution and monitoring* finally cover all activities that are carried out during a process conducted by actors in a specific organizational situation. Monitoring is especially relevant for identifying deviations from expected process outcomes and triggering activity bundles to compensate these deviations.

The activity bundles of S-BPM can be linked to the KLC at several points. Integrating S-BPM with the KLC leads to an instance of the KLC that is methodologically augmented to handle business process knowledge. The activity bundles of S-BPM will be put into relationship with the building blocks of the KLC in Fig. 5.16, which provides an overview of the integrated framework.

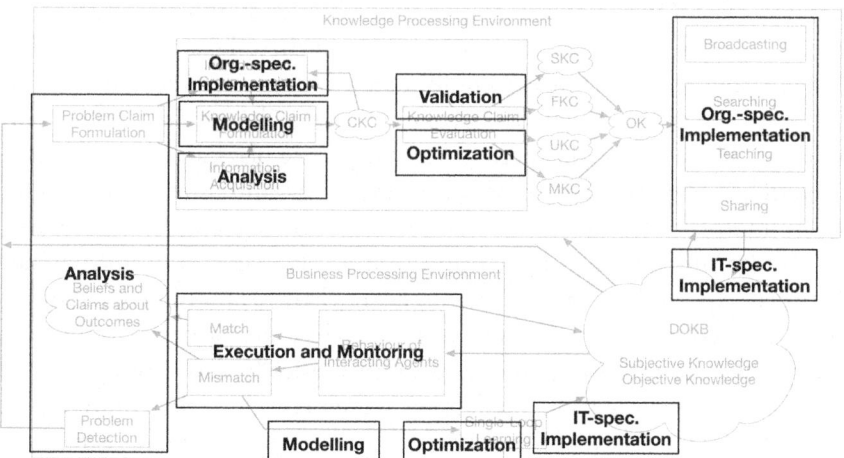

Fig. 5.16 Integration of the KLC with S-BPM activity bundles

5.3.1 S-BPM Activity Bundles in the Business Processing Environment

Execution and Monitoring can be located at the core of the business processing environment in the KLC. It covers actual work being performed by actors (termed "behavior of interacting agents" in the KLC) and the identification of deviations between expected and actual outcome of work (which refers to the "monitoring"-part of this activity bundle). It is important to note that "behavior of interacting agents" does not necessarily refer to the execution of a complete work process (i.e., working until the aim of the work should basically have been reached). The identification of mismatches is an ongoing, parallel activity that can trigger compensation activities, whenever contingencies or unclear situations arise.

The activities during execution are selected and motivated by the contents of the distributed organizational knowledge base (DOKB), which can be codified in IT-support systems and thus be propagated to the business process environment (as, for example, is the case if workflow management systems are used). In this case, the behavior of the actors is directed and coordinated by IT system. In a less IT-supported system, information about how to perform work under particular circumstances may be codified in documents and be provided to the actors for reference. Finally, the DOKB also covers personal experience and expertise of all members of the organization, which also can drive how work is carried out. In real-world settings, a combination of any of these information types will be encountered. The proposed approach especially focuses on settings, where personal experience and expertise plays an important role (i.e., in knowledge-intense processes or expert-organizations, respectively) in an interplay with one of the first two DOKB-sourced drivers of work.

Monitoring covers the identification of matches and mismatches between expected and actual outcomes. In S-BPM, monitoring is originally limited to technical measures to collect performance indicators (e.g., from a workflow engine) and mathematically derive deviations of performance figures from predefined reference values. In combination with the KLC, this understanding of monitoring has to be extended. People involved in the work process can identify mismatches of outcome, contingencies, or unclear situations based on their knowledge and thus are also triggers for compensation activities. In terms of S-BPM roles, not only *actors* can trigger this behavior from directly within the work process but also *governors* might be able to identify problems or mismatches from their ongoing monitoring activities.

Independently of how the problematic situation or outcome deviation has been identified, the choice of whether to compensate directly (entering the single-loop learning cycle) or trigger a more fundamental examination (may lead to double-loop learning) is a decision to be made by humans (made by *actors* themselves or *governors*). In the case of actors identifying a problem and compensating it directly within the work process, the whole sequence of problem identification, deciding what to do, and performing the compensation activities is not necessarily a conscious process. Still, altering one's behavior due to perceptions of the environment that deviate from what was expected alters how an individual will react in future situations and thus is considered learning (cf. Chap. 1 for a more elaborate argumentation on this claim). If the problem is consciously recognized, actors or governors might still decide not to question the way in which work is performed fundamentally, but to immediately compensate the problems that have occurred (actually, this will be the far more common case).

If immediate compensation is chosen, that is the single-loop learning cycle of the KLC is followed, the business processing environment is not left. The KLC here does not give any clues on which activities are happening in this case. When integrating the KLC with the S-BPM activity bundles, there are several candidates for activities that can be performed in this case. They should, however, be considered optional, as their application depends on the situation the problem was triggered in and the organizational context in which the work was performed (e.g., it makes a difference whether the problem occurred in work process which has been codified in a workflow engine or solely was driven by personal expertise without any external guidance).

Modeling the alternative steps that were taken to compensate the problem can be a first step to persist in learning about the specific situation and potentially make it available to other people in the organizations. Modeling here does not necessarily refer to building a formally correct and complete business process model but to any form of conceptual externalization of which activities were set in reaction to the particular situation that occurred.

Optimization is the core step of the single-loop learning sequence and covers the activities that adjust aspects of the work process to fit the given situation. This goes beyond the original S-BPM-understanding of this activity bundle that mainly covers measures to increase process efficiency. If a codified model of the work process exists, either being already available in the DOKB or being created on the fly during an instance of the modeling activity bundle, optimization can be considered tinkering the process parameters, such as which and how many people are involved in a certain role in the work process or altering execution and communication patterns to compensate the problems that have occurred. Again, optimization might also be an "invisible" activity, being performed by actors unconsciously.

Finally, if IT-support has been involved in the work process, the altered way of performing work might require adaptation in the IT-systems, which is covered by the S-BPM activity bundle "IT-specific implementation." This covers changes to workflow definitions (e.g., introducing alternative activities and changing sequences) but also configurations of groupware systems, such as granting additional actors access to communication facilities in which the respective work process is coordinated.

Whatever compensation activities have been taken and whether they lead to externalized, codified results or solely new behavioral patterns of the involved actors, the changes become part of the DOKB and influence future executions of the work process in similar situations.

5.3.2 S-BPM Activity Bundles in the Knowledge Processing Environment

The double-loop learning cycle that involves activities in the knowledge processing environment is triggered either by a problem that cannot be resolved immediately by compensation activities in the business processing environment, or deliberately by actors or governors to revise a work process, that regularly causes problems that can be compensated but cause overhead and hamper effective and/or efficient execution of work. Triggering the knowledge processing environment in the latter case should be considered an asynchronous activity that does not prevent further instances of the particular work process to be executed in the business processing environment until the double-loop learning cycle is completed.

The block "problem detection" in the KLC refers to identifying problems in the DOKB that lead to the mismatch in outcomes or cause the problems that have been observed. Using the S-BPM activity bundles as a frame of reference, this block already belongs to "Analysis" and leads the way towards leaving the business processing environment and entering the knowledge processing environment. In the course of the analysis activities, the problem in the DOKB is not only identified but being codified in a "problem claim." A problem claim describes which element in the DOKB is likely to cause the problem, how the problem manifests in the actual execution of work, and under which situation (i.e., conditions in the work environment) it occurs. An element in the DOKB can be anything that bears or codifies knowledge, that is, a set of actors, documents, and workflow definitions.

Based upon the problem claim, the people to be initially involved in the resolution of the problem (referred to as "knowledge production" in the KLC) are identified. This activity and all remaining activities in the knowledge processing environment are driven by a person taking the S-BPM role of a *facilitator*. This person might have been involved in the original work process as an *actor* or a *governor* but can also be a dedicated organizational role or be recruited from outside the organization. Identifying the initially involved actors of the knowledge production process also brings in the people referred to as *experts* in S-BPM, that is, people that have not necessarily been involved in the work process that has triggered the double-loop learning process, but who have expertise that is expected to be relevant for the resolution of the problem.

Activities in the knowledge production process of the KLC are highly iterative and also involve several S-BPM activity bundles. The main goal of these activities is to produce a codified knowledge claim. A knowledge claim, in contrast to the problem claim, describes how the occurrence of the problem can be avoided and eventually has to be integrated in the

DOKB to guide future behavior in similar situations occurring in the business processing environment. The knowledge claim has to be codified to allow for evaluation and distribution across the organization.

The evolution of a knowledge claim is an iterative process involving the KLC activities "information acquisition", "knowledge claim formulation", "individual and group learning", and eventually "knowledge claim evaluation". In terms of S-BPM activity bundles, this potentially involves all bundles except IT-specific implementation. In the following, the first three KLC activities concerned with the formulation of the knowledge claim will be put into relationship to S-BPM. In a second step, knowledge claim evaluation will be brought together with the S-BPM activities.

The codified knowledge claim is produced in an interplay of explicitly formulating the knowledge claim, developing new ideas as an individual, and transferring knowledge within the group of involved people as well as performing further research to gain more information necessary to develop the knowledge claim.

Knowledge claim formulation is linked with the S-BPM activity bundle of modeling. As mentioned, modeling does not necessarily refer to producing business process models that strictly adhere to the syntax and semantics or to a specific language but should describe how work and interaction are performed in a given situation. Having said that, it might still be necessary to produce formally correct models to allow for simulation, validation, and IT-specific implementation of the model—but that is not a requirement in the early stages of knowledge production. Modeling should be considered a group activity here, as the codified knowledge claim evolves through a cooperative process involving all actors of the knowledge production process (Nolte and Prilla 2012). Individual contributions still can be externalized separately and eventually be brought together in a group process.

Information acquisition instantiates a second iteration through the S-BPM analysis activity bundle. After the formulation of the problem claim or in the course of knowledge claim formulation, the need for further information might become evident. Acquisition activities can include identification of knowledge that led to successful work in similar situations, the involvement of further *experts*, or research activities in resources stemming from outside the organization (such as scientific or professional literature and case studies).

Individual and group learning finally involve activities to reflect and revise one's own and the whole group's assumptions of how the problem can be prevented to not occur again. This block is not fully covered by any of the S-BPM activity bundles, as they omit the learning perspective on business process management. The activity bundle termed "organization-specific implementation," however, might be triggered in the course of individual and group learning, as the learning processes inherently change how the group of people involved in the knowledge production environment interact. This directly affects future activities in the business processing environment, as, following the S-BPM paradigm, the actors executing work processes are also part of the group performing knowledge processing activities. As long as the actors of the work processes currently being processed are involved, individual and group learning thus will directly contribute to organization-specific implementation of the work process.

A codified knowledge claim that appears to be appropriate to at least some of the involved people participating in the knowledge production process is evaluated in the next step. Knowledge claim evaluation covers all activities that can be performed to justify that the new knowledge claim will appropriately solve or avoid the problematic situation in the business processing environment. This is where S-BPM can contribute most to the activities in the knowledge processing environment—both the activity bundle "Validation" and "Optimization" can contribute to this step in the KLC. Validation refers to activities that check the effectiveness of a process before it is put to practice. Optimization, though not being relevant here as a whole, includes activities to evaluate efficiency of a process and thus asses the quality of a knowledge claim before putting it to practice. Activities include acting out the process and interacting with the other roles in the process as if it were executed in a real-world setting or simulating process execution by using statistical models. In both cases, ineffective, inefficient, or simply malfunctioning parts of knowledge claims can be identified. Simulation and IT-supported validation are two of the cases that were referred to earlier to require models adhering to formal syntactical and semantic rules. Modeling activities thus have to include means to guide the involved people to appropriately represent their models.

If evaluation activities identify shortcomings in or inappropriateness of the knowledge claim, this again triggers learning activities which, in turn, lead to revised knowledge claims. If evaluation is finished successfully or further iterations are not considered reasonable, knowledge production ends. The resulting knowledge claim is considered a "surviving knowledge claim" if evaluation was successful. If the evaluation still showed that the problem is not been resolved or evaluation did not lead to unambiguous results and further iterations were not made, the knowledge claim is considered "falsified" or "undecided", respectively. Still, both of the latter cases are valid outcomes of the knowledge production process as they at least augment the DOKB in terms of solutions that are likely to not work (and thus do not need to be tried in real-world occurrences of the problem).

Together with information about how the knowledge claim has been produced (e.g., who was involved, which information was built upon, how many and which revisions have been made), the knowledge claim needs to be integrated in the DOKB. Again, this step is more detailed in the KLC than it is in the S-BPM activity bundles. Cross-leveling a knowledge claim can be performed by activities like sharing, teaching, searching, and broadcasting (as described in the KLC). All these activities alter how the organization will in future react to the occurrence of the problem that the knowledge claim is intended to solve. They are thus part of what S-BPM refers to as organization-specific implementation. Knowledge claims might also involve adaptations of an organization's IT-infrastructure (e.g., workflow descriptions, as mentioned earlier). While not an explicit part of the KLC knowledge integration activities, such IT-specific implementations (as referred to in S-BPM) are also part of the activities that lead to integration of the knowledge claim in the DOKB.

The (re-)integration of a knowledge claim in the DOKB ends the double-loop learning process and closes the knowledge lifecycle. S-BPM activity bundles can be found in all steps of the KLC and augment various aspects of it with a more concrete approach on how to implement the steps. In turn, several steps of the KLC, especially those directly concerned with learning and knowledge transfer, offer a more detailed concept of how to implement those activities than S-BPM does. S-BPM and KLC thus augment each other on a conceptual level and offer a starting point to populate their building blocks with instruments that aid their implementation. Both, the KLC steps and the S-BPM activity bundles, however, do not leave the organizational level when describing process development and knowledge development and knowledge application. To implement the actor-centric approach outlined by S-BPM, a conceptual bridge between the super-individual phenomena described here and actually applicable instruments on an individual or group level needs to be constructed. Articulation Work and Mental Model Theory are candidates to provide the foundations for this bridge and will be reviewed in the following section before an attempt is made to put them into the context of the KLC.

5.3.3 Tool Support

The execution of subject-oriented representation schemes can be supported by an appropriate workflow system (Krenn and Stary 2016). In the following example, we show how the workflow system is used to execute an application for vacation using a generic communication scheme.

Figure 5.17 shows a generic subject-oriented specification scheme with three involved parties. It fits to the holiday application process, as the three subjects are employee (Subject 1), HR department (Subject 2), and manager (Subject 3). Each of the parties exchange messages with another party.

Each subject starting message exchange is marked with a small white triangle (Subject 1).

Each subject can send messages with the name Message to any other subject any time. Figure 5.18 shows the behavior of the subject with the name Subject 1. Since Subject 1 is the subject who starts a process, its start state is the state select. The start state is marked with a thick frame. The state "start" and the transitions to the state select will be never executed in the start subject. This state is the start state in all the other

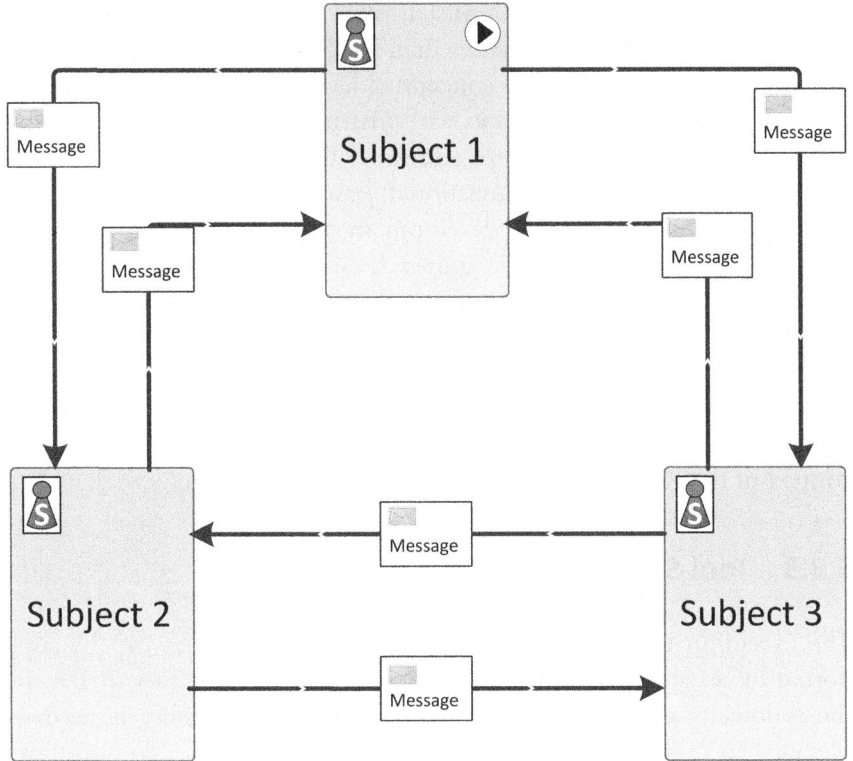

Fig. 5.17 Subject-oriented representation schema for three-party process

subjects. All the other subjects are waiting for a message from all the other subjects.

In this way all subjects that are not start subjects have to receive at least one message before they can start to send messages. The start subject sends a message to any other subject. The receiving subject can now reach the state select. In that state, any subject can decide upon its next action without restriction. A subject which is in state select can send a message to other subjects which are still in the state start. Now these subjects can also reach the select state and can send messages. Finally, all subjects are in the state select and can communicate when addressed.

In the "select" state, the start subject decides whether it wants to send or to receive a message. In order to start a workflow, it does not make sense to receive a message because the other subjects are waiting for mes-

Acting on Work Designs: Providing Support for Validation... 229

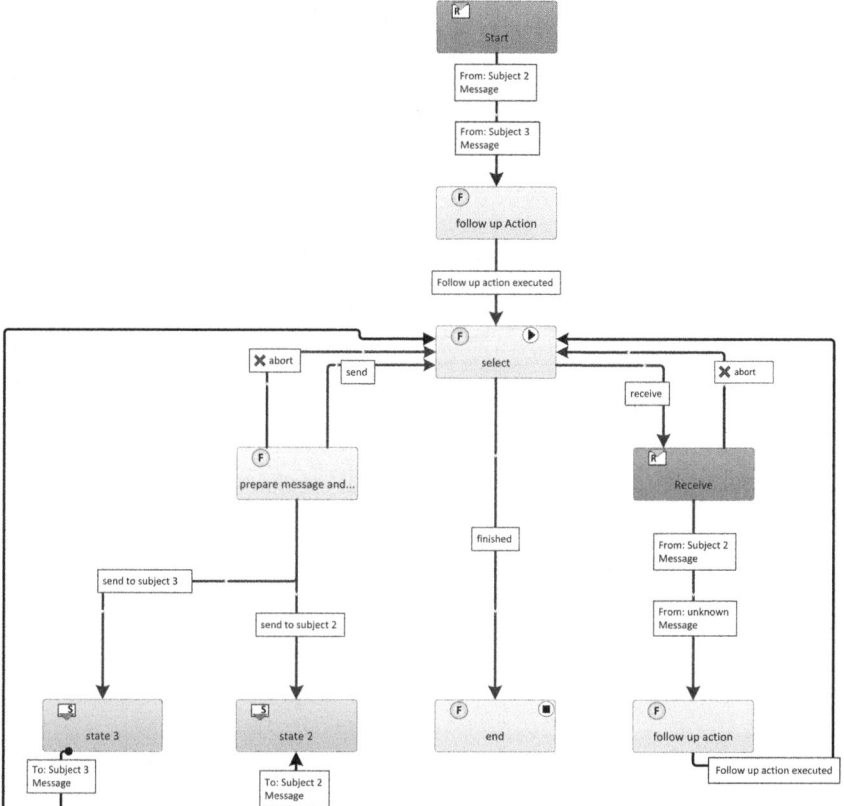

Fig. 5.18 Generic behavior of the start subject "Subject 1"

sages. All the other subjects are in the state start which is a receive state (cf. Fig. 5.19). This means the start subject will start with sending messages. Now the message exchange can begin. In the select state, a subject decides to use the send transition. In the state "prepare message and select address," the subject fills out the business object that is transmitted by the message "message." After that, a subject decides to which subject the message with the business object as content will be sent.

In the select state, a subject can also decide whether it wants to receive a message. If a message from the expected subject is available, the message can be accepted, and a follow-up action can be executed. It is not specified what the follow-up action is. This is like receiving an e-mail. The

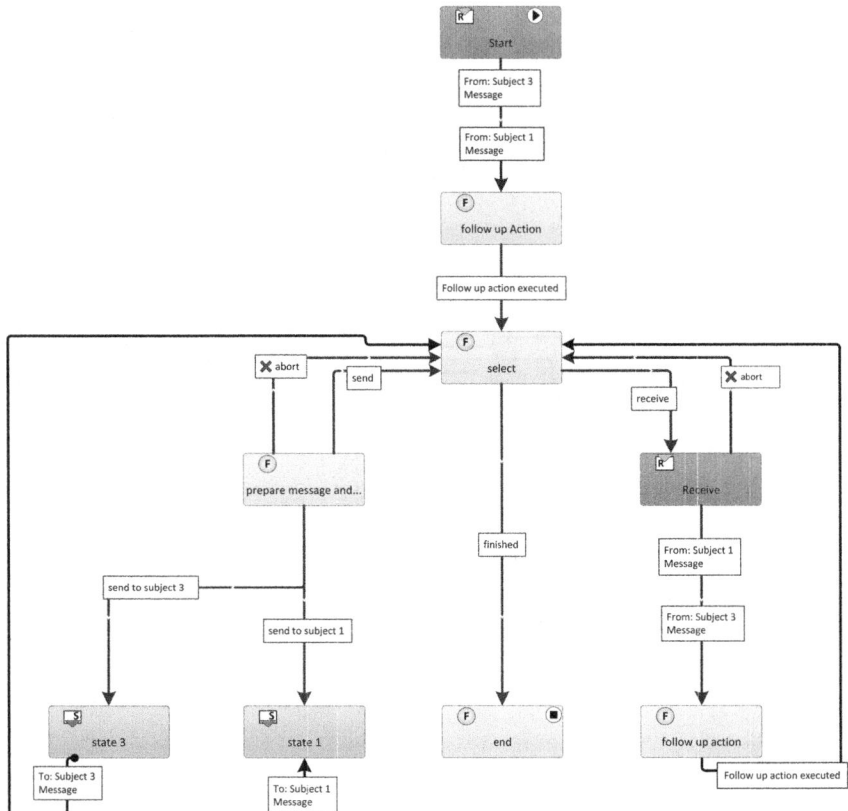

Fig. 5.19 Generic behavior of "Subject 2"

receiver can interpret the content of an e-mail and knows what the corresponding follow-up action is. The abort transitions back to the select state enable to step back in case a subject has made the wrong choice.

With the message "Message," a corresponding business object is sent. The structure of this business object corresponds to the structure of a mail with some extensions like keyword and signature. Figure 5.20 shows the specification of the business object message in an XSD notation.

Whenever a message "Message" is sent, such a business object is sent. The values for the components of the business message object correspond to the content of a traditional mail.

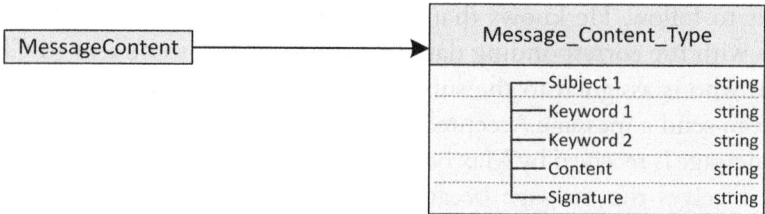

Fig. 5.20 Generic structure of the business object "Mail"

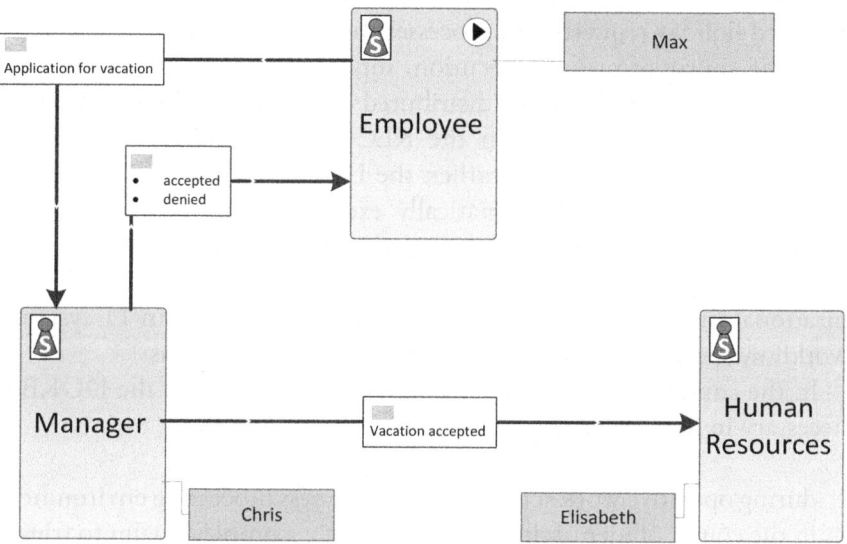

Fig. 5.21 Instantiating a process scheme

For the specification of an actual workflow, the various subjects of a process must be assigned to existing roles and persons or agents. The example shown in Fig. 5.21 demonstrates such an assignment to the three-party process scheme.

The workflow support system is configured in a way, that the actors Max, Helge, and Josi can be assigned to subject "employee." Since these actors are assigned to the start subject, all of them can start the process. For instance, Max creates a process instance and is then guided through the process. He is asked by the workflow system about the transition he wants to follow. He knows that he has to fill out the business message form with the corresponding data and that form has to be sent to Chris. Chris who is assigned to the subject "manager" can accept that message and can send a message Accepted or Denied back to subject employee—the message is received by Max because he is assigned to subject employee. Max receives the message because in his environment or context, the process is started. If another person assigned to subject employee starts a process, this process instance is executed in his or her environment. Elisabeth from HR (Human Resource department) receives already accepted holiday requests and processes them accordingly.

In the course of process execution, support for disseminating knowledge that is represented in the distributed organizational knowledge base (DOKB) as conceptualized in the KLC is also required for effective knowledge use. As described earlier, the DOKB does not solely contain technically represented, automatically executable knowledge claims. It rather conceptually includes all knowledge of the members of an organization, their experiences and expertise, as well as all individual and organizational procedures and facts that have been codified in IT-systems, workflow specification, or other organizational descriptions.

In the context of the KLC, dissemination of content of the DOKB is necessary in four different cases, namely

- during operative work activities in the business processing environment
- in the course of identifying and formulating a problem claim to trigger activities in the knowledge processing environment
- during knowledge production
- during knowledge integration

During operative work activities, the actors need to be able to access the DOKB in order to decide on how to react on the currently perceived situation in the work environment. In terms of accessing the explicitly codified part of the DOKB, this can be realized by providing in situ scaffolding, that is, providing suggestions on how to continue to actors based either on their own previous task implementations or on other's task implementations ("what can I do now/next?"). Alternatively, actors also need to be able to assess whom to approach, if they cannot decide on how to proceed ("whom can I ask?").

In the course of formulating a problem claim, access to documentation of previously executed processes of the same process class is necessary. This allows identifying potential sources of the problematic situation and reflecting upon the different variants on work execution that might provide deeper insights in how the addressed problematic situation has developed.

Activities in knowledge production again require access to historic process execution information and transactive knowledge of which information has been codified, under which circumstances, and by whom. Sharing of instance knowledge between the involved actors is required to facilitate alignment of meaning as well as for development and evaluation of the knowledge claim.

Knowledge integration aims at disseminating a new knowledge claim to the organization, eventually making it part of the DOKB. For all activities carried out in this step of the KLC, it is necessary to put the new knowledge claim into the context of the current state of knowledge of the targeted actors. Knowledge dissemination thus not only has to be tailored to the specific situation during a work process but also has to allow actors to put the new knowledge claim in the context of their own mental model during dedicated learning activities in knowledge integration.

Summarizing, support for knowledge dissemination should support the following scenarios for dissemination knowledge about work processes:

- In situ scaffolding during operative work processes (what can I do now/next)

    – based on previous own task implementations
    – based on others' task implementations

- Learning from other actors who previously took the same role in a task bundle or are experts in the area during operative work processes or in knowledge integration (whom can I ask)

- Reflection on previous own and others' task implementations as a part of problem claim identification and knowledge production (what have we done)

    - individually
    - as a group

- Aligning with other actors who take other roles in a task bundle during knowledge integration (how can we work)

These requirements can be methodologically and technically approached using instruments for self-directed learning. Based on the learning support platform (Auinger and Stary 2005), concepts have been developed to extend learning support to organizational learning settings (Neubauer et al. 2011, 2013). We will re-address these concepts in the digital work design framework developed in Chap. 6.

5.4 Synthesis

Table 5.1 gives a structured overview of the presented techniques of this chapter. It structures each approach according to its

- focus revealing its objective
- essential support features
- means of representation, in order to validate and process documented work knowledge
- procedure to follow for putting process knowledge to work practice

Validation and enactment of work models can be facilitated by scaffolds and subject-oriented processing. Role-specific behavior representations on various levels of detail play a crucial role in stakeholder-driven reflection of the (future) organization of work tasks. Scaffolds open up for additional articulation while execution enables interactive experience of models in terms of actual process support of individual work practices in the selected business process (Table 5.2).

Table 5.1 Processing work models for validation and enactment

	Virtual enactment through scaffolding	S-BPM-based validation and execution
Focus	Sharing simulated process-model walk-throughs in a role-conform way	Building interactive process experience enabling role-specific semantic behavior checks for a specific work process
Essential support feature	Adjusted scaffolds guiding validation	Software engine supporting the execution of interacting behavior encapsulations in a role-specific way with a dual view on models and user interactions
Means of representation	Workflow DiagramInteraction Relations	Subject Interaction Diagram (SID)Subject Behavior Diagram (SBD)
Validation and implementation procedure	1. Scaffold preparation 2. Participatory simulation preparation 3. Detailing specifications 4. Collective reflection	1. Provision of Subject Interaction Diagram 2. Provision of Subject Behavior Diagrams corresponding to 1 3. Preparation of execution environment 4. Execution of behavior diagrams 5. Reflection of interactive process experience(s)

From a procedural perspective, scaffolding for virtual enactment comprises several phases:

1. It involves preparation of the setting, actors, and instruments, comprising (i) determining the type of scaffolds and scaffolds that guide the implementation activities, (ii) instrument support including digital media support as execution environment, (iii) actors willing to take responsibility for validating role-specific behavior and rethinking, (iv) a facilitator to guide the validation procedure and use of enactment environment.

Table 5.2 Elicitation requirements and scaffolding-based validation and virtual enactment

Elicitation requirement	Scaffolded virtual enactment
Awareness on role(s) and their management	Walk-throughs in the course of validation and alignment are performed along role-specific behavior representations including their interactions with other roles. Depending on the type of scaffold, further articulation of work knowledge could be required and it needs to be aligned among role carriers. As such, the approach can be considered as active role management—determining the various role behaviors develop over time and can be re-arranged on the fly using the corresponding digital tools.
Situation awareness	Validation is framed by situation information as defined at the core of role behaviors. The participatory and multifaceted scaffolding approach helps creating situation awareness in the participating group of stakeholders.
Conceptual understanding of complex systems	Complexity of systems can be addressed by the variety of scaffolding approaches that can be applied for the role-specific walk-throughs. Rather than prescribing linear (for the roles) and networked (by interaction between roles) validation and enactment, work knowledge can be represented according to the capacities of the participating stakeholders. Complex system specification can develop step by step. A further facilitator is the switching between views, referring to actors or the flow of work in the course of virtual enactment.
Creating a reflective practice for situations-to-be	Future actor behavior can be compared with originally articulated and previously aligned representations via the models that are created through scaffolding and virtual enactment support. Again, the participating stakeholders can switch between views, either referring to actors or the flow of work.
Focusing while utilizing multiple perspectives	On the one hand, the different types of scaffolds enable various perspectives on work processes with respect to the completeness or prescriptiveness of specifications. On the other hand, the virtual enactment support allows taking various views on processes, namely, view per actor, overall actor-centric view, and overall flow-oriented view, and is augmented with information about the current instance, such as the currently available activities and the path through the process.

(*continued*)

Table 5.2 (continued)

Elicitation requirement	Scaffolded virtual enactment
Articulating intangible assets	Based on the view per actor, overall actor-centric view, and overall flow-oriented view, and the information about the current instance, visualizations are created dynamically. Hence, becoming aware of intangible assets relevant for the work process at hand could easily be embodied in terms of encoded process elements.
Engage in alignment for collective intelligence	Although the focus is on role-specific behavior, the overall flow-perspective allows all stakeholders to engage in experiencing models by walk-throughs and share the documented knowledge, guided by the prompting mechanism when probing novel organizational behavior.

2. Situation-sensitive articulation is supported through the scaffolds and the enacting environment when aiming to implement specifications of work processes.
3. Facilitation is required (i) to set the stage involving stakeholders as role carriers, (ii) to ensure the usefulness and effectiveness of the selected scaffolds, and (iii) to support the use of the virtual enactment tool.
4. Representational alignment might need to be facilitated when the participants aim to consolidate their findings into a shared representation in the course of enactment.
5. Organizational implementation needs to be documented when the participants validate how work processes could become part of future workplace designs.

We now discuss the requirements with respect to subject-oriented validation and execution based on S-BPM (Table 5.3).

From a procedural perspective, S-BPM-based validation and execution needs to take into account the following sets of activities:

1. The preparation of validation and execution requires (i) determining the models within the scope of implementation, that is, some business process, (ii) providing the models and the probing actors as role carriers, (iii) configuring the tool for validation and execution of the role-specific models, (iv) providing a facilitator to effectuate the procedure including articulation of additional work knowledge when probing.

Table 5.3 Elicitation requirements and S-BPM-based validation and execution

Elicitation requirement	S-BPM-based execution
Awareness on role(s) and their management	Roles and role-specific interactions are constitutive elements of subject-oriented representation that lay ground for validation and implementation. Automated execution of Subject Behavior Diagrams helps to experience role-specific behaviors, and thus manage roles based on their function flows and interaction patterns.
Situation awareness	The selection of roles and thus subjects depends on the situation to be addressed in the course of articulation, alignment, and implementation. Consequently, a situation is grasped and modeled implicitly through the resulting interaction pattern between roles and their behavior specifications.
Conceptual understanding of complex systems	A system can be validated as being composed of an arbitrary set of entities. The ultimate criterion is whether the constellation of subjects encapsulates intended work behavior. Validation can be achieved by specifying subject behaviors and their interactions, and by executing these representations by choreographic workflow engines. According to S-BPM methodological approach, several facilitating organizational roles can be identified for validation and execution. They are either governing or acting throughout implementation, aiming to build up digital work experiences based on validated process models.
Creating a reflective practice for situations-to-be	The baseline in S-BPM can either be a situation as-it-is, or a situation to-be, or even a mixture of existing and envisioned patterns of work. It depends on the purpose of validating knowledge and the status of implementing models.
Focusing while utilizing multiple perspectives	Each subject pursues a specific role perspective in the course of validation and execution. The overall picture becomes visible through interaction between the subjects. When executing the models by digital means, the overall process flow, as well as the individual models of the role-specific behavior, is available. Hence, for probing and interactive process experience, a dual view (design and runtime) can be provided. In addition, the procedure of validation and execution is supervised by additional organizational perspectives, in particular through the governor and facilitator role.

(*continued*)

Table 5.3 (continued)

Elicitation requirement	S-BPM-based execution
Articulating intangible assets	Subject-oriented validation is based on explicit work models, in particular, it allows to explore how work task are accomplished when collaborating in a certain role with other roles. Whenever a participating stakeholder becomes aware of work knowledge not externalized so far, it can be articulated in the course of model validation and represented to become part of the subject representations that can be executed.
Engage in alignment for collective intelligence	A subject-oriented business process consists of collaborating subjects. For validation and implementation, all involved stakeholders can experience how models could be implemented in work practice. Hence, it is advised from a project management perspective to involve the actual takers of each role to check whether the models provide a balanced perspective on individual mental models of work. Otherwise, the impact of implemented work processes on the collective intelligence could only be evaluated after probing the models.

2. Situation-sensitive validation and execution features are modeling and tool functions referring to the notation (e.g., editor) and execution support (e.g., bootstrapping through subject-wise execution), as the subject constellation refers to the situation to be targeted. It is captured through a functional and interactional perspective.
3. *Situation-as-is versus situations-to-be*: Both can be addressed by models when validating and executing them as being constructed. Hence, also the transformation from as-it-is to as-it-could-be can be experienced interactively, which is particularly helpful when stakeholders start revisiting existing work patterns, role labels, and work assignments, and try to generate novel structures of work. Our practical experiences show a strong preference of stakeholders of this middle-out approach.
4. *Representational validation*: The specification of work processes is complete with respect to the information required for executing models without further transformation. Hence, stakeholders are enabled to change patterns of behavior back and forth in the course of valida-

tion and execution, depending on the addressed situation (to-be or as-it-is).
5. Organizational validation and execution is best achieved involving role carriers when probing models through executing them. It fosters a lively engagement when aiming to find organization-wide consensus along interactive experiences of process models.

Cross-checking the presented validation and implementation techniques, each of the presented technique focus on role-specific behavior and their interaction patterns. Scaffolding extends the variability of how to align and validate knowledge for an organizational consolidation and, finally, implementations. Opening up for innovative or novel process design is considered particularly useful in case of highly complex situations or conflicting perspectives on workflows. Scaffolds also support middle-out development of organizational structures, as they enable multilayered perspectives and bootstrap alignment processes. The latter is supported only implicitly by going back and forth in subject-oriented modeling coupled with interactive prototyping of executable process models.

References

Abdu, Rotem, Baruch Schwarz, and Manolis Mavrikis. 2015. Whole-Class Scaffolding for Learning to Solve Mathematics Problems Together in a Computer-Supported Environment. *Zdm* 47 (7): 1163–1178 (Springer).

Arias, E.G., and G. Fischer. 2000. Boundary Objects: Their Role in Articulating the Task at Hand and Making Information Relevant to It. In *Proceedings of the International ICSC Symposium on Interactive and Collaborative Computing (ICC'2000)*.

Arias, Ernesto, H. Eden, Gerhard Fischer, A. Gorman, and E. Scharff. 2000. Transcending the Individual Human Mind—Creating Shared Understanding Through Collaborative Design. *ACM Transactions on Computer-Human Interaction (TOCHI)* 7 (1): 84–113 (New York: ACM Press).

Auinger, Andreas, and C. Stary. 2005. *Didaktikgeleiteter Wissenstransfer. Interaktive Informationsräume Für Lern-Gemeinschaften Im Web*. Deutscher Universitätsverlag.

Berztiss, Alfs. 1996. Business Process Prototyping. In *Software Methods for Business Reengineering*, 122–125. Springer.

Beyer, H., and K. Holtzblatt. 1997. *Contextual Design: Defining Customer-Centered Systems*. Morgan Kaufmann.

Briggs, Robert O., G.L. Kolfschoten, Gert Jan De Vreede, Stephan Lukosch, and Conan C. Albrecht. 2013. Facilitator-in-a-Box: Process Support Applications to Help Practitioners Realize the Potential of Collaboration Technology. *Journal of Management Information Systems* 29 (4): 159–194 (Taylor & Francis).

Bruno, G., F. Dengler, B. Jennings, R. Khalaf, S. Nurcan, M. Prilla, Marcello Sarini, R. Schmidt, and R. Silva. 2011. Key Challenges for Enabling Agile BPM with Social Software. *Journal of Software Maintenance and Evolution: Research and Practice* 23 (4): 297–326 (Wiley Online Library).

Bulu, Saniye Tugba, and Susan Pedersen. 2010. Scaffolding Middle School Students' Content Knowledge and Ill-Structured Problem Solving in a Problem-Based Hypermedia Learning Environment. *Educational Technology Research and Development* 58 (5): 507–529 (Springer).

Caporale, T. 2016. A Tool for Natural Language Oriented Business Process Modeling. In *Proceedings of ZEUS 2016*, 49. ceur-ws.org.

Chin, Christine, and Jonathan Osborne. 2010. Students' Questions and Discursive Interaction: Their Impact on Argumentation During Collaborative Group Discussions in Science. *Journal of Research in Science Teaching* 47 (7): 883–908 (Wiley Online Library).

Convertino, Gregorio, Helena M. Mentis, Mary Beth Rosson, John M. Carroll, Aleksandra Slavkovic, and Craig H. Ganoe. 2008. Articulating Common Ground in Cooperative Work: Content and Process. In *CHI '08: Proceeding of the Twenty-Sixth Annual SIGCHI Conference on Human Factors in Computing Systems*, 1637–1646. New York: ACM. https://doi.org/10.1145/1357054.1357310.

Dalmaris, Peter, Eric Tsui, Bill Hall, and Bob Smith. 2007. A Framework for the Improvement of Knowledge-Intensive Business Processes. *Business Process Management Journal* 13 (2): 279–305 (Emerald).

Davies, Islay, Peter Green, Michael Rosemann, Marta Indulska, and Stan Gallo. 2006. How Do Practitioners Use Conceptual Modeling in Practice? *Data & Knowledge Engineering* 58 (3): 358–380.

Dennen, Vanessa Paz. 2004. Cognitive Apprenticeship in Educational Practice: Research on Scaffolding, Modeling, Mentoring, and Coaching as Instructional Strategies. In *Handbook of Research on Educational Communications and Technology*, 2nd ed., 813–828. Mahwah, NJ: Lawrence Erlbaum Associates.

Dittmar, A., P. Forbrig, S. Heftberger, and C. Stary. 2004. Support for Task Modeling—A "Constructive" Exploration. In *International Workshop on Design, Specification, and Verification of Interactive Systems*, 59–76. Berlin, Heidelberg: Springer.

Ellson, J., E. Gansner, L. Koutsofios, S.C. North, and G. Woodhull. 2001. Graphviz—Open Source Graph Drawing Tools. In *International Symposium on Graph Drawing*, 483–484. Berlin, Heidelberg: Springer.

Firestone, J.M., and M.W. McElroy. 2003. *Key Issues in the New Knowledge Management*. Butterworth-Heinemann.

Fleischmann, Albert, Werner Schmidt, C. Stary, Stefan Obermeier, and Egon Börger. 2012. *Subject-Oriented Business Process Management*. Springer.

Floch, Jacqueline, Svein Hallsteinsen, Erlend Stav, Frank Eliassen, Ketil Lund, and Eli Gjorven. 2006. Using Architecture Models for Runtime Adaptability. *IEEE Software* 23 (2): 62–70 (IEEE).

Floyd, Christiane. 1984. A Systematic Look at Prototyping. In *Approaches to Prototyping*, 1–18. Springer.

Forster, Simon, Jakob Pinggera, and Barbara Weber. 2013. Toward an Understanding of the Collaborative Process of Process Modeling. *CAiSE Forum*, 98–105.

Franco, L. Alberto. 2013. Rethinking Soft or Interventions: Models as Boundary Objects. *European Journal of Operational Research* 231 (3): 720–733 (Elsevier).

Franco, L. Alberto, and Etiënne A.J.A. Rouwette. 2011. Decision Development in Facilitated Modelling Workshops. *European Journal of Operational Research* 212 (1): 164–178. https://doi.org/10.1016/j.ejor.2011.01.039.

Frederiks, P.J.M., and Th.P. van der Weide. 2006. Information Modeling: The Process and the Required Competencies of Its Participants. *Data & Knowledge Engineering* 58 (1): 4–20. https://doi.org/10.1016/j.datak.2005.05.007.

Front, A., D. Rieu, M. Santorum, and F. Movahedian. 2017. A Participative End-User Method for Multi-Perspective Business Process Elicitation and Improvement. *Software & Systems Modeling* 16 (3): 691–714.

Ge, Xun, and Susan M. Land. 2004. A Conceptual Framework for Scaffolding Ill-Structured Problem-Solving Processes Using Question Prompts and Peer Interactions. *Educational Technology Research and Development* 52 (2): 5–22 (Springer).

Hartmann, B., L. Yu, A. Allison, Y. Yang, and S.R. Klemmer. 2008. Design as Exploration: Creating Interface Alternatives Through Parallel Authoring and Runtime Tuning. In *Proceedings of the 21st Annual ACM Symposium on User Interface Software and Technology*, 91–100. ACM.

Herrmann, T., and K.-U. Loser. 2013. Facilitating and Prompting of Collaborative Reflection of Process Models. In *Proceedings of MoRoCo@ECSCW 2013*, 17–24. ceur-ws.org.

Herrmann, Thomas, G. Kunau, K.U. Loser, and N. Menold. 2004. Socio-Technical Walkthrough: Designing Technology Along Work Processes. In *Artful Integration: Interweaving Media, Materials and Practices. Proceedings of the Eighth Conference on Participatory Design*, 132–141. New York: ACM Press.

Herrmann, Thomas, K.U. Loser, and I. Jahnke. 2007. Sociotechnical Walkthrough: A Means for Knowledge Integration. *The Learning Organization* 14 (5): 450–464.

Herrmann, Thomas, M. Hoffmann, G. Kunau, and K.U. Loser. 2002. Modelling Cooperative Work: Chances and Risks of Structuring. In *Cooperative Systems Design, a Challenge of the Mobility Age. Proceedings of COOP 2002*, 53–70. IOS Press.

Hjalmarsson, Anders, Jan C. Recker, Michael Rosemann, and Mikael Lind. 2015. Understanding the Behavior of Workshop Facilitators in Systems Analysis and Design Projects: Developing Theory from Process Modeling Projects. *Communications of the AIS* 36 (22): 421–447.

Holbrook, H. 1990. A Scenario-Based Methodology for Conducting Requirements Elicitation. *ACM SIGSOFT Software Engineering Notes* 15 (1): 95–104. https://doi.org/10.1145/382294.382725.

Jumaat, N.F., and Z. Tasir. 2014, April. Instructional Scaffolding in Online Learning Environment: A Meta-analysis. In *2014 International Conference on Teaching and Learning in Computing and Engineering*, 74–77. IEEE.

Kensing, Finn, and Jeanette Blomberg. 1998. Participatory Design: Issues and Concerns. *Computer Supported Cooperative Work (CSCW)* 7 (3–4): 167–185 (Springer).

King, Alison, and Barak Rosenshine. 1993. Effects of Guided Cooperative Questioning on Children's Knowledge Construction. *The Journal of Experimental Education* 61 (2): 127–148 (Taylor & Francis).

Kobbe, Lars, Armin Weinberger, Pierre Dillenbourg, Andreas Harrer, Raija Hämäläinen, Päivi Häkkinen, and Frank Fischer. 2007. Specifying Computer-Supported Collaboration Scripts. *International Journal of Computer-Supported Collaborative Learning* 2 (2–3): 211–224 (Springer).

Krenn, Florian, and Christian Stary. 2016. Exploring the Potential of Dynamic Perspective Taking on Business Processes. *Complex Systems Informatics and Modeling Quarterly*, no. 8: 15–27. https://csimq-journals.rtu.lv/issue/view/87.

Land, Susan M., and Carla Zembal-Saul. 2003. Scaffolding Reflection and Articulation of Scientific Explanations in a Data-Rich, Project-Based

Learning Environment: An Investigation of Progress Portfolio. *Educational Technology Research and Development* 51 (4): 65–84 (Springer).
Mayer, R.E. 1989. Models for Understanding. *Review of Educational Research* 59 (1): 43–64.
Minor, M., A. Tartakovski, and D. Schmalen. 2008. Agile Workflow Technology and Case-Based Change Reuse for Long-Term Processes. *International Journal of Intelligent Information Technologies (IJIIT)* 4 (1): 80–98 (IGI Global).
Monsalve, C., A. April, and A. Abran. 2010. Representing Unique Stakeholder Perspectives in BPM Notations. In *2010 Eighth ACIS International Conference on Software Engineering Research, Management and Applications (SERA)*, 42–49.
Mori, G., F. Paternò, and C. Santoro. 2002. CTTE: Support for Developing and Analyzing Task Models for Interactive System Design. *IEEE Transactions on Software Engineering* 28 (9): 797–813.
Morris, W.T. 1967. On the Art of Modeling. *Management Science* 13 (12): B–707–B–717. http://pubsonline.informs.org/doi/abs/10.1287/mnsc.13.12.B707.
Neubauer, Matthias, Stefan Oppl, C. Stary, and Georg Weichhart. 2013. Facilitating Knowledge Transfer in IANES—A Transactive Memory Approach. In *Innovation Through Knowledge Transfer 2012*, Smart Innovation, Systems and Technologies, vol. 18, ed. R. Howlett, B. Gabrys, K. Musial-Gabrys, and J. Roach, 39–50. Berlin and Heidelberg: Springer. https://doi.org/10.1007/978-3-642-34219-6_5.
Neubauer, Matthias, C. Stary, and Stefan Oppl. 2011. Polymorph Navigation Utilizing Domain-Specific Metadata: Experienced Benefits for E-Learners. In *Proceedings of the 29th European Conference on Cognitive Ergonomics (ECCE 2011)*, 45–52. ACM Press.
Nielsen, Jakob. 1993. Iterative User-Interface Design. *Computer* 26 (11): 32–41 (IEEE).
Nolte, Alexander, and M. Prilla. 2012. Normal Users Cooperating on Process Models: Is It Possible at All? In *Collaboration and Technology*, 57–72. Berlin: Springer.
Oppl, Stefan. 2015. Articulation of Subject-Oriented Business Process Models. In *Proceedings of S-BPM ONE 2015*, 1–11. New York: ACM Press. https://doi.org/10.1145/2723839.2723841.
———. 2016. Articulation of Work Process Models for Organizational Alignment and Informed Information System Design. *Information & Management* 53 (5): 591–608. https://doi.org/10.1016/j.im.2016.01.004.
Oppl, Stefan, and Nancy Alexopoulou. 2016. Linking Natural Modeling to Techno-Centric Modeling for the Active Involvement of Process Participants in Business Process Design. *International Journal of Information System*

Modeling and Design 7 (2): 1–30. https://doi.org/10.4018/IJISMD.2016040101.

Ozmantar, M.F., and T. Roper. 2004. Mathematical Abstraction Through Scaffolding. In *Proceedings of the 28th Conference of the International Group for the Psychology of Mathematics Education*, vol. 3, 481–488.

Pinelle, D., and C. Gutwin. 2002, April. Groupware Walkthrough: Adding Context to Groupware Usability Evaluation. In *Proceedings of the SIGCHI Conference on Human Factors in Computing Systems*, 455–462. ACM.

Pirnay-Dummer, Pablo N., and A. Lachner. 2008. Towards Model Based Knowledge Management. A New Approach to the Assessment and Development of Organizational Knowledge. In *Annual Proceedings of the AECT 2008*, ed. M. Simonson, 178–118.

Polson, Peter G., Clayton Lewis, John Rieman, and Cathleen Wharton. 1992. Cognitive Walkthroughs: A Method for Theory-Based Evaluation of User Interfaces. *International Journal of Man-Machine Studies* 36 (5): 741–773 (Elsevier).

Powell, S.G., and T. R. Willemain. 2007. How Novices Formulate Models. Part I: Qualitative Insights and Implications for Teaching. *Journal of the Operational Research Society*, 58: 983–995 (JSTOR).

Prilla, Michael. 2015. Supporting Collaborative Reflection at Work: A Socio-Technical Analysis. *AIS Transactions on Human-Computer Interaction* 7 (1): 1–17.

Prilla, M., and Alexander Nolte. 2012. Integrating Ordinary Users into Process Management: Towards Implementing Bottom-Up, People-Centric BPM. In *Enterprise, Business-Process and Information Systems Modeling*, 182–194. Springer.

Prusak, Naomi, Rina Hershkowitz, and Baruch B. Schwarz. 2012. From Visual Reasoning to Logical Necessity Through Argumentative Design. *Educational Studies in Mathematics* 79 (1): 19–40 (Springer).

Recker, J.C., Norizan Safrudin, and Michael Rosemann. 2012. How Novices Design Business Processes. *Information Systems* 37 (6): 557–573 (Elsevier Science Ltd).

Reichert, Manfred, and Peter Dadam. 2009. Enabling Adaptive Process-Aware Information Systems with ADEPT2. In *Handbook of Research on Business Process Modeling*, 173–203. IGI Global.

Roberts, Andrew. 2009. Encouraging Reflective Practice in Periods of Professional Workplace Experience: The Development of a Conceptual Model. *Reflective Practice* 10 (5): 633–644 (Taylor & Francis).

Sandkuhl, Kurt, Janis Stirna, Anne Persson, and Matthias Wißotzki. 2014. Elicitation Approaches in Enterprise Modeling. In *Enterprise Modeling*, The

Enterprise Engineering Series, 39–51. Berlin and Heidelberg: Springer Berlin Heidelberg. https://doi.org/10.1007/978-3-662-43725-4_4.

Santoro, Flávia Maria, Marcos R.S. Borges, and José A. Pino. 2010. Acquiring Knowledge on Business Processes from Stakeholders' Stories. *Advanced Engineering Informatics* 24 (2): 138–148. https://doi.org/10.1016/j.aei.2009.07.002.

Schiffner, S., T. Rothschädl, and N. Meyer. 2014, September. Towards a Subject-Oriented Evolutionary Business Information System. In *2014 IEEE 18th International Enterprise Distributed Object Computing Conference Workshops and Demonstrations*, 381–388. IEEE.

Smeds, Riitta, and Jukka Alvesalo. 2003. Global Business Process Development in a Virtual Community of Practice. *Production Planning & Control* 14 (4): 361–371 (Taylor & Francis).

Sousa, Kênia, Hildeberto Mendonça, Amandine Lievyns, and Jean Vanderdonckt. 2011. Getting Users Involved in Aligning Their Needs with Business Processes Models and Systems. *Business Process Management Journal* 17 (5): 748–786 (Emerald Group Publishing Limited).

Stender, Peter, and Gabriele Kaiser. 2015. Scaffolding in Complex Modelling Situations. *Zdm* 47 (7): 1255–1267 (Springer).

Su, J.-M. 2015. A Self-regulated Learning Tutor to Adaptively Scaffold the Personalized Learning: A Study on Learning Outcome for Grade 8 Mathematics. In *2015 8th International Conference*—Presented at the *Ubi-Media Computing (UMEDIA)*, 376–380.

Sukaviriya, N., V. Sinha, T. Ramachandra, S. Mani, and M. Stolze. 2007, September. User-Centered Design and Business Process Modeling: Cross Road in Rapid Prototyping Tools. In *IFIP Conference on Human-Computer Interaction*, 165–178. Berlin, Heidelberg: Springer.

Tavella, Elena, and Thanos Papadopoulos. 2014. Expert and Novice Facilitated Modelling: A Case of a Viable System Model Workshop in a Local Food Network. *Journal of the Operational Research Society* 66 (2): 247–264 (Palgrave Macmillan).

Van de Pol, Janneke, Monique Volman, and Jos Beishuizen. 2010. Scaffolding in Teacher—Student Interaction: A Decade of Research. *Educational Psychology Review* 22 (3): 271–296 (Springer).

Vogel, F., C. Wecker, I. Kollar, and F. Fischer. 2017. Socio-Cognitive Scaffolding with Computer-Supported Collaboration Scripts: A Meta-analysis. *Educational Psychology Review* 29 (3): 477–511.

Vygotsky, L.S. 1978. *Mind in Society*. Harvard University Press.

Wachholder, Dominik, and Stefan Oppl. 2012. Stakeholder-Driven Collaborative Modeling of Subject-Oriented Business Processes. In *S-BPM*

ONE—Scientific Research, ed. C. Stary, 145–162. Berlin and Heidelberg: Springer. https://doi.org/10.1007/978-3-642-29133-3_10.

———. 2014. Interactive Coupling of Process Models: A Distributed Tabletop Approach to Collaborative Modeling. In *ECCE '14: Proceedings of the 2014 European Conference on Cognitive Ergonomics*, September, 1–8. New York: ACM Request Permissions. https://doi.org/10.1145/2637248.2637262.

Walls, Joseph G., George R. Widmeyer, and Omar A. El Sawy. 1992. Building an Information System Design Theory for Vigilant EIS. *Information Systems Research* 3 (1): 36–59 (Informs).

Weske, Mathias. 2010. *Business Process Management: Concepts, Languages, Architectures*. Springer.

Willemain, Thomas R. 1995. Model Formulation: What Experts Think About and When. *Operations Research* 43 (6): 916–932 (Informs).

Wood, David, Jerome S. Bruner, and Gail Ross. 1976. The Role of Tutoring in Problem Solving. *Journal of Child Psychology and Psychiatry* 17 (2): 89–100 (Wiley Online Library).

Open Access This chapter is licensed under the terms of the Creative Commons Attribution 4.0 International License (http://creativecommons.org/licenses/by/4.0/), which permits use, sharing, adaptation, distribution and reproduction in any medium or format, as long as you give appropriate credit to the original author(s) and the source, provide a link to the Creative Commons licence and indicate if changes were made.

The images or other third party material in this chapter are included in the chapter's Creative Commons licence, unless indicated otherwise in a credit line to the material. If material is not included in the chapter's Creative Commons licence and your intended use is not permitted by statutory regulation or exceeds the permitted use, you will need to obtain permission directly from the copyright holder.

6

Enabling Emergent Workplace Design

This chapter offers a synthesis of the conceptual and methodological considerations of the former chapters. It brings together the lines of argumentation offered in Chaps. 3, 4, and 5 and provides a unifying framework that guides the design of organizational interventions that enable emergence of novel digital workplace designs and work practices.

Process-oriented organizational learning (OL) approaches assume the existence of commonly agreed upon or prescribed work processes (Wargitsch and Wewers 1997; Abecker et al. 2001; Diefenbruch et al. 2002; Hinkelmann et al. 2002). OL support systems then augment these processes with work-relevant knowledge during execution of the workflow (Abecker et al. 1998). These systems, however, do not explicitly allow for or even consider deviations from the prescribed work process. Such deviations, however, happen regularly due to contingencies that arise during execution of a workflow. Another reason of process deviation is individual expertise that allows for shortcuts or emphasis on certain aspects of the workflow depending on the actual set of people executing it. In knowledge-intensive business environments (Dalmaris et al. 2007), both triggers for deviations are rather a standard case than an exception (Marjanovic and Freeze 2011).

Business process support in knowledge-intense environments requires three dimensions to be brought together:

- *Business processes* describe how actors work together and perform their work in an organization to pursue a common goal
- *Knowledge* is required to perform the business processes and enables actors to make their decisions on how to continue work based on their perceptions of the environment
- *Actors* are those entities in an organization who actually carry out business processes based on their knowledge

In the light of these dimensions, the two major goals for support measures can be defined as follows:

- *Agility* is the ability of allowing actors to deviate from a given business process based on their knowledge about their work and their perceptions of the environment
- *Actor-awareness* is the ability of a support system to adapt its behavior to a specific actor being active in a business process and provide the actor with any information that is necessary to build up knowledge and make informed decisions

Any two of the three dimensions have already been brought together in earlier research. The frameworks resulting from this earlier works, however, always omit the third dimension:

- The *Knowledge Lifecycle* (KLC) of Firestone and McElroy (2003) brings together business process management and knowledge management. It, however, does not explicitly consider actors and how they perform the activities described in the KLC.
- *Subject-oriented Business Process Management* (S-BPM), as developed by Fleischmann et al. (2012), explicitly considers the role of actors during process design and execution. The primary elements of structuring are subjects that separate a process along who performs work in the process. How to deal with knowledge, however, is not explicitly taken into account in this approach.

- *Mental Model Theory* (MMT), as initially described by (Johnson-Laird 1981), provides a way of understanding how people make decisions based upon their perception of their current work situation and their prior knowledge. Being a generic approach, the theory is not specific to people's behavior in business processes and thus does not explicitly consider this dimension.

The KLC and S-BPM augment each other. The KLC focuses on the identification of knowledge needs, knowledge production, and knowledge distribution. S-BPM focuses on business process execution and monitoring. Both, however, provide overlapping aspects that act as docking points to intertwine both frameworks. Mental Model Theory, however, does not easily integrate with the other two frameworks. MMT is a psychological approach focusing on the cognitive processes of individuals. The KLC and S-BPM, however, take an organizational perspective and consider individuals as atomic entities in the organizational knowledge base (KLC) or interacting with each other (S-BPM). The remaining gap can be bridged by the sociological theory of *Articulation Work* (Strauss 1993). Articulation Work describes the phenomenon of resolving work situations that are considered problematic by the participating individuals. Problematic work situations occur whenever the application of the current mental model of any participant does not lead to the desired outcome (Pirnay-Dummer 2006). The resolution of such problematic work situations leads to changes in the workflow among the involved actors and eventually can be a trigger for learning processes in an individual and inter-individual (i.e., organizational level).

In the following sections, we outline how methodological inputs from MMT and Articulation Work help to inform learning practice to finally develop an integrated framework.

6.1 Articulation Work and Mental Models

Work is an inherently cooperative phenomenon (Helmberger and Hoos 1962). Whenever people work, they have interfaces to others, either cooperating directly to perform a task or mediated via artifacts of work, which they share (Strauss 1985). The cooperative nature of work and its support with social and technical means have been subject to research for decades now (Schmidt and Bannon 1992).

Cooperative work requires that participating parties have a common understanding of the nature of their cooperation. This includes dimensions such as when, how, and with whom to cooperate using certain means. The mutual understanding of cooperation has to be developed when cooperative work starts and has to be maintained over time, as changing environment factors may influence cooperation (Fujimura 1987). All activities concerned with setting up and maintaining cooperative work are summarized using the term "Articulation Work" (Strauss 1985). Articulation Work mostly happens implicitly and is triggered during the actual productive work activities whenever contingencies arise (Gerson and Star 1986). Cooperative practices are established without a conscious act of negotiation in "implicit" Articulation Work, relying on social norms and observation to form a mutually accepted form of working together (Strauss 1988).

Implicit Articulation Work, however, is not sufficient when cooperative work situations are perceived to be "problematic" or "complex" by at least one of the involved parties (Strauss 1993). The terms "problematic" and "complex" here explicitly refer to individual perceptions and are intrinsically subjective. As such, they cannot be detailed from an outsider's perspective. Consequently, implicit Articulation Work can influence cooperation substantially. Different understandings of the same work situation impact the way of accomplishing tasks and the quality of work results, once Articulation Work remains on an implicit level.

The act of negotiation and development of a common understanding of the cooperative work processes has to be carried out deliberately and consciously in such cases. This act has been termed "explicit" Articulation Work by Strauss (1988). It has not been detailed methodologically initially, but rather omitted deliberately (Strauss 1993, p. 131). However, explicit Articulation Work has to be carried out whenever problematic or complex work situations arise. Its expected outcome is to enable involved stakeholders starting or continuing their cooperative work towards a shared goal. The roles and activities of stakeholders involved in explicit Articulation Work need to be clarified, as they go beyond implicit Articulation Work and prevention of "problematic" (as termed by Strauss) situations. Existing studies largely focus on the support of implicit Articulation Work and the prevention of "problematic" situations (e.g., Schmidt and Bannon 1992; Grinter 1996; Sarini and Simone 2002a). The findings do not explicitly address the individual dimension of explicit Articulation Work (e.g., Schmidt and Simone 1996 or Herrmann et al. 2002), nor remain on a conceptual level without deriving implications for supporting explicit Articulation Work (e.g. Fjuk and Dirckinck-Holmfeld 1997).

Conducting Articulation Work facilitates the alignment of individual views about collaborative work. Strauss (1993) argues that these individual views (termed as "thought processes" and "mental activities") affect human work and direct individual action. Research in the field of Articulation Work and its methodological support has hardly ever addressed the roles of the involved individuals in the alignment process. Both Herrmann et al. (2002) and Jørgensen (2004) present approaches that state that explicitly considering the individuals' views on work is crucial for successful Articulation Work, but it does not explicitly consider complex work situations. Consequently, a common understanding of the concepts used for describing work cannot be taken for granted (Sarini and Simone 2002b) and should be subject to alignment itself (Sarini and Simone 2002a). For problematic or complex work situations in particular, where social means of alignment (Wenger 2000) might not be sufficient and even a common understanding of the used terms cannot be expected (Sarini and Simone 2002a), a closer look at the individuals' understandings of their and others' work is of development interest. It should support designers to provide effective support measures. From how "thought processes" are described by Strauss (1993), they correspond to instances of the concepts of "schemes" and "mental models" in cognitive sciences (Johnson-Laird 1981).

6.2 Mental Models Theory and Articulation Work for Organizational Learning

Both mental model theory and Articulation Work can be linked to different steps in the organizational learning approach taken by the KLC. Together with the S-BPM, a comprehensive picture is taken of how an actor-centric, process-oriented approach to organizational learning can be supported methodologically and technically.

Figure 6.1 gives an overview about where mental model theory and Articulation Work play a role in the KLC. Starting with the Articulation Work, the mapping of the different concepts to the KLC is straightforward. All activities that are carried out to the reach the goal of work are referred to as "Production Work." As soon as a problem arises (i.e., some outcomes do not meet the expectations of any of the involved people), Articulation Work is triggered. How these problems are identified cannot be explained using Articulation Work; Mental Model theory will provide support here. If a problem can be compensated without any dedicated engagement in revising the work process, "implicit Articulation Work" happens. This is equivalent to what is referred to as single-loop learning in the KLC. Following the approach of "implicit Articulation Work", the modes for performing single-loop learning are extended in reference to the S-BPM instantiation of single-loop learning. More informal, even completely unobservable alignment activities among the workers might compensate the problem and affect the distributed organizational knowledge base (DOKB) only in terms of the actor's knowledge that has assimilated the new information of which contingencies can occur in a particular process and how it can be resolved.

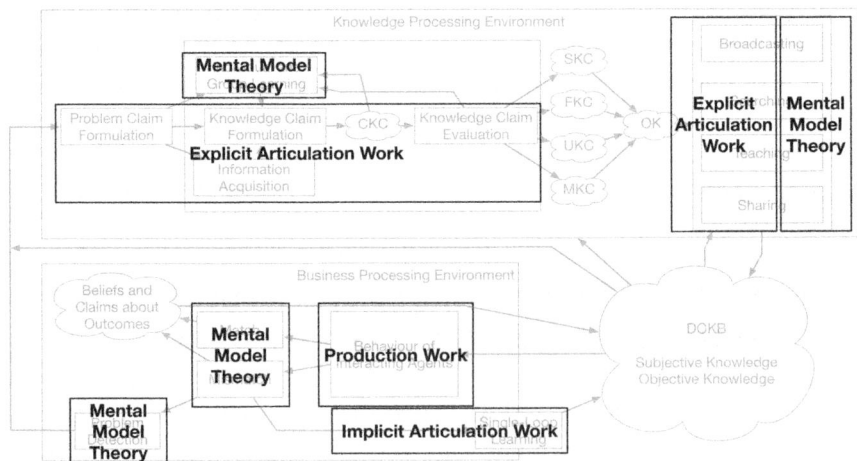

Fig. 6.1 Mental model theory and Articulation Work in the KLC

If a situation appears to be too problematic to be resolved without any interruptions in the work process (i.e., stepping out of the business processing environment), "explicit Articulation Work" is triggered. Explicit Articulation Work refers to dedicated negotiation and alignment activities that are carried out to enable people to effectively perform their work. Both, the activities included in knowledge production as well as those in knowledge integration, can be considered to be part of explicit Articulation Work. With this mapping of the KLC knowledge processing environment to explicit Articulation Work, the spectrum of methodological support for the latter is opened up for implementing organizational learning support.

Mental model theory augments the KLC in two specific situations: It can be used to explain how matches and mismatches in outcomes are identified by actors, and it provides a frame of reference for learning and behavioral change processes on an individual level.

People involved in a work process perform their activities according to their perceptions of the current situation and their schemes and mental models (i.e., their beliefs about how the environment will react to their activities). As soon as a scheme does not provide a starting point on how to determine what to do next or if the environments' response to a certain activity does not match what has been expected, a problematic situation occurs (this is equivalent to the match/mismatch section within the KLC). People try to compensate this problem using their mental models. This might or might not involve other people participating in the work process but will always lead to learning in the sense of the KLC. The situation is considered too problematic to be compensated, if the existing mental models of the involved people do not allow for compensation of the problem. In these cases, explicit Articulation Work, that is, activity in the knowledge processing environment is triggered.

During different phases of the knowledge processing environment, the mental models of different groups of people are altered. Those involved in developing the knowledge claim accommodate the new theories about how to resolve or avoid the problematic situation during the individual and group learning steps. In the course of integrating the new knowledge claim in the DOKB, the mental models of other people for whom the claim can be important during work need to accommodate it in their mental models. Instruments supporting model-based learning activities can potentially be of use here.

An alternative to immediate distribution of a new knowledge claim throughout an organization is to delay delivery to the affected people until they are confronted with the problematic situation. Following the theory of model-based learning, people are able to accommodate new information in their mental models more easily if the accommodation process is triggered by an actual situation (Ifenthaler 2006). This, in turn, requires an appropriate form of representation for the knowledge claim that allows for individualized and situated delivery of the information necessary to resolve a particular problematic situation. The requirements on this representation and a conceptual approach on how to implement it are presented in the next section.

A support system making use of such a representation is able to provide agile—that is, appropriately situated—and actor-aware—that is, personalized and individually adaptable—process implementation and improvement support, and in this way realizes the foundation for implementing an integrated learning framework.

6.3 Towards an Integrated Framework

In order to establish a common framework for the design of work processes in digitally augmented organization, we outline a framework on how to describe work in such organizations in the following. It should allow for situation-specific agreements on work carried out by a group of organizational actors that dynamically match their skills and work processes to reach a shared organizational aim. Work processes are not considered to be static flows of activities and interaction throughout a whole organization in this approach but vary in their implementation depending on the current situation and the actors carrying out the process.

This approach requires introducing concepts to describe such processes and their variations as well as the situations that determine which variation of a process is actually carried out. The relevant concepts are printed in italics in the following section and are put into mutual context in Fig. 6.2.

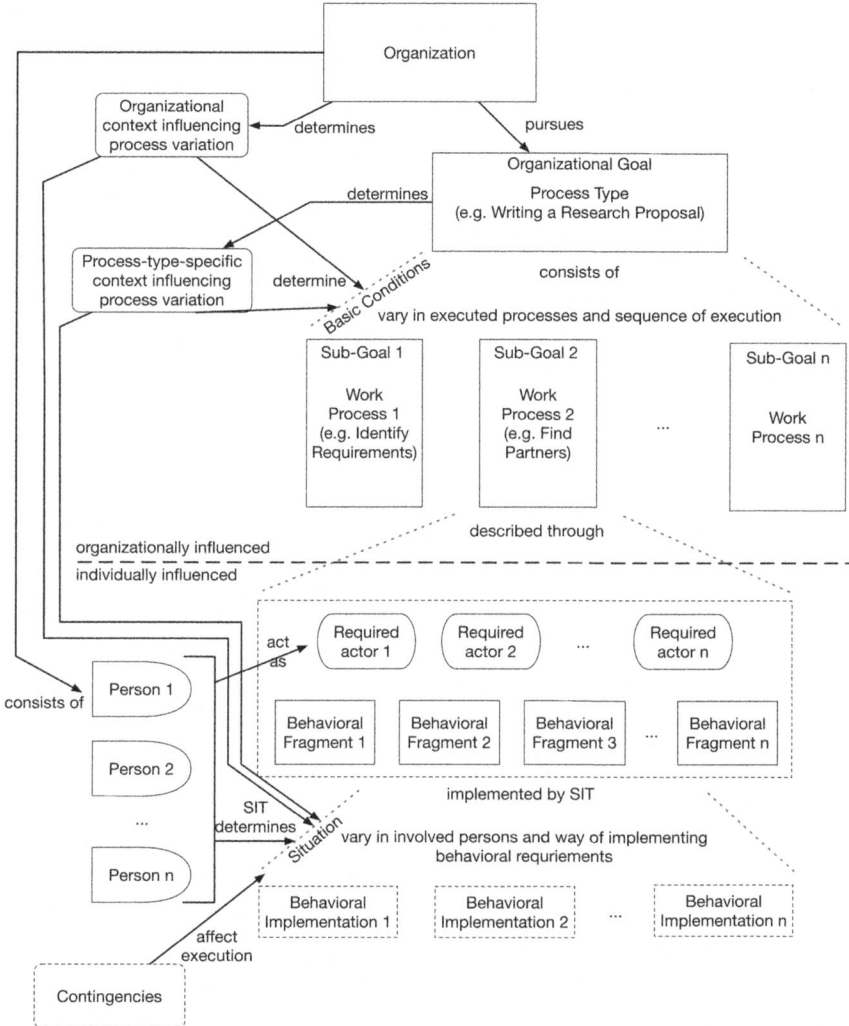

Fig. 6.2 Conceptual framework

6.3.1 Relevant Concepts

Organizations are the frame of reference for work processes. An organization here not necessarily corresponds to a single company but might span across any number of legal or administrative entities. An organization is determined by the *persons* that participate in it. Persons are active entities that carry out work within the organization. Persons are the carriers of knowledge and determine with their expertise how work is actually carried out.

An organization pursues one or many *organizational goals*. Such goals determine the scope of work of the organization and are specified by the entities being responsible for the organization. Each *organizational goal* is pursued by a particular type of process. A *process type* is not an actual process but a means of structuring the activities of the organization. Process types are specified independently of any influence factors except the organizational goal to be pursued. A process type thus contains a number of actual work processes that are determined by the *basic conditions* under which an organizational goal is pursued.

The basic conditions for process execution are determined by the *context influencing the process variation*. This context is determined by two major sources. *Organization-specific context* refers to the general conditions within the organization under which a process is carried out. This includes financial and administrative guidelines as well as business rules to be adhered. Organization-specific context is identical throughout the whole organization and equally affects all process types. *Process-type-specific context* refers to aspects that are only relevant for pursuing one specific organizational goal. This includes varying legal circumstances, potentially changing partners from outside the organization, or requirements specified by internal or external entities such as superiors or customers, respectively. The actual values these factors take in a particular situation make up the basic conditions under which a process of a particular process class is executed.

A process type consists of a set of *work processes* that allow reaching the organizational goal. Work processes are distinguished by goals that can be pursued separately and finally lead to archiving the overall organizational goal. The sequence of work processes and the question whether a particular work process is executed at all are determined by the basic conditions. A work process might not be relevant under certain conditions and thus is omitted. The sequence of work processes might be of importance under certain conditions, whereas it might not matter otherwise.

6.3.2 Implementation of Work Processes

Work processes are described by determining the *required actors*, who interact based on *behavioral fragments*, which determine the tasks to be completed and interaction to be carried out in any case. A required actor is described by the set of tasks it is responsible for (as will be elaborated on in more detail later). The set of required actors and behavioral fragments is not static and predefined for a particular work process. Which actors are actually required, and which behavioral fragments have to be implemented, are determined by the context factors described earlier as well as by the *persons* who act as members of a *situation-specific interdisciplinary team (SIT)* in a particular work process instance. As noted, persons are the carriers of knowledge and implement processes according to their expertise and the situation they perceive while performing their work.

The *situation* a process is implemented in consequently is determined by the organization-specific and process-class-specific context factors as well as the persons that are involved in the process. Furthermore, unforeseen *contingencies* can affect process execution, as ad hoc adaptations and workarounds might become necessary. Contingencies are a part of the situation a work process is carried out in but cannot be described at the time the execution of the work process starts. They are thus specific to a particular execution of the process, in contrast to the other influence factors that remain stable for identical situations (given that such identical situations would occur).

6.3.3 Responsibilities and Skills

Based on the aforementioned description, *work processes* can be considered collections of areas of responsibility, as visualized in Fig. 6.3.

The notion of *team member* refers to a set of activities that are completed by a single person in a *work process*. Team members are required to interact with each other as part of a SIT to achieve the aim of the work process. In the context of the work process, these team members play different roles when carrying out the work process.

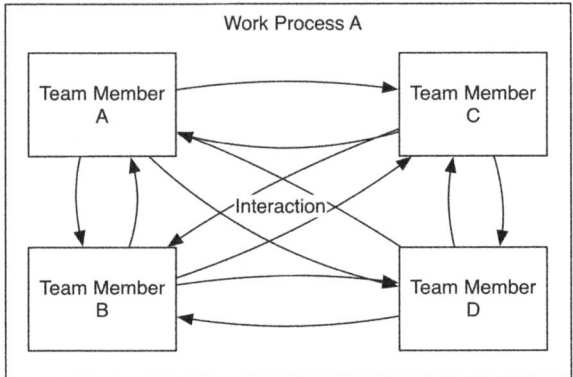

Fig. 6.3 Work processes and areas of responsibility

Persons can take these areas of responsibility as visualized in Fig. 6.4.

Persons are active entities in an organization that are able to perform tasks. A person can act as an actor in one or more teams within a work process and can even take over the responsibilities of several team members.

Organizational roles are used to cluster areas of responsibilities across work processes. Persons can take organizational roles and still take over areas of responsibility not included in this specific role, as visualized in Fig. 6.5.

The notion of *organizational role* refers to an area of responsibility within an organization. Organizational roles are specified by the set of team memberships they comprise. Persons take a specific role in a defined part of an organization. The role a person takes designates the person's formally necessary competences, that is, which team memberships the person has to take in the work processes it is involved in. The skills of a person (i.e., the set of team memberships a person is able to take in general) might go beyond those required by the person's organizational role.

Team Members (areas of responsibility) are specified regarding their interface towards other subjects and the behavior required to fulfill the responsibilities, as visualized in Fig. 6.6.

A *team member* is described using a set of requirements that specifies it expected behavior towards other team members and guides its actual behavior in a specific work setting. Vague behavior blocks specify the

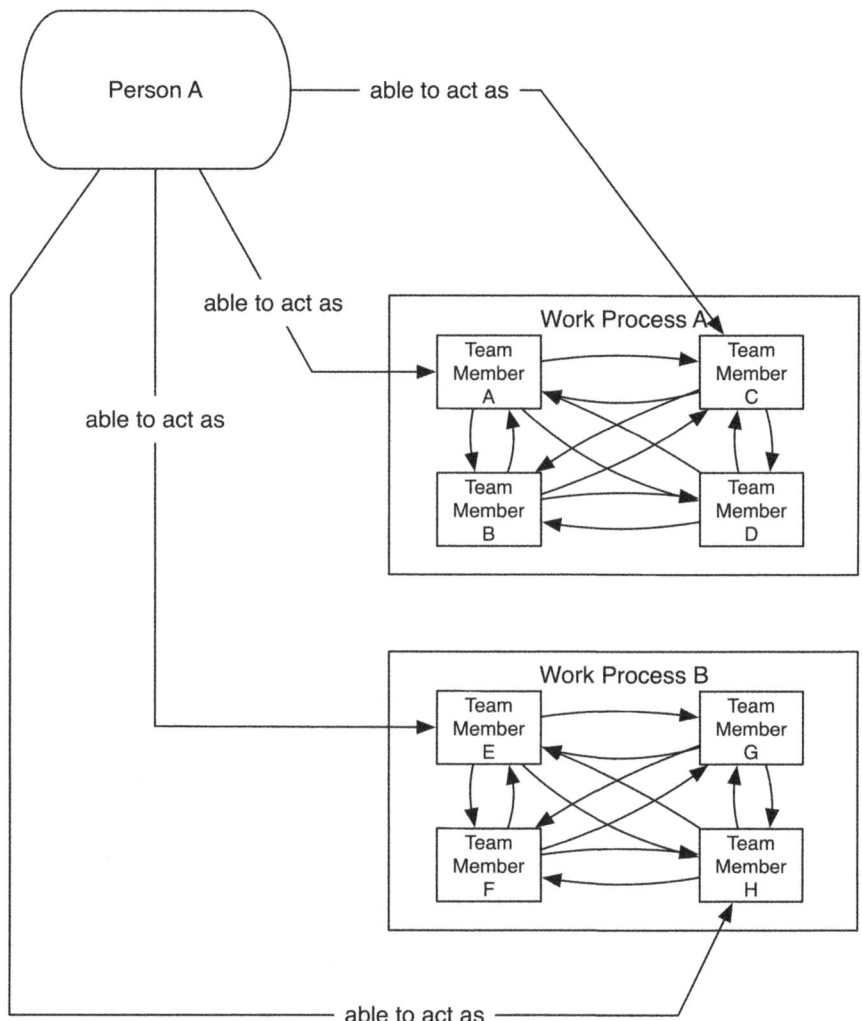

Fig. 6.4 Persons and areas of responsibility

minimum requirements on expected behaviors, whereas behavior block implementations show the actual behavior variants for different variants of a work process.

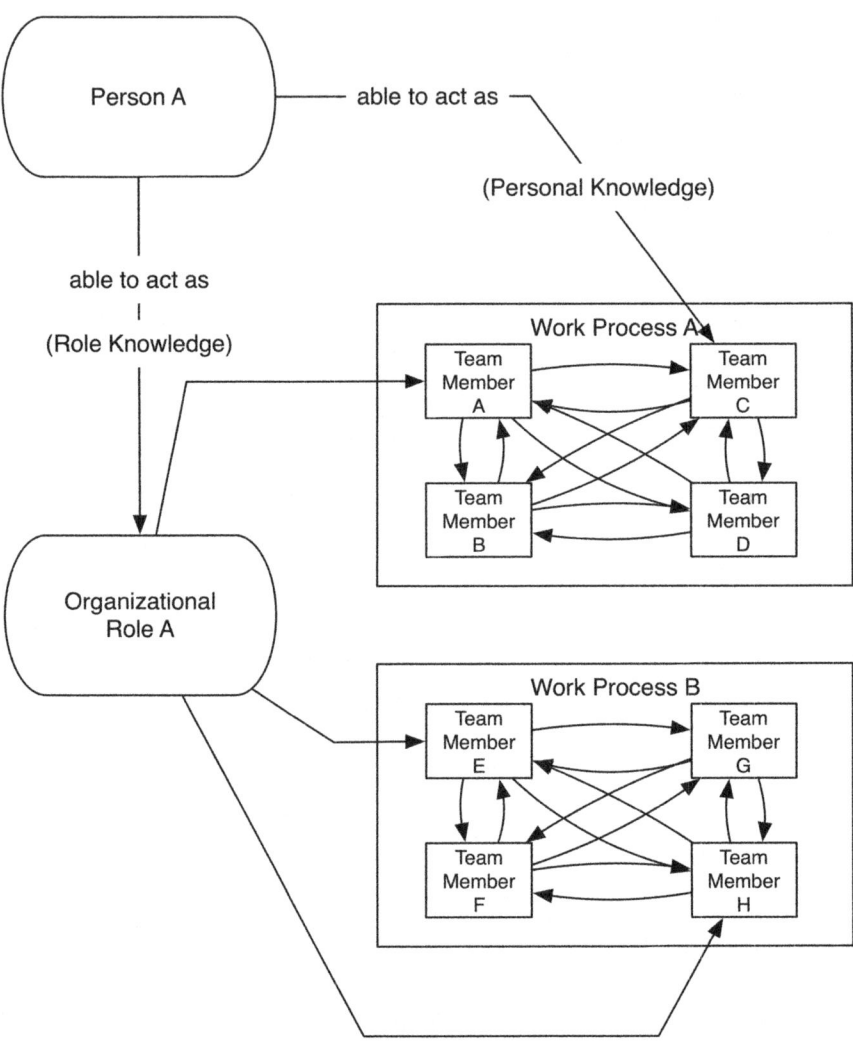

Fig. 6.5 Organizational roles clustering areas of responsibility in different work processes

Enabling Emergent Workplace Design 263

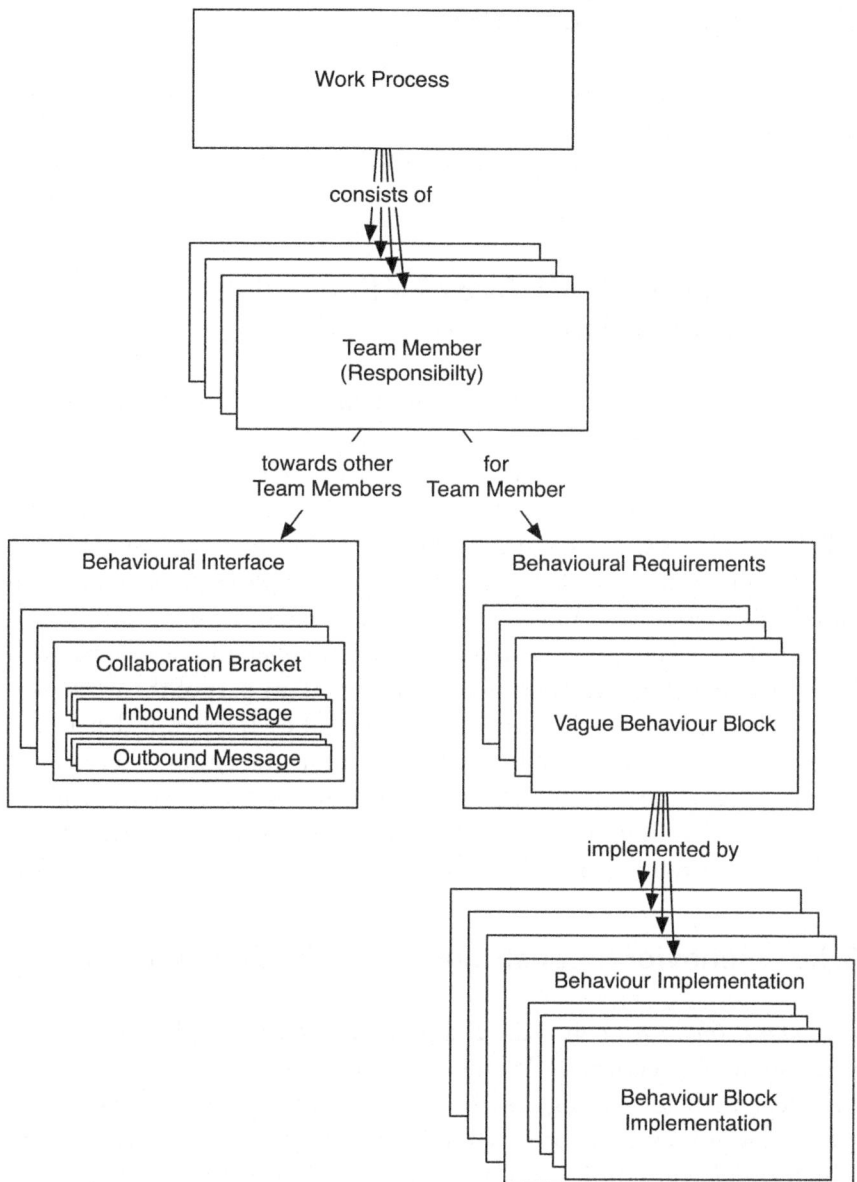

Fig. 6.6 Interfaces and behaviors of team members

6.3.4 Towards Instantiation

For single behavioral fragments, this leads to a three-tier conceptual structure, as shown in Fig. 6.7.

Areas of responsibility can be interlinked by matching their behavioral interfaces, as shown in Fig. 6.8.

6.3.5 Behavioral Interfaces for Interaction Coordination

Expected behavior towards other team members is described in a *behavioral interface*. The behavioral interface contains all messages a team member is able to receive and provide. If particular messages are interdependent of each other within the work process, they are grouped using *collaboration brackets*. Messages inside a collaboration bracket follow a specified order, where order specification can include sequences, optional parts, and alternatives. A behavioral interface can comprise multiple collaboration brackets, thus enabling a team member to collaborate with different other team members while completing its part of the work process. Collaboration brackets are independent of each other. If dependencies exist, they are to be handled within the implementation of the subject and thus are reflected in the behavioral requirements of a team member.

Behavior fulfilling an area of responsibility can be implemented by executing differently refined behavioral requirements as shown in Fig. 6.9.

6.3.6 Behavioral Constraints for Individual Actions

Behavioral fragments constrain and guide the actual implementation of team member activities. Behavioral requirements refer to orders or interdependencies among collaboration brackets and constraints and guidelines of how an actor may implement a subject's activities (e.g., in which order communication with different subjects has to be carried out or which activities have to be performed in any case before a message is sent).

Enabling Emergent Workplace Design 265

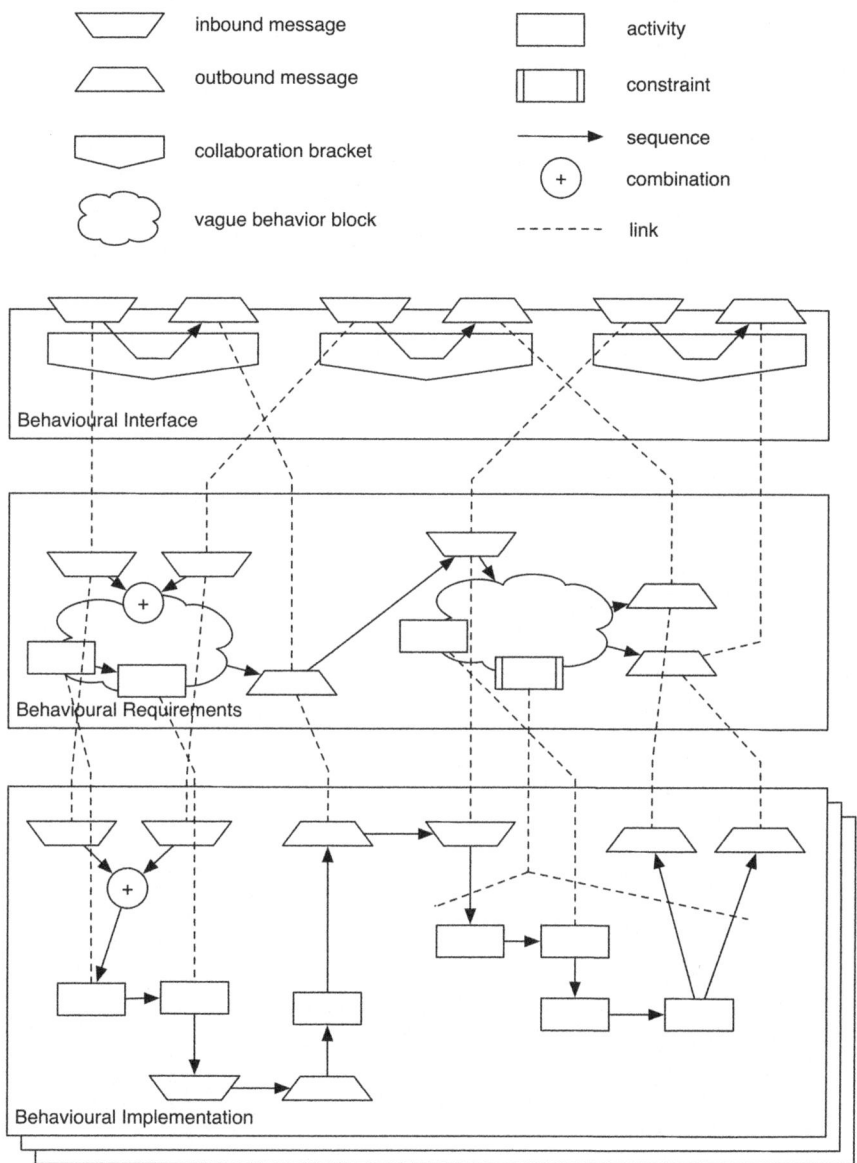

Fig. 6.7 Instantiation of behavior fragment

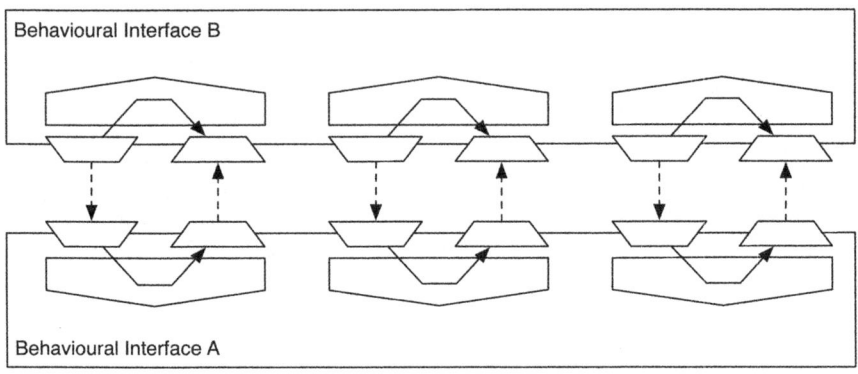

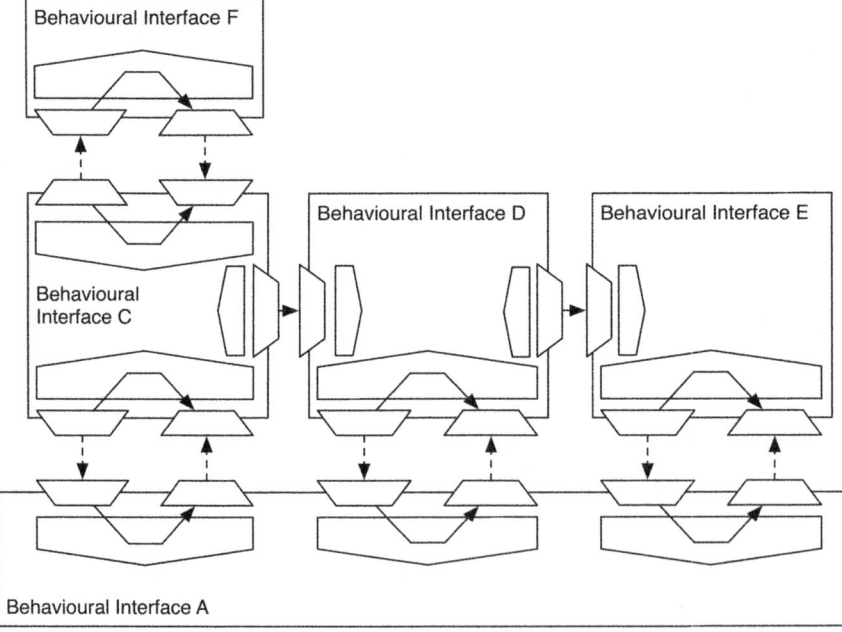

Fig. 6.8 Linking behavioral interfaces

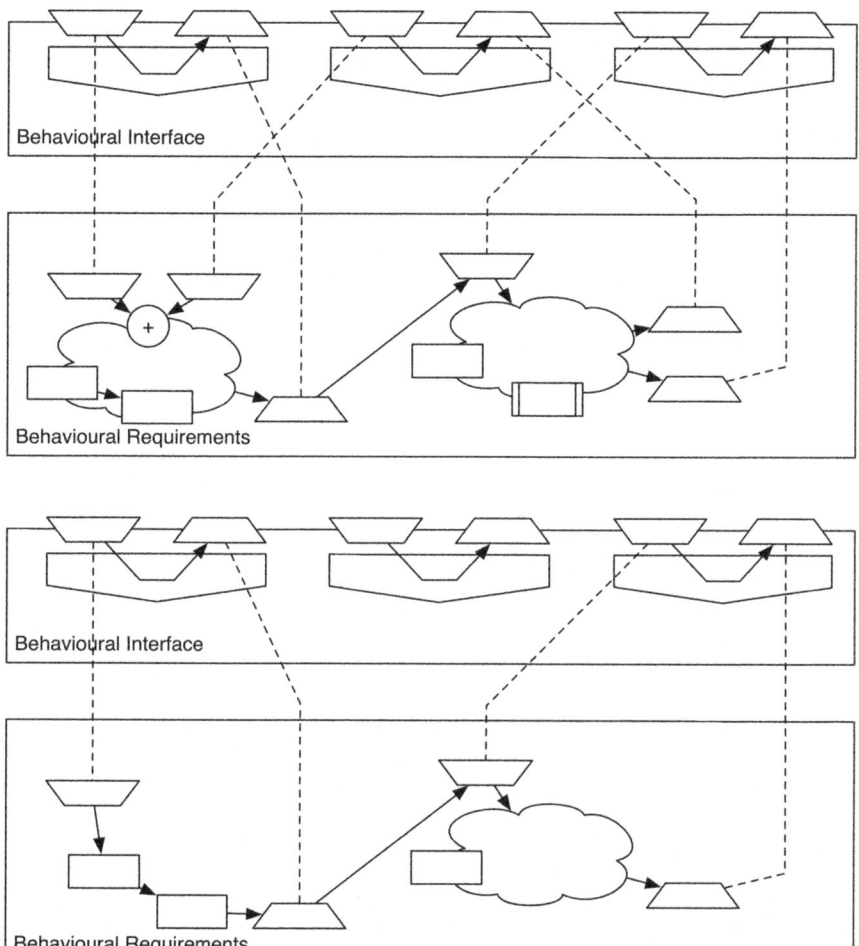

Fig. 6.9 Different behavioral requirements for a single behavioral interface

Behavioral fragments can be specified in a procedural way. They remain vague where no constraints apply, but are specified in more detail where particular activities or collaboration sequences are required.

Each set of behavioral fragments can be realized in different behavioral implementations, as shown in Fig. 6.10.

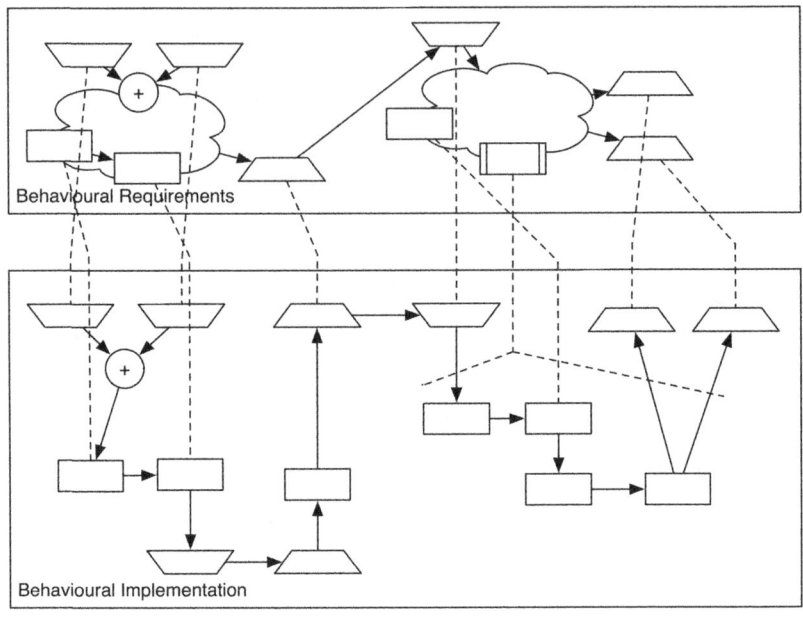

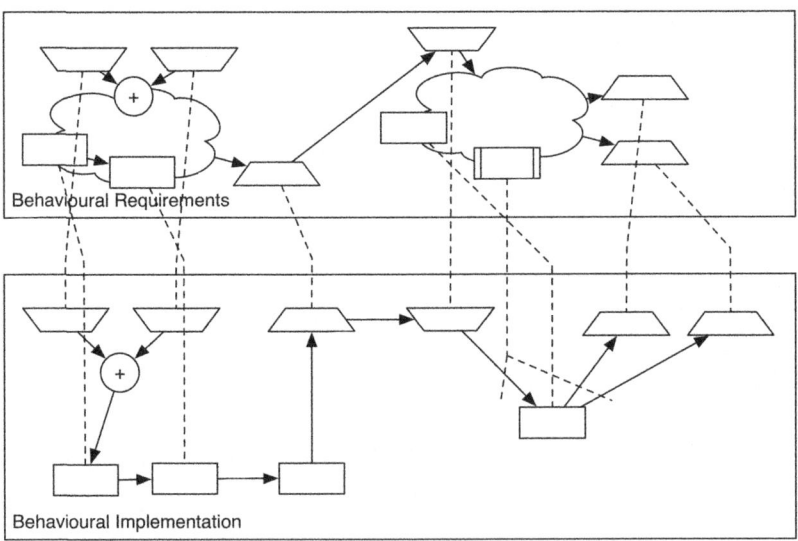

Fig. 6.10 Meeting behavioral requirements through different behavioral implementations

Collaboration requirements set on a subject separate different *vague behavior blocks* from each other. Each of the vague behavior blocks can be constrained by additional requirements. The actual implementation of the vague blocks is specific for a particular situation the business process is carried out in. The overall implementation is referred to as *behavior implementation*.

6.3.7 Varying Degrees of Freedom in Individual Activity

This structure can be put in the context of the theory developed earlier and allows implementing organizational work support for SITs, as shown in Fig. 6.11.

The parameters relevant to describe a situation involving self-organizing actors are organization- and process-specific, and additionally include the set of persons participating in the SIT.

A particular team membership thus might be implemented by different behavior implementations. Given potentially specified collaboration requirements, the granularity of behavior implementation is brought down to the level of vague behavior blocks. A vague behavior block is delimited by specified incoming and outgoing messages and might be constrained by requirements on the activities that are carried out during its implementation.

6.4 Articulation Engineered for Organizational Learning

In this section, we provide a conceptual architecture for developing learning support for situated team members and organizational actors. We ground the component-based approach on the concepts detailed in this chapter so far, in particular focusing on (i) mental model elicitation and articulation, (ii) continuous documenting of organizational knowledge (creation), and (iii) coupling execution with modeling facilities for actor-centric prototyping and probing of work processes. Hence, any technical

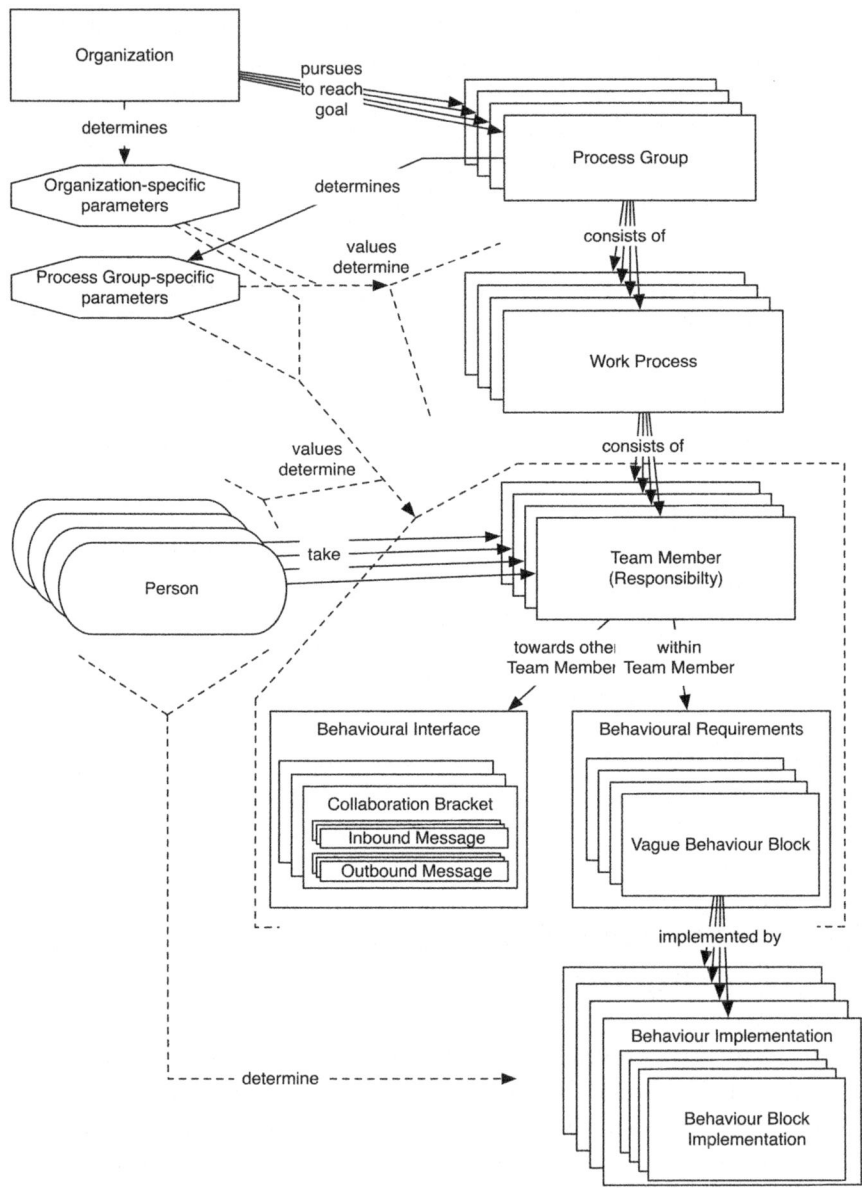

Fig. 6.11 Conceptual framework for situation-specific interdisciplinary teams

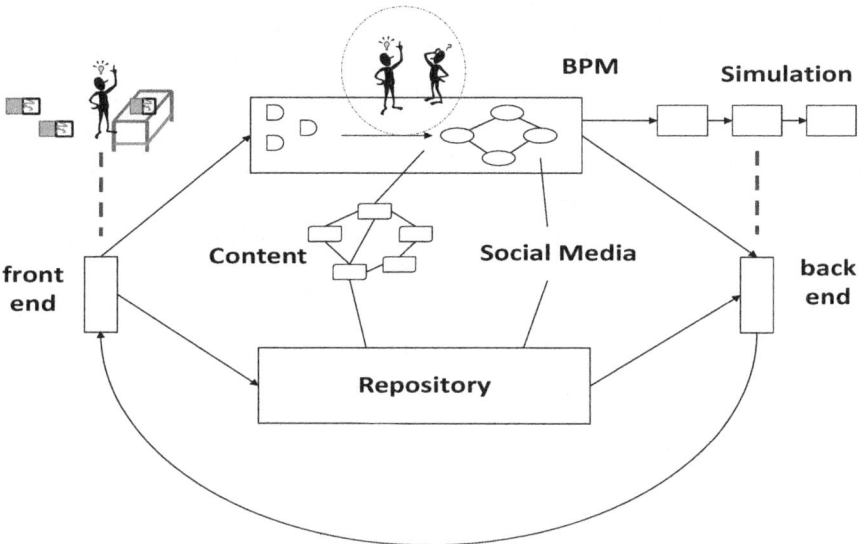

Fig. 6.12 Articulation engineered for organizational learning (Chris Stary 2014)

support system for elicitation, articulation, and exploration of knowledge needs some kind of shared repository for storing all different types of (created) content, including social media entries—thus prepared material. Preferably, it also contains information about the learning process itself, including social media interaction, in order to trace learning steps and model construction processes. Such a component can be based on intelligent content management and networked social media to inform project management (including contracting with responsible persons) and execution—see Fig. 6.12.

Since learning is not only an intertwined cognitive and social endeavor, front- and back-end components help strengthening stakeholder commitment allowing deep enquiry and sustainable capacity building. The component framework abstracts from concrete applications in terms of generic learning support systems and consists of:

- A set of front-end technologies for *articulation support*: It comprises state-of-the art devices, such as tablets with touchscreen structure elaboration as well as on-the-edge to market devices, such as tabletop systems for concept mapping and process modeling (e.g., Metasonic.de/touch, Comprehand Oppl and Stary 2014; Wachholder and Oppl 2014). These technologies serve as means to structure, share, and reflect on individual and collective mental models, either in the beginning of a project (idea development), in between (orientation), or when completing a project (reflection of achievements, check for completeness).

- A set of back-end technologies for *process execution*: Similar to the set of front-end components, it comprises state-of-the art devices for simulating or prototyping, for example, process management triggering production lines (process execution) (cf. www.so-pc-pro.eu). These technologies serve as means to implement production processes as projected by planning and monitoring, and finally to produce goods.
- A set of capacity building and project management technologies for *learning management and support*: It comprises state-of-the art systems, in particular intelligent content management and state-of-the-art social media, either well established, such as Facebook, or on-the-edge to market, such as for video annotation. It is crucial that capacity building is supported by intertwining features from social media, with didactically prepared content, as well as contract and project management, to organize learning steps.
- A storage technology as a *memory support*: It includes text, diagrammatic information, video material, or hypermedia, and supports storing articulated items or process representations such as interactive structure elaboration of critical incidents. It also provides all prepared material and authoring functionality to create work content, up to executable process models. All information, either stemming from preparation, planning, learning, execution, or simulation, can be kept in that component and reused by searching via metadata and changing the context of use.

The concept supports coupling existing support systems, that is, the concept of federated systems becomes important. Each system of a federated system can then be operated autonomously, while at the same time being an interoperable part of a larger system (Wachholder and Stary 2014). Hence, dedicated effort has to be taken for ensuring the interoperability of the involved technologies.

6.4.1 Featuring OL Processes

In this section, we outline how the different steps in the framework can be supported methodologically and technically. Seven different areas of support have been in line with the KLC and its adjustment with the S-BPM activity bundles:

- *Support for repository access* is necessary to provide means for input, output, and representation of the codified parts of the repository throughout the whole learning cycle
- *Delivering actor- and situation-aware process support* aids the execution of a work process by providing access to knowledge about work when needed and also providing the underpinnings of activities during knowledge integration
- *In-situ process collection, adaptation, and refinement* are necessary when a problem is encountered during work and can be compensated from directly within the work process
- *Identification of the need for explicit acquisition and alignment activities* is triggered when in-situ compensation is not possible, and a problem claim needs to be formulated for further processing
- *Process knowledge collection, reflection, and alignment* cover all activities concerned with the social interactions during knowledge production
- *Process validation and simulation for reflection and alignment* are necessary to evaluate knowledge claims and check their feasibility
- *Process visualization for elicitation, reflection, and sharing* is a means of support for all activities in the knowledge processing environment that require a codified form of the knowledge claim that is produced and eventually integrated into the repository.

6.4.2 Support for Repository Access

The repository does not solely contain technically represented, automatically executable knowledge claims. It conceptually rather covers all knowledge of the members of an organization, their experiences and expertise, as well as all individual and organizational procedures and facts that have been codified in IT-systems, workflow specification, or other organizational descriptions.

In dealing with this manifold of potentially relevant knowledge while still keeping maintainable complexity of the support tools, we follow a transactive memory approach proposed by Wegner (1987). Transactive memory is a conceptual type of memory (i.e., stored knowledge) that augments the (individual) internal memories and (codified) external memories. It refers to "a set of individual memory systems in combination with the communication that takes place between individuals." Following this definition, knowledge transfer here is bound to the opportunity of direct interaction among the involved people. The challenge of building transactive memory support systems to facilitate knowledge sharing beyond a directly interacting group level has initially been addressed by Nevo and Wand (2005).

Nevo and Wand (2005) claim that transactive memory in distributed settings can be supported by an organizational memory system providing access to:

- *Role knowledge*—knowledge that is required by definition to take a certain role (e.g., knowledge about how to write program code in a specific language for application developers).
- *Instance knowledge*—knowledge a person has but which would not be required by his or her formal role (e.g., experiences in supporting international research projects for a secretary).
- *Transactive knowledge*—knowledge about how to effectively extend one's knowledge by interacting with others. This includes:

    – Conceptual meta-knowledge (ontological concepts needed to describe a knowledge domain).

- Descriptive meta-knowledge (information about role or instance knowledge, like author, scope, format, or creation date).
- Cognitive meta-knowledge (knowledge about one's own knowledge and abilities).
- Persuasive meta-knowledge (knowledge about the credibility and expertise of the source).

The conceptual distinction between role and instance knowledge is equivalent to the approach of process groups and processes as described earlier, which has been introduced to handle variants of process execution due to personal expertise and situated influence factors. The concept of transactive knowledge allows providing support for distributing knowledge that is not codified in a technically executable way. In providing actors with knowledge about how to interact with other member of the organization, their ability to extend their knowledge in a self-directed way is supported. The concept of transactive memory maps to the meta-information about knowledge claims that need to be integrated in the repository according to the KLC (Neubauer et al. 2013).

The state of the art in organizational memory systems research is to enable situated knowledge delivery by offering technical support systems that monitor the current state of a work process and provide access to information relevant to the current work step (Mühlburger et al. 2017). We go beyond this approach in two respects: (a) it not only focuses on providing support during the operative work process but supports activities throughout the whole KLC (indicated by the arrows reaching out from the repository to different phases in the KLC, and (b) it does not assume the existence of a standardized (i.e., unique) way of carrying out a work process in a particular situation but explicitly also considers the configuration of involved actors and their individual experiences and expertise.

Following this actor-centric approach, the use cases visualized on the right margin of Fig. 6.13 can be identified. According to the terms of Firestone and McElroy (2003), the different types of descriptions representing knowledge can be considered to cover the explicitly codified part of the repository (in contrast to non-codified subjective knowledge

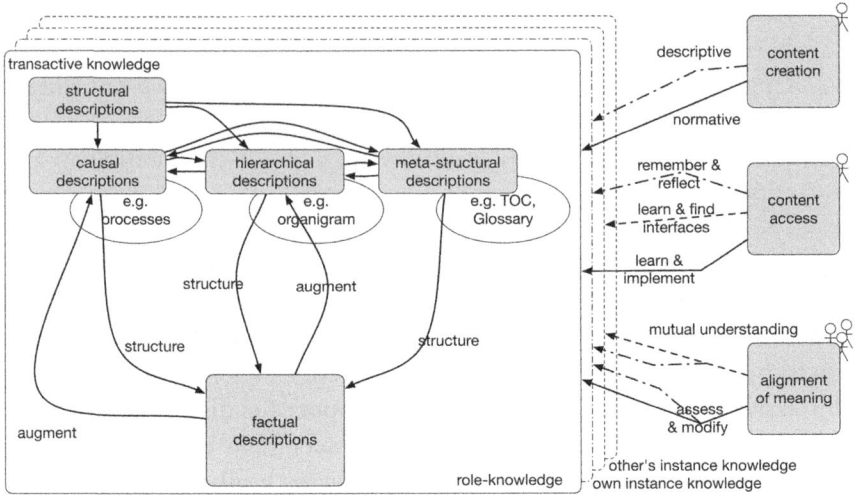

Fig. 6.13 Transactive memory concept used for the codified part of the repository (according to Neubauer et al. 2013)

and knowledge implicitly codified in technology-based work support systems).

Role knowledge is considered that part of organizational information that is mutually agreed upon or considered an organizational guideline of how a certain role in a particular work process should act. That corresponds to the task bundles with their proposed roles and tasks described earlier. In addition to role knowledge, each actor has *instance knowledge* that is different from the formally agreed upon view, goes beyond it, and potentially leads to different behavior in particular situations (dashed frames in Fig. 6.13).

6.4.3 Process Knowledge Elicitation and Knowledge Claim Development

Following the actor-centric approach, process knowledge has to be collected from the people actually performing the work processes. While there might be an organization-wide defined way of carrying out work for a particular class of processes, such models cannot cover all specifics of process execution that come with experience and expertise in knowledge-intense processes. These specifics, however, have to be collected, if individual or—even more—organizational learning processes should be enabled. This includes

- individual reflection of how work processes have been performed historically by oneself, and
- inter-individual alignment of understanding of how a cooperative process is performed introducing actors to how they can implement a role for a particular class of processes under given basic conditions or even in a particular situation

Fundamentally, process knowledge elicitation is a two-step process, where the first step is optional: (i) In-situ process collection, adaptation, and refinement, and (ii) process knowledge collection, reflection, and alignment.

6.4.3.1 In-situ Process Collection, Adaptation, and Refinement

In-situ process collection, adaptation, and refinement refer to activities that are conducted during and as an integral part of the work process ('in-situ').

Process collection refers to activities that build representations of process knowledge from directly within the work process. The information necessary to do this can be collected technically or using methods from social and cognitive sciences. On the technical side, information can be collected from workflow support systems or groupware that technologically mediates the work processes and can keep track of the activities performed with its help. If technological means are not available or their use is not appropriate (e.g., due to generated overhead or privacy reasons), methods to diagnose how people perform their work (such as "storytelling", Santoro et al. 2010) and to diagnose their motivations and reasons for acting in a certain way (such as "thinking aloud", Van Someren et al. 1994, or "structure elaboration techniques", Groeben and Scheele

2001) are used, both being recognized methods to learn about an actor's mental models.

Process adaptation and refinement is a requirement specific to actor-centric process modeling, as deviations from potentially pre-specified task bundles have to be possible and can be kept track of technologically. Even specifying a task-bundle variation from scratch during its execution has to be possible—the virtual enactment instrument described in Chap. 5 allows for such scenarios of operation.

6.4.3.2 Process Knowledge Collection, Reflection, and Alignment

Process knowledge collection refers to dedicated modeling activities that are conducted before or after the actual work process execution. Using the KLC as a frame of reference, this step is located in the knowledge processing environment.

This step can be carried out based upon data collected in situ throughout step 1, can build upon previously built process or task bundle models (e.g., for reflective purposes), or can start from scratch if process knowledge is to be represented for the first time.

Methodological support in performing these activities is even more important here than it was in step 1. While some information in step 1 could be collected technologically without any direct involvement of the actors, knowledge externalization in step 2 can only be performed directly by the actors. The process of externalization has to be guided methodologically in order to, on the one hand, not restrict users in expressing their perceptions on their work contributions, and, on the other hand, capturing this knowledge in a communicable form, codified for further processing. Technical tools can be used as support measures; however, their use should be informed and guided by methodological considerations.

Promising methodological approaches for process collection and refinement are structure elaboration techniques and concept mapping approaches. Both have proven to aid externalization of mental models (as noted earlier, these mental models guide individual task implementations) and support the development of a common view on cooperative task bundles.

The second aspect of methodological support, to allow for capturing of knowledge in a communicable form to be further processed, requires to guide actors from their initial, unrestricted form of process visualization (e.g., created using open structure elicitation techniques or concept mapping) towards a more structured, refined version that uses a defined language for representation and is sufficiently specified to be mapped to micro-processes, that, in turn, allow for delivering process support during task execution. While still open at this point in time, this aspect will be approached with a scaffolding approach (as known from individual learning support) that provides guidance for actors in refining their models towards the desired target format.

6.4.4 Process Visualization for Elicitation and Reflection

Situation-specific process representations, as introduced earlier, are not an appropriate means of process visualization for actors refining, reflecting upon, or aligning their views on their work. A more accessible presentation of the model has to allow for in-depth visualization of individual contributions and interactions while still maintaining the overall view on the whole process, or at least the task bundle variant. Additionally, the influencing factors have to be visible and configurable from within this presentation.

6.4.5 Process Validation and Simulation for Reflection and Alignment

Knowledge Claim Evaluation requires checking whether a newly proposed knowledge claim appropriately solves the problematic situation and will work in organizational reality when applied in the business processing environment. For knowledge about work processes, a common approach to evaluation is to conduct validation and simulation activities.

In both cases, the process is enacted in an artificial environment to examine its properties during execution and draw conclusions for its real-world implementation.

Validation refers to activities that involve actors in process evaluation and allows them to directly enact the process in an artificial environment. They can uncover problems and inefficiencies in the process in this way and directly apply their experiences to the knowledge claim in the course of the "individual and group learning" feedback cycle in the KLC knowledge production step. Validation can be implemented in different ways, varying in the depth the artificial environment enactment happens in, that is, how closely the model of the environment matches the real-world work environment and how many parameters characterizing the current work situation can be accounted for.

Simulation in contrast to validation does not directly involve actors in the execution of a process for evaluation purposes. Simulation rather applies statistical models to represent environmental parameters (e.g., how often a process is triggered, how long it takes an actor to complete a task) and uses IT-based instruments to automatically enact a process. Process models in general, and the flexible process representation approach proposed here in particular, have a large number of potential instantiations due to the choices that can be made during execution. While during validation only some cases can be evaluated, simulation can be used to check the process as a whole and uncover problems in any of the potential instantiations.

The subject-oriented approach requires simulation to explicitly consider the work being executed in parallel by different actors and being aligned by acts of communication among these actors. The ASM-approach to S-BPM proposed by Börger (2012) provides a starting point here for being able to formally verify a process (e.g., identifying deadlocks) and simulation with an ASM-implementation based on multi-agent system (cf. Lerchner 2015; Lerchner and Stary 2016).

6.5 Conclusion

Starting out with organizational development in an actor-centered way requires a value chain in digital work design that keeps actors involved. Hence, in this chapter, we have integrated the methods and instruments presented in the former chapters in a coherent framework and have identified design space elements for various tool chains supporting stakeholder-based agile work design. The integration process, however, was challenged by the contextual elicitation and alignment activities to be coordinated for stakeholder-centered learning and process development—the conceptual and empirical findings provided a rich set of design elements, such as tasks, business objects, various relationships, and levels of abstractions, and indicated intertwined learning loops which are indirectly synchronized by a design repository (Documented Knowledge Base).

The framework for practical design support proposed in this chapter is based on four elementary components, namely, (i) a front-end for articulation and elicitation of work knowledge, (ii) a learning environment coupling content with communication features, (iii) a container facility for multiple content types, and (iv) a processing component for automatically executing business processes. They allow reflecting on existing work practice and formulating knowledge claims that need to be negotiated before fed back to a Documented Knowledge Base interfacing the operational (business processing) environment. In this way, digital work design becomes a traceable and transparent way for stakeholders.

References

Abecker, A., G.N. Mentzas, M. Legal, S. Ntioudis, and G. Papavassiliou. 2001. Business-Process Oriented Delivery of Knowledge Through Domain Ontologies. In *Proceedings of the 12th International Workshop on Database and Expert Systems Applications*, 442–446.

Abecker, Andreas, Stefan Decker, and Otto Kühn. 1998. Organizational Memory. *Informatik-Spektrum* 21 (4): 213–214.

Börger, E. 2012. The Subject-Oriented Approach to Software Design and the Abstract State Machines Method. In *Proceedings of S-BPM ONE 2012*, ed. C. Stary, 1–21. Springer.

Dalmaris, Peter, Eric Tsui, Bill Hall, and Bob Smith. 2007. A Framework for the Improvement of Knowledge-Intensive Business Processes. *Business Process Management Journal* 13 (2): 279–305 (Emerald).

Diefenbruch, M., T. Goesmann, Thomas Herrmann, and M. Hoffmann. 2002. KontextNavigator Und ExperKnowledge—Zwei Wege Zur Unterstützung Des Prozesswissens in Unternehmen. In *Geschäftsprozessorientiertes Wissensmanagement*, ed. A. Abecker, K. Hinkelmann, H. Maus, and H.J. Müller, 275–292. Berlin: Springer.

Firestone, J.M., and M.W. McElroy. 2003. *Key Issues in the New Knowledge Management*. Oxford: Butterworth-Heinemann.

Fjuk, A, and L. Dirckinck-Holmfeld. 1997. Articulation of Actions in Distributed Collaborative Learning. *Scandinavian Journal of Information Systems* 9 (2): 3–24 (Unknown).

Fleischmann, Albert, Werner Schmidt, C. Stary, Stefan Obermeier, and Egon Börger. 2012. *Subject-Oriented Business Process Management*. New York: Springer.

Fujimura, J H. 1987. Constructing 'Do-Able' Problems in Cancer Research: Articulating Alignment. *Social Studies of Science* 17 (2): 257–293 (SAGE Publications).

Gerson, E.M., and Susan Leigh Star. 1986. Analyzing Due Process in the Workplace. *ACM Transactions on Information Systems (TOIS)* 4 (3): 257–270 (ACM Press).

Grinter, R.E. 1996. Supporting Articulation Work Using Software Configuration Management Systems. *Computer Supported Cooperative Work (CSCW)* 5 (4): 447–465 (Springer).

Groeben, Norbert, and Brigitte Scheele. 2001. Dialogue-Hermeneutic Method and the "Research Program Subjective Theories". *Forum Qualitative Sozialforschung/Forum: Qualitative Social Research* 1 (2): 1–10. http://nbn-resolving.de/urn:nbn:de:0114-fqs0002105.

Helmberger, P., and S. Hoos. 1962. Cooperative Enterprise and Organization Theory. *American Journal of Agricultural Economics* 44 (2): 275.

Herrmann, Thomas, M. Hoffmann, G. Kunau, and K.U. Loser. 2002. Modelling Cooperative Work: Chances and Risks of Structuring. In *Cooperative Systems Design, a Challenge of the Mobility Age. Proceedings of COOP 2002*, 53–70. IOS Press.

Hinkelmann, K., D. Karagiannis, and R. Telesko. 2002. PROMOTE—Methodologie Und Werkzeug Für Geschäftsprozessorientiertes Wissensmanagement. In *Geschäftsprozessorientiertes Wissensmanagement*, ed. A. Abecker, K. Hinkelmann, H. Maus, and H.J. Müller, 65–90. Berlin: Springer.

Ifenthaler, D. 2006. Diagnose Lernabhängiger Veränderung Mentaler Modelle—Entwicklung Der SMD-Technologie Als Methodologisches Verfahren Zur Relationalen, Strukturellen Und Semantischen Analyse Individueller Modellkonstruktionen. University of Freiburg.

Johnson-Laird, P.N. 1981. Mental Models in Cognitive Science. *Cognitive Science* 4 (1): 71–115 (Elsevier).

Jørgensen, H.D. 2004. Interactive Process Models. Department of Computer and Information Sciences, Norwegian University of Science and Technology Trondheim.

Lerchner, Harald. 2015. An Abstract State Machine Interpreter for S-BPM. In *S-BPM in the Wild*, 219–233. Springer.

Lerchner, H., and C. Stary. 2016. Model While You Work: Towards Effective and Playful Acquisition of Stakeholder Processes. In *Proceedings of the 8th International Conference on Subject-Oriented Business Process Management*, 1. ACM.

Marjanovic, Olivera, and R Freeze. 2011. Knowledge Intensive Business Processes: Theoretical Foundations and Research Challenges. In *Proceedings of the 44th Hawaii International Conference on System Sciences (HICSS), 2011*, January, 1–10. IEEE. https://doi.org/10.1109/HICSS.2011.271.

Matthias, Neubauer, Stefan Oppl, C. Stary, and Georg Weichhart. 2013. Facilitating Knowledge Transfer in IANES—A Transactive Memory Approach. In *Innovation Through Knowledge Transfer 2012*, Smart Innovation, Systems and Technologies, ed. R. Howlett, B. Gabrys, K. Musial-Gabrys, and J. Roach, vol. 18, 39–50. Berlin and Heidelberg: Springer. https://doi.org/10.1007/978-3-642-34219-6_5.

Mühlburger, M., S. Oppl, and C. Stary. 2017. KMS Re-contextualization—Recognizing Learnings from OMIS Research. *VINE Journal of Information and Knowledge Management Systems* 47 (3): 302–318.

Nevo, D., and Y. Wand. 2005. Organizational Memory Information Systems: A Transactive Memory Approach. *Decision Support Systems* 39 (4): 549–562. https://doi.org/10.1016/j.dss.2004.03.002.

Oppl, Stefan, and C. Stary. 2014. Facilitating Shared Understanding of Work Situations Using a Tangible Tabletop Interface. *Behaviour & Information Technology* 33 (6): 619–635. https://doi.org/10.1080/0144929X.2013.833293.

Pirnay-Dummer, Pablo N. 2006. Expertise Und Modellbildung—MITOCAR. University of Freiburg.

Santoro, Flávia Maria, Marcos R.S. Borges, and José A. Pino. 2010. Acquiring Knowledge on Business Processes From Stakeholders' Stories. *Advanced*

Engineering Informatics 24 (2): 138–148. https://doi.org/10.1016/j.aei.2009.07.002.

Sarini, Marcello, and C. Simone. 2002a. The Reconciler: Supporting Actors in Meaning Negotiation. In *Proceedings of the Workshop on Meaning Negotiation*.

——— 2002b. Recursive Articulation Work in Ariadne: The Alignment of Meanings. In *Proceedings of COOP 2002*, 191–206.

Schmidt, K., and C. Simone. 1996. Coordination Mechanisms: Towards a Conceptual Foundation of CSCW Systems Design. *Computer Supported Cooperative Work* 5 (2–3): 155–200. https://doi.org/10.1007/BF00133655.

Schmidt, K., and L. Bannon. 1992. Taking CSCW Seriously: Supporting Articulation Work. *Computer Supported Cooperative Work (CSCW)* 1 (1): 7–40 (Springer).

Stary, C. 2014. From e-Learning to Organizational Learning—A Knowledge Life Cycle Perspective on Operational Designs. In *Proceedings of the 10th International Conference on Knowledge Management*. Antalya.

Strauss, A. 1985. Work and the Division of Labor. *The Sociological Quarterly* 26 (1): 1–19 (Blackwell Publishing Ltd).

———. 1988. The Articulation of Project Work: An Organizational Process. *The Sociological Quarterly* 29 (2): 163–178.

———. 1993. *Continual Permutations of Action*. New York: Aldine de Gruyter.

Van Someren, M.W., Y.F. Barnard, and J.A.C. Sandberg. 1994. *The Think Aloud Method: A Practical Guide to Modelling Cognitive Processes*. London: Academic Press.

Wachholder, D., and C. Stary. 2014. Bigraph-Ensured Interoperability for System (-of-Systems) Emergence. In *OTM Confederated International Conferences "On the Move to Meaningful Internet Systems"*, 241–254. Berlin, Heidelberg: Springer.

Wachholder, Dominik, and Stefan Oppl. 2014. Interactive Coupling of Process Models: A Distributed Tabletop Approach to Collaborative Modeling. In *ECCE '14: Proceedings of the 2014 European Conference on Cognitive Ergonomics*, September, 1–8. New York: ACM Request Permissions. https://doi.org/10.1145/2637248.2637262.

Wargitsch, C., and T. Wewers. 1997. WorkBrain: Merging Organizational Memory and Workflow Management Systems. In *Workshop of Knowledge-Based Systems for Knowledge Management in Enterprises at the 21st Annual German Conference on AI (KI-97)*.

Wegner, D.M. 1987. Transactive Memory: A Contemporary Analysis of the Group Mind. *Theories of Group Behavior* 185: 208.

Wenger, E. 2000. Communities of Practice and Social Learning Systems. *Organization* 7 (2): 225–246.

Open Access This chapter is licensed under the terms of the Creative Commons Attribution 4.0 International License (http://creativecommons.org/licenses/by/4.0/), which permits use, sharing, adaptation, distribution and reproduction in any medium or format, as long as you give appropriate credit to the original author(s) and the source, provide a link to the Creative Commons licence and indicate if changes were made.

The images or other third party material in this chapter are included in the chapter's Creative Commons licence, unless indicated otherwise in a credit line to the material. If material is not included in the chapter's Creative Commons licence and your intended use is not permitted by statutory regulation or exceeds the permitted use, you will need to obtain permission directly from the copyright holder.

7

Putting the Framework to Operation: Enabling Organizational Development Through Learning

This chapter picks up the framework developed in Chap. 6 and shows how it can be put to operation using instruments that have been successfully deployed in practice (cf. Chap. 8). These instruments enable articulation and alignment of work process knowledge, allow its representation and transfer within organizations, and facilitate acting on these representations for validation and implementation in diverse organizational settings. We here adopt the organizational learning perspective, already proposed in Chap. 1, and situate the presented instruments along a multi-perspective learning chain informed by the components of the framework presented in Chap. 6. This allows us to offer an integrated view in Sect. 7.5, which shows how the framework can be instantiated in organizational practice.

Organizational learning comprises learning on two layers:

1. Individual level
2. Collective level within an organization (e.g., CoP) or beyond (e.g., company network)

According to the Knowledge Lifecycle (KLC) (Firestone and McElroy 2005), it also can occur within running business operation and beyond. However, learning outputs need to become concrete and visible, in particular when implemented in the business operation. Single loops target towards an envisioned optimum of operational procedures, for example, zero-tolerance with respect to quality of products. Learning is thus related to business processes and their management. Learning is less concrete when questioning assumptions and background of business operations. Double loops allow claiming and evaluating change proposals that could affect business operation. Deutero-learning reflects double-loop learning. It structures reflection on operation and learning procedures (knowledge processing), keeping multiple levels of development activities apart with dedicated point of coupling, triggering redesign of processes (Table 7.1).

With respect to handling knowledge according to an organizational learning architecture (see earlier in the chapter), reflection, referring to as-the-organization-is, and some prospective organization of work need to be supported and may occur intertwined. Hence, stakeholders need to be able to

- *Express* themselves in terms what they know, in order to document starting points of change
- *Reflect* on articulated knowledge, either alone, with peers, or other groups
- *Represent* and *manipulate* codified knowledge, forming baselines for further steps
- *Store* to avoid loss of information and process know-how ('again-invented-here')
- *Process* knowledge to evaluate or establish adjunct or resulting operational procedures
- *Share* knowledge by distributing content to put it to operation

Summing up, the development of support (chains of) technologies or enablers is a multi-dimensional endeavor, as it needs to address

- individual/group learning
- single/double/deutero-loop learning

Table 7.1 Learning/design dimensions, activities, and tools

Levels, loops, and enablers / Learning chain elements	Individual layer Features	Collective layer Features	Learning loops: single/double/deutero Constituents (what it is about)	Enablers: cognitive/social/emotional (m = method, t = tool) m also includes tool-specific feature set
Articulation and elicitation	Express (model), prepare for sharing, Reflect on shared knowledge	Active sharing (express and negotiate)	Knowledge about operation (processes), content/assumptions, claims, change proposals/reflected learning processes	Concept mapping (m), VNA (m), S-BPM modeling (m), structure elaboration (m), (rural) Comprehand (t), Faraw (ARCOR) (m)/social media (forum, chat, blog, Facebook, Twitter, LinkedIn) (m + t)/social media (m + t)
Representation	Navigate, search, store, trace, Versioning	Navigate, search, store, trace, Versioning	Business process models (business processing)/Knowledge (change) claims (knowledge processing)	Filter (m), topic map (m), annotations (m), Nymphaea-content management, social media support and intertwining (t), Metasonic Build (t)/annotations (m), Nymphaea view management (t)/annotations (m), Nymphaea view management (t)
Manipulation	Search, retrieve, develop perspective Edit content, prepare sharing perspective	Search, retrieve, develop group perspective, edit content, perspective management,	Perspective on constituents	Tagging (m), tracing (history) (m), editor (t), Comprehand(t)/like, follow, exchange annotations (m) social media (t), Comprehand (t)/like, endorse, follow (m) social media (t)
Processing	Explore	Explore impact on organization level	Evaluated knowledge claim	Validation and processing of BP (m) BP-execution engine (t), Simulating BP (m) Simulation engine (t)

- cognitive (content, domain)/social processes
- knowledge elicitation, representation, visualization, presentation and communication, sharing, processing

Support technologies require respective features, may stem from different kinds of applications:

- articulation, elicitation, elaboration/modeling tools
- content management systems
- social media
- business process management suites
- CSCW systems

Support technologies are of different kinds, with respect to mobility and chaining:

- Stationary/mobile
- Articulation—learning—processing

Support for explication, exploration, and distribution should capture both, scenarios for individual and collaborate development and deployment:

- In-situ scaffolding during operative work processes (what can I do now/next)

    - based on previous own task implementations
    - based on others' task implementations

- Learning from other actors who previously took the same role in a task bundle or are experts in the area during operative work processes or in knowledge integration (whom can I ask)
- Reflection on previous own and others' task implementations as a part of problem claim identification and knowledge production (what have we done)

- individually
- as a group

* Aligning with other actors who take other roles in a task bundles during knowledge integration (how can we work)

These requirements can be methodologically and technically approached using instruments for self-directed learning. In the following, we revisit the set of tools described throughout the former chapters and put them into mutual context using the framework developed above.

7.1 Sample Actor-Centric Tool Support for Articulation and Elicitation

The aim of the articulation and elicitation phase is to methodologically and technically allow for modeling of organizational phenomena whenever the need arises, especially of that directly situated in the actual work context. The approach exemplified here follows a physical card placement paradigm, for example, as proposed for structure elaboration techniques in the field of mental model externalization and alignment (Dann 1992) and, for example, adopted in the field of BPM by Luebbe and Weske (2011). The semantics of the used cards as well as their spatial arrangement is predetermined to be interpretable as actor-centric, communication-oriented business processes.

A modeling approach, which serves the goal of capturing a comprehensive representation of the overall business process, needs to take into account all individual contributions and facilitate identifying and making visible different mental models of how the collaborative aspects are performed (Fischer and Mandl 2005). In order to be able to identify different perceptions of how collaborative work is carried out, the individual mental models of the collaborating contributors need to be made accessible for alignment (Engelmann and Hesse 2010). Consequently, an approach for collaborative modeling of work should profit from a stage during which the participants individually externalize their mental model of the business process in the form of a conceptual model.

The results of these individual modeling activities provide a foundation for argumentative co-construction of a shared understanding about the business process. Argumentative co-construction can again be facilitated by conceptual models that serve as a shared artifact (Fischer and Mandl 2005). A conceptual modeling approach supporting this process should allow expressing individual claims and collaboratively putting them in the context of other claims for referral in the argumentative chain.

7.1.1 Comprehand Cards

Comprehand Cards (Oppl 2017; Oppl et al. 2017) enable creating models of work without the need for any dedicated technical infrastructure. The aim of this component is to allow for collaborative modeling whenever the need arises, especially of that which is directly situated in the actual work context. The Comprehand Cards approach (cf. Fig. 7.1) follows a physical card placement paradigm, for example, as proposed for structure elaboration techniques in the field of mental model externalization and alignment (Dann 1992). The semantics of the different card types is

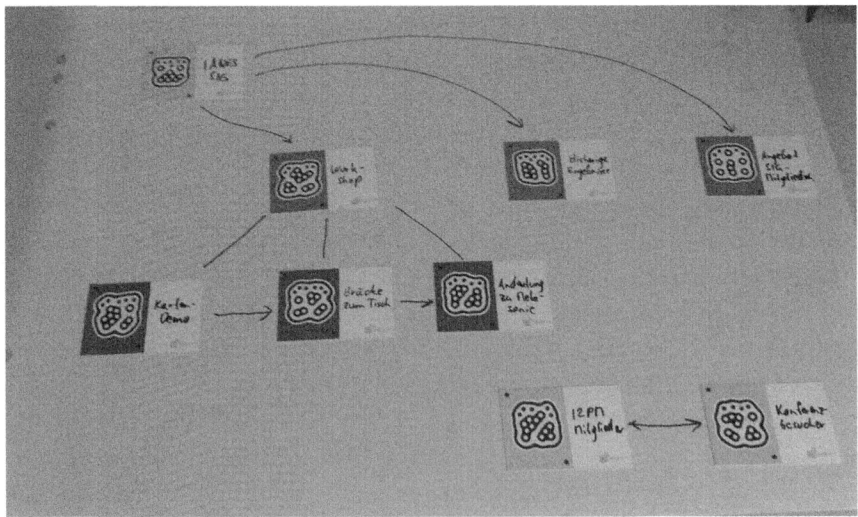

Fig. 7.1 Sample model created with modeling cards

determined by the use case they are deployed in—the system can be used for semantically open modeling (e.g., for concept mapping) as well as procedural modeling adhering fixed semantics as described in Sect. 5.1. What sets apart Comprehand Cards from other physical card modeling approaches (e.g., Decker and Weske 2009) is an additional software component, which allows extracting the conceptual model from any picture taken of the card-based model. The model extraction algorithms recognize concepts (i.e., cards), concept types (i.e., card types), and relationships between concepts (i.e., connections drawn between cards). Labels written on cards or besides connections are also extracted and provided as scaled and rectified images.

The recognition engine is designed to be used with pictures taken by smartphone cameras without any strict constraints on image angles and lighting conditions (Oppl et al. 2017). Pictures of a model are uploaded to an online platform that acts as a front-end for model extraction. If a model is too large to be depicted on one image in sufficient detail, multiple pictures covering distinct but overlapping areas of the model can be uploaded, which are then automatically processed by the system to improve recognition quality. For recognition of the cards, an adapted version of ReacTIVision (Kaltenbrunner and Bencina 2007) is used. All cards thus bear optical markers that make them uniquely identifiable by the system. Connections are traced using image recognition algorithms adapted from the study by Jiang et al. (2011). The extracted model information is represented in a configurable XML-based format that allows for further processing of the model in any compatible tool (cf. next sections).

7.1.2 Comprehand Table

The *Comprehand modeling table* (Oppl 2006; Oppl and Stary 2009; Oppl and Rothschädl 2014) is an interactive collaborative modeling environment that used graspable modeling elements to support externalization activities and to allow for equal access to the model for multiple modelers at the same time. The Comprehand Table aims at supporting in-depth, potentially controversial, modeling situations, where the need for flexibly altering the model conflicts with the constraints of card-based models and their hand-drawn, hardly changeable connections. The table (cf. Fig. 7.2) thus focuses on features supporting the modeling process, such as easy connection creation and deletion, altering the model layout without changing its conceptual structure, and tracking the modeling history, thus allowing to revert the modeling process to earlier stages. A detailed description of the features and their mode of operation is provided in Oppl and Stary (2014).

Technologically, the modeling table is an implementation of a tangible tabletop interface (Ishii and Ullmer 1997) that uses back-projection on the table surface to blend the physical modeling elements with digital information. Tangible interfaces (Ishii and Ullmer 1997) are an approach

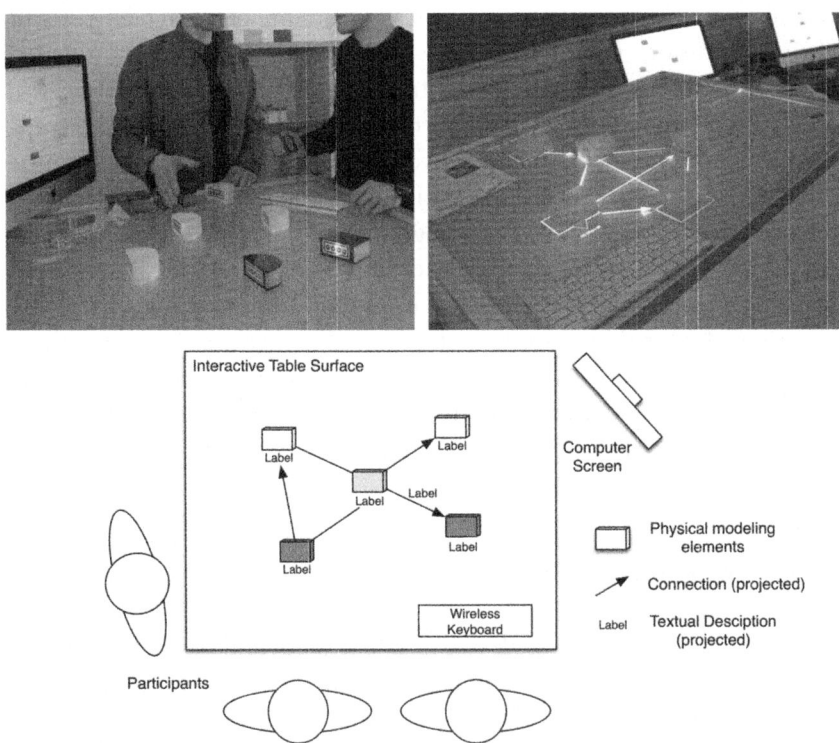

Fig. 7.2 Comprehand Table overview (top-left: interaction on table surface; top-right: modeling tokens with projected connections; bottom: schematic bird's eye view of tabletop)

to HCI that aims at bridging the gap between artifacts in the physical world and digital information conceptually belonging to those artifacts. Information augments the physical artifacts, is accessible through them, and can also be manipulated directly using the artifacts. These properties make tangible interfaces a candidate for integration of the superficially opposing requirements of physical immediacy and computer support during externalization of mental models.

Tangible interfaces are not only favorable from an externalization point of view. Previous research has shown that tangible interfaces also facilitate cooperation (Hornecker 2001) and learning (Resnick et al. 1998; Zuckerman et al. 2005). Hornecker (2001) examines the effects of tangible interfaces on human cooperation (with a focus on tabletop interface, which are built upon a common physical surface used for interaction with the system). Based on a review of existing literature and validated empirically, she identifies four social effects of tangible interfaces that facilitate cooperation: they act as enabler for (1) intuitive and simultaneous manipulation, (2) focusing, (3) awareness of gestures and performatives of actions, and (4) are facilitator for externalization and role as boundary object. Regarding effect 1, Hornecker (2004) claims that tangible interfaces lower the barrier for initial usage and allow for performing the actual task of coordination instead of investing effort in handling the tool. Tangible interfaces are also claimed to facilitate focusing on the topic of coordination for all involved individuals (effect 2) by making the physical representation a spatially focal point of interaction and in this way, creating a transactional space. The co-located setting around the interface also facilitates communication of non-verbal signals and performatives of actions of other individuals (effect 3). The physical representation also enriches communication beyond verbal expression, for example, by allowing gestural referencing of aspects of the shared information. Finally, the physically shared representation acts as a boundary object, providing an anchor for the development of shared understanding among the involved individuals (effect 4).

Comprehand has been implemented using an interactive table and uses physical tokens for cooperative structure elaboration. The semi-transparent table surface is back-projected from below to display additional information like connections or captions (cf. Fig. 7.2, bottom image). Labels of modeling elements as well as connections between them are part of the projected information, which allows for easy rearrangement of models. The recognition of modeling elements and manipulation tools (such as a connection removal tool) is again based upon the ReacTIVision system (Kaltenbrunner and Bencina 2007), which makes the Comprehand Table compatible to the Comprehand Cards system in terms of recognizable model elements. While the cards could be used on the table, the standard configuration uses 3D modeling blocks, which are graspable more easily and additionally can be opened physically to embed smaller modeling elements that can be bound to additional contextual work information, such as documents, forms, or other, already existing models. This approach allows mutually linking models, and more explicitly situates them in their context of work.

Interaction with the system has been designed based upon real-world activity metaphors (Fishkin 2004), which improve learnability of the interaction with the system (Oppl and Stary 2011). The interaction features are described in more detail in the following:

User-defined representational semantics. Models of individual perceptions of work have to allow arbitrary model element types to avoid misrepresentation (Oppl 2018) or loss of information due to lacking support of what people want to express (Goguen 1993; Sarini and Simone 2002). Typically, the semantics of a representation also evolves in the course of identifying/putting nodes on the tabletop and creating links. As more than one person may be part of a mapping session, essential nodes and relations can be shared and stored as meaningful information for groups, including the generation of variants with respect to a certain issue (Rentsch et al. 2010). Typical variants of course designs are subject-specific lectures for different curricula, for example, computer science and business information systems, involving educators with different intentions and learners with heterogeneous backgrounds.

The tool provides various types of tokens for modeling, and their respective meaning has to be assigned by the user(s) through labeling. An arbitrary number of token types is supported, and the shapes available as hardware tokens can be configured dynamically in the software system. Various categories of tokens enable the flexible specification of concept classes and the flexible assignment of meaning in the course of structure and/or behavior modeling.

At the same time, semantics can also be constrained to the pre-specified elements of existing modeling languages such as Subject-oriented Business Process Management (S-BPM). In this case, the interactive modeling support still provides the features listed in the following but is used analogous to the card-based modeling approach described in Sect. 3.2.

Labeling and associating. Work process modeling relies on the ability to assign names to model elements, to define associations between them, and—in case of clarification—to attach annotations to objects. In traditional, software-based modeling tools, these interactions are performed using mouse and keyboard. Using traditional input devices in a physical modeling environment requires switching between media. It might distract users from their original modeling task.

Accordingly, the interface has been designed to avoid input devices like mouse or keyboard. Several tools can be used to manipulate the model directly on the surface: Tokens are associated by putting them into close proximity and then placing them back in their original position (like linking them with a rubber band). Directed connections can be created using an arrow-tip-shaped tool that is put onto the connection near the intended endpoint. A rubber-shaped token enables users to delete connections. In case of multiple connections between two tokens, these connections are dynamically spread to avoid overlapping. Token types (e.g., red, blue, and yellow elements) are not semantically predefined. Users can assign meaning to them in the course of using Comprehand, as described earlier.

Labeling (i.e., assigning designators to concepts, cf. Fig. 7.3) is performed by using the keyboard for naming tokens or connections. The input text is assigned to the most recently added element (token or association), thus avoiding explicit selection of the target. A pen-shaped selection token enables the explicit selection of an element to rename it.

Abstraction support. Features like zooming or the selective display of concepts allow reducing the complexity of visualizations. They are, however, restricted to the computer-based desktop or multi-touch tabletops without any tangible elements. The tokens act as containers in order to overcome this limitation and to reduce complexity in physical models, too (cf. Fig. 7.4). They represent either an arbitrary digital resource (file), or a model state captured previously. The latter information type enables users generating parts of a work representation separately and connecting these parts on a higher level of abstraction. In this way, the common

Fig. 7.3 Labeling and associating

Fig. 7.4 Users can open a token and put additional information into it. Additional information is bound to smaller tokens

modeling concept of abstraction through subsuming or detailing representations is mapped to the physical world. In the case of S-BPM modeling, the container metaphor is used to link individual behavior models in subjects, which are then used for interaction modeling.

The ratio between token size and the size of the table surface allows about 10-15 tokens to be placed on the physical surface simultaneously. For complex modeling tasks, this number of elements might be too small. The container feature is designed to overcome this deficiency according to its purpose of nesting and embodying.

History and reconstruction. The traceability of the modeling process is ensured in Comprehand through capturing the modeling history. This feature also facilitates the understanding of a representation (Klemmer et al. 2002), in particular for cooperative endeavors. The modeling history enables participants to recapitulate and reflect the modeling steps made so far, even when they join a session later on, or in case they have to continue working on a model generated by different individuals.

The tool captures the modeling history by taking snapshots automatically. Whenever the model has not changed for a couple of seconds, the system takes a snapshot of the current state. In addition, a dedicated camera-shaped token enables users to take snapshots on demand. It allows explicit capturing and storing a certain model state using the back-end system for later retrieval. The users can navigate back and forth in the modeling process using the stored information.

The history mode (i.e., recalling former model states) can be activated using a clock-shaped token. It can be rotated counterclockwise or clockwise to go back and forth in time, respectively. When the users switch to the history mode, the computer screen displays a graphical visualization of the currently selected model state along with a status bar, indicating the point in time when the state has been captured.

Additionally, the modeling history enables support for rolling back changes of the model. This is necessary when encouraging the exploration of potential model elements (concepts) and associations. Experimental changes need to be reversible. Such a requirement can be implemented in a straightforward way for desktop applications, but hard to accomplish in a physical modeling environment. The reconstruction feature built upon the history navigation mode supports the physical reconstruction of a previously selected model state. When triggered, Comprehand guides the users step by step and indicates visually which physical tokens need to be (re)moved and/or added according to the differences between current and requested model state, in order to complete a reconstruction of the model state on the table surface. When reconstruction is triggered, a new sequence of modeling states is forked from the original strain of model development, as proposed by (Klemmer et al. 2002). In this way, not only the temporal evolution of the model but also its conceptual development and alternative ways of model representation become accessible. These complex model histories, however, are not accessible via the tabletop but only via the desktop system in order to keep interaction during modeling simple.

Figure 7.5 shows the set of tabletop elements and the toolset for:

- Selecting elements (node or link) of a tabletop map going to be manipulated
- Marking a link as directed relationship, for example, indicating a procedure (chain)

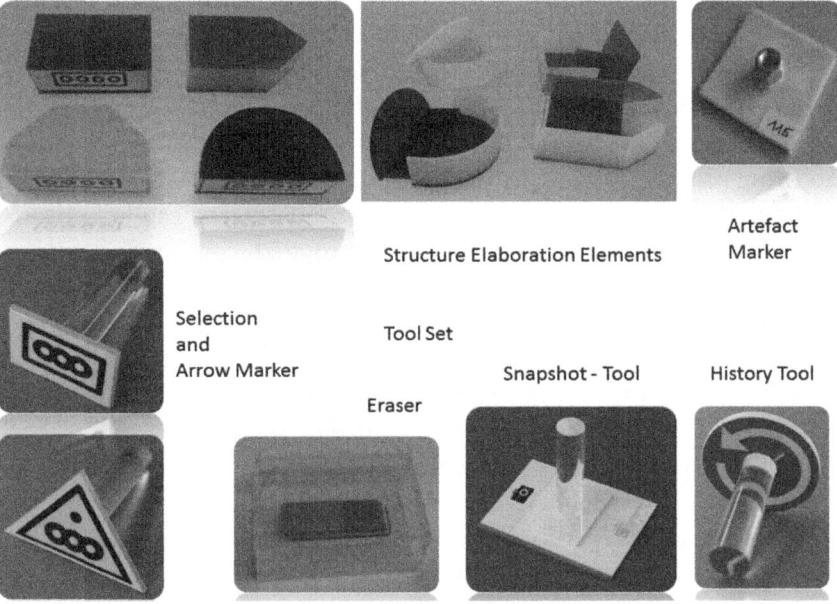

Fig. 7.5 Elements and tools for tabletop concept mapping

- Removing a link or text label of a node or link (eraser tool)
- Storing the current state of the map as a snapshot in a repository for later use (snapshot tool)
- Step back in time showing previous snapshots (history tool)

7.1.3 Collaborative Model Articulation and Exploration

Comprehand is generally used by individuals for reflection and articulation purposes, although it can be applied in multi-user settings as well (Furtmüller and Oppl 2007; Wachholder and Oppl 2012; Oppl 2013). In a multi-user setting, the participants gather around the table. From a methodological point of view, they need to agree on their modeling task in the form of a focus question (Novak and Canas 2006) or a topic of interest (Dann 1992) that targets at the cooperative work process or contingency at hand. The remainder to the instrument's application is a self-moderated process, with interventions only happening in case technical issues arise.

The tabletop allows for simultaneous access to the modeling surface, allowing all participants to freely place and associate physical tokens. They start by placing a token on the surface and assign a designator describing the meaning of the concept represented by the token. Tokens are available in three different shapes and colors. They are, however, semantically not predefined. When using a certain kind of tokens, the participants cooperatively specify their meaning and thus a class of concepts relevant to the topic to be modeled. This meta-information is also captured by Comprehand, which provides means of textual specification of the concept type whenever a new kind of token is used.

As the model evolves, more tokens are placed on the table surface. They are again labeled and can be put into explicit mutual relationship by briefly moving them into close proximity. Projected lines between the associated tokens visualize relationships. These associations also can be labeled, if considered necessary by the participants. All labeling processes are performed using a wireless keyboard, which is passed among the participants as required.

The model on the table surface eventually represents the agreed-upon view all involved participants have on the topic of modeling. During the modeling process, however, the individual views might be incomplete, complementary, or even controversial. Exploration of the modeling space (in terms of different concepts and associations among them) facilitates the articulation process among the involved participants and allows resolving open issues. The system keeps track of the model evolution and separates different strains of development in case of temporary experimental changes. This allows recapturing the modeling process at a later point in time but also enables active exploration of different model variants and their documentation, as model changes become traceable and can be undone. Comprehand supports reconstruction of prior model states on the tabletop surface, as described earlier.

When the participants agree that they have come to an end and have found a common model representing their views on the modeling topic, they are able to make persistent both, the final model and the modeling process. Using the semantically flexible form of representation via Topic Maps, as mentioned earlier, the model, its semantics and its history can be captured in a single, self-contained file using standardized means of representation. The modeling process can be reproduced along its time line and can be resumed on the table surface whenever requested. Comprehand Tables in addition can be linked with each other to allow for spatially (Oppl 2011) or conceptually distributed collaborative modeling (Oppl and Rothschädl 2014; Wachholder and Oppl 2012).

The Comprehand Table provides all these features in order to support conceptual modeling with a focus to facilitate explicit articulation and alignment of mental models through cooperative modeling rather than in a generic way. The tabletop acts as an enabler for communication between the involved persons in the process of performing Articulation Work, requiring the elaboration of models only in so far as they can serve as common point of reference. Model completeness and correctness with regards to formal semantics are not relevant in this case. The evaluation of the instrument consequently has focused on the usability of the toolset (i.e., not hampering users in their alignment activities), its ability to support concept mapping as a means to support Articulation Work, and the effects of its application on actual cooperative work processes.

Models created with both, the table and the card-based instrument, are represented on a conceptual level and—together with their creation and revision history—become part of the distributed organizational knowledge base (DOKB) as knowledge claims to be processed further in the course of double-loop learning processes, according to the KLC.

7.2 Sample Actor-Centric Tool Support for Representation

An OL platform needs to aim at putting people seeking for or being able to provide knowledge in control of the transfer process. It also allows situation-specific communication among the parties involved in the transfer process. In the following sections, we review the feature-set of an OL platform clustered along the different knowledge types described earlier.

7.2.1 Representing Role Knowledge and Descriptive Meta-knowledge

Role knowledge is represented in the OL platform using fine-grain content objects. A content object is a conceptual building block within the knowledge to be represented, such as a definition, an example, or an explanation. Instead of using a document-centric approach to provide information, content is split in its fundamental didactical elements. These elements can be flexibly arranged and reused to form representations of role knowledge.

Descriptive meta-knowledge is codified in the navigation structures of developed demonstrator Nymphaea (Neubauer et al. 2013; Weichhart et al. 2018) as well as directly anchored on content objects (Neubauer et al. 2009, 2011). The Nymphaea learning environment provides different "workspaces" for users, each one containing the relevant knowledge representations, necessary for a certain task. It provides meta-knowledge about the domain-specific scope of knowledge, its author, and creation date. Content within a certain workspace comprises of modules and hierarchically structured content objects. These content objects are enriched with educational meta-knowledge such as 'definition', 'motivation', 'background information', 'directive', 'example', or 'self-test'. This didactic meta-knowledge is displayed on the right side, on top of each content element (cf. Fig. 7.3), and can be used in the course of individualization when filtering content according to metadata.

Besides structuring content according to educational and domain-specific metadata, Nymphaea provides means to structure content according to level of details, allowing actors to retrieve content in the desired granularity.

7.2.2 Representing Conceptual Meta-knowledge

The OL platform provides an alternative navigation design focusing on domain-inherent structures that can be used complementary to hierarchical navigation. The navigation design represents and organizes conceptual meta-knowledge in a graphical concept map.

Concept maps (Novak 1995) are established means to organize and represent knowledge. They can be used to support the process of eliciting, structuring, and sharing knowledge and aim to enable meaningful learning (Chabeli 2010; Stoyanova and Kommers 2002; Steiner et al. 2007). Concept maps use concepts as entity to structure items of interest. Concepts might be central terms, expressions, or metaphors, as they represent a unit of information for the person(s) using it. Those items are put into mutual context, leading to a network of concepts. Persons express the items of interest and the relationships by means of language constructs. Per se, there are no restrictions in the naming of concepts or relationships.

Compared to the traditional navigation design, the concept map navigation enables domain-specific and cross-border relationships. Knowledge acquisition paths can considerably differ when using the concept map approach. Instead of implicit learning paths—via hierarchies of modules, content units, blocks, via internal/external links—learning paths using a concept map are oriented towards explicit structural relationships beyond hierarchies and domains.

Figure 7.6 depicts a part of a cross-disciplinary concept map for codified knowledge about 'Enterprise Architecting'. It can be used for navigating content.

Putting the Framework to Operation: Enabling Organizational...

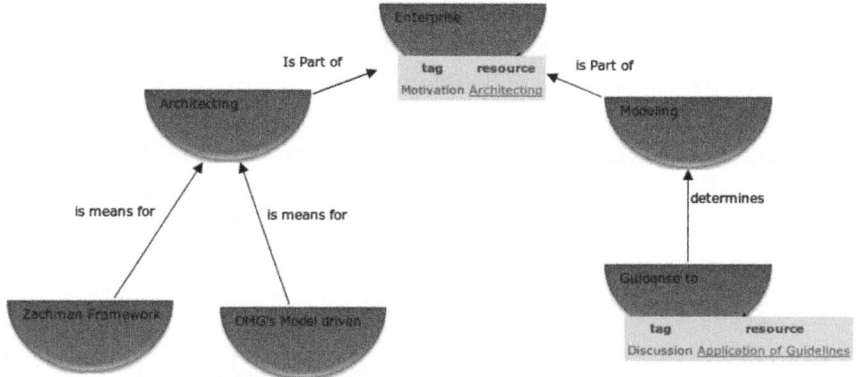

Fig. 7.6 Exemplifying CMap navigation and content links

Within the map, domain-specific associations are used for relating concepts. Furthermore, descriptive meta-knowledge (such as motivation and discussion—see Fig. 7.6) is used to semantically describe links between concepts and information resources. Hence, the associative navigation provides additional structural information that shapes learning paths and can guide the individual exploration of content.

7.2.3 Enabling the Assessment of Cognitive Meta-knowledge

Cognitive meta-knowledge (i.e., knowledge about one's own knowledge) and persuasive meta-knowledge (i.e., knowledge about the credibility and expertise of the knowledge source) are not directly represented in content structures in learning support systems, such as Nymphaea. However, they rather provide means to enable people seeking for knowledge to assess whether they need to access and acquire certain content. A specific instrument termed "Intelligibility Catcher" aids the acquisition of meta-knowledge (Chris Stary 2007). Intelligibility Catchers (ICs) are grounded in reformist-pedagogic and constructivist didactics in general, and the Dalton Plan approach in particular (Parkhurst 1922; Chris Stary 2007; Eichelberger et al. 2008).

The main objectives of the Dalton Plan are to provide freedom to learners and enable them to follow individual learning paths, involving group interaction and collaboration (Parkhurst 1922; Chris Stary 2007). The main vehicle guiding learners through larger and complex (group) learning tasks is the learning contract. It provides a specific structure which can be implemented through ICs. Based on the pedagogic foundation of Parkhurst's assignments, ICs are additionally tailored to be methodically and functionally integrated in online knowledge transfer environments. The IC's structure is as follows (clustered along transactive memory dimensions):

- Elements aiding the assessment of cognitive meta-knowledge

    - *Preface (Orientation section)*: The preface section provides the context motivating the knowledge acquisition tasks.
    - *Topic/objectives*: This section clearly states the central idea of the subject to be addressed. This helps learners to stay focused and reflect about their own work on/about this topic.
    - *Problems and tasks*: This section includes all tasks learners work within the frame of the current contract. It is advisable to state here which problems are to be solved individually by each learner and which problems are to be solved by a group of learners.

- Elements aiding the creation of instance and transactive knowledge

    - *Written work*: This section identifies the documentation to be provided by learners. When finished, the involved people discuss written work within a meeting/conference (see later in the chapter). ICs particularly should include references to functionality provided by a learning support system platform to effectively structure and support the learning process.
    - *Memory work*: In this part of the assignment, the intellectual and cognitive work is described. It comprises the intellectual effort to be spent when exploring content and apply it in a reflective way for problem solving.
    - *Conferences/meetings*: While learners are required to manage their own (learning) time, it is often advisable to schedule (online) meetings and check the intermediate progress.

- Elements containing descriptive meta-knowledge

    - *References*: All additional content for which it is advisable to be read, are referenced here.
    - *Equivalents*: The estimated effort (in hours of work) for the assignment is provided here.
    - *Bulletin Board*: A forum dedicated to discussions related to the assignment is provided.

Using this structure, guidance on how to use and interact based upon content for knowledge acquisition is provided. Following a self-regulated learning paradigm (Oppl et al. 2010), cognitive meta-knowledge is not directly captured and represented explicitly in a platform; however, users are enabled to self-assess their existing knowledge and learning requirements. Elements specifically targeting at didactically reasonable interaction among learners and knowledge providers facilitate building transactive knowledge and allow making use of instance knowledge within and external to the platform.

7.3 Sample Actor-Centric Tool Support for Intelligent Content Manipulation

Support for individualization of content can be provided in an OL platform with respect to capture instance knowledge (Fürlinger et al. 2004). Annotations enable individuals to (i) annotate or alter a specific content element, (ii) post questions, answers, or comments directly anchored on content, and (iii) additionally link the contribution to a discussion theme from the system's global discussion board. The latter link (being part of navigation) guides users to the adjacent discussion of the learning material. In case of real-time online connections, such as chats, the questions and answers can pop up immediately on the displays of all connected users (available in a buddy list). In addition, the content elements referred to can be displayed at the same time.

Annotation support for content is realized using a view concept. As soon as provided content is displayed, a view is generated like an overlay transparency. The view is kept for further access and reloaded when the content is accessed again. Within a certain view, users can (i) highlight, (ii) link, and (iii) add remarks to content elements. The features for view management (add view layer, delete view layer, share view layer, show available views) as well as those for annotations are located in the ribbon-bar at top, whereas the selection of a certain view is provided at the right-hand top of the content area (cf. "MyView" in Fig. 7.6).

While annotating content, users can add internal and external references to content items. Internal references are links between content and communication items, such as entries in the discussion forum or Infoboard, which support context-sensitive discussions. Furthermore, internal links might refer to other elements within the same or a different module. The corresponding features have been included into the annotation icon bar (cf. Fig. 7.6 'Link'). Editing internal links requires marking a position in the text that should represent the link. After evoking the respective function located in the ribbon bar at the top, a tree with the node of the currently addressed module is displayed. It allows users to select the target of the link (e.g., a forum entry or another content item).

Coupling content and communication is the core concept in the learning support instrument to foster sharing of instance knowledge. Features supporting sharing are integrated with the individualization features to comprise the possibility to contextualizing individual interactions by directly anchoring them on content elements. Sharing of individual views or creation of shared views, as suggested by (Shi-Kuo Chang et al. 1998), is enabled in the system.

The system allows linking content elements to forum and discussion entries, and vice versa. Sharing these links in a group enables the group to discuss the provided content in context. This feature is particularly useful when users are not only "passive" recipients of content but also actively provide or augment role or instance knowledge in the workspace. Having the discussion documented in the forum provides new users with justifications and background information that has led to previous revisions of content (Weichhart and Stary 2009).

7.4 Sample Actor-Centric Tool Support for Processing Work Models

Validation and refinement of work models by simulated enactment are supported by a workflow management system (WfMS) that has been adapted to allow for dynamic reconfiguration of processes during runtime. The WfMS is based on the S-BPM paradigm (Fleischmann et al. 2012) and integrates with a computer-based modeling environment via a shared model repository, which is part of the DOKB. One of the methodological requirements is to enable refinement of the process models during the simulated enactment whenever an issue is recognized. To avoid losing the context of the enacted work process (i.e., losing all entered data or information about already made decisions), these model changes must not require a restart of the simulated enactment. The need to restart workflow execution in the case of model changes is a technical constraint of most currently available workflow execution engines (Rothschädl 2012). The execution engine has been functionally extended in the course of adapting it to the needs of the methodology presented here and now supports deviations from a currently executed process model during runtime without the need for restarting the process and losing the execution context (Rothschädl 2012).

The overall architecture of the system is outlined in Fig. 7.7. A central model repository is used to store process models. The model importer

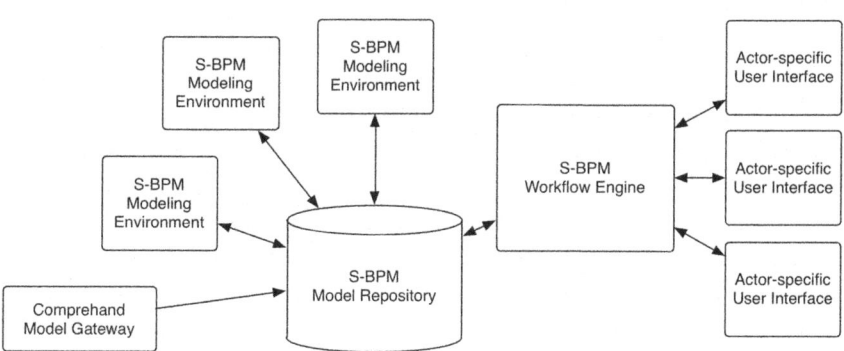

Fig. 7.7 Architecture of process enactment environment

provides the gateway to the articulation tools described earlier and uses the repository as a target for the resulting S-BPM models. The repository is used by the workflow engine to retrieve information about currently instantiated processes. Running process instances are made accessible via a web interface that offers individualized task lists for all actors. Process models stored in the repository can be altered at any time by accessing them via a modeling environment. The architecture visualized in Fig. 7.6 of the repository allows for synchronously altering models from different modeling environments.

7.5 Towards Seamless Tool Support—A Showcase

This section demonstrates the methodology and the use of the toolset based on a simple organizational setting. The aim of this section is to visualize the potential interplay between the distinct areas of support described in the former section. The same set of tools and methods as presented there, is used here for reasons of consistency.

The organizational setting used as a showcase here is not complex, either content-wise or in terms of required collaboration. Its simplicity, however, allows visualizing the intended effect of the proposed methods and tools in a straightforward way. More complex situations would require a more comprehensive description of the organizational setting, which would be beyond the scope of this paper.

The sample process is applying for a vacation. As a starting situation, we assume that the process of vacation application has been handled informally so far in our sample organization. The aim now is to establish a process that is agreed upon throughout the organization and support it by means of IT. In this context, informally established routines should be made visible, questioned, and examined for potential improvements (cf. double-loop learning in the KLC). Three organizational roles are originally involved in the process: an employee, the secretary, and the manager. For the initial articulation and elicitation activities, representatives of these roles are brought together for a workshop.

7.5.1 Articulation and Elicitation

The front-end for articulation and elicitation of knowledge about work processes is based upon the concept of structure elaboration techniques. In the initial workshop, card-based models are created to establish a shared understanding of the process across all involved roles and identify potential differences in how the necessary interaction is perceived.

In a first step, the individual views on one's own contributions to the work process are articulated by each involved person separately. In a second step, those individual views are consolidated in a common model. In the process of consolidation (cf. Oppl 2015), different views are made explicit and are required to be resolved. The card-based model here acts as a mediator between the involved people and always reflects the current state of shared understanding (cf. Fig. 7.8, left).

The interactive tabletop surface can also be used for modeling (cf. Fig. 7.8, right). In collaborative settings, it is used for concept mapping (Oppl and Stary 2009) to reconcile different understandings of fundamental elements of a work situation, such as artifacts, documents, or involved roles and their relationships. In the showcase, this would involve agreeing upon the relevant information to be submitted with a vacation application, identifying used artifacts such as a calendar and reflecting on which organizational roles are actually involved.

The interactive tabletop can also be used individually and collaboratively to reflect upon the procedural aspects of work. It here replaces the card-based system. While the tabletop is not as flexible with regards to

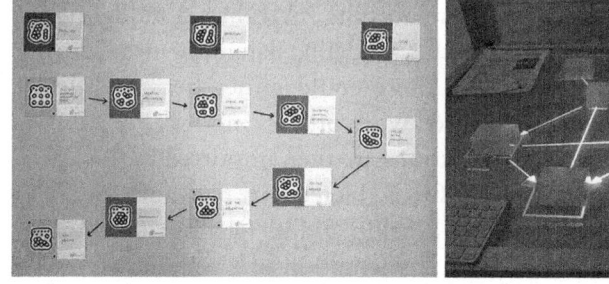
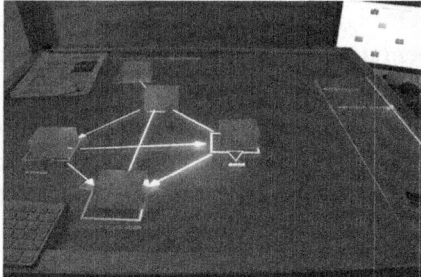

Fig. 7.8 Card-based model (left), interactive surface modeling (right)

spatial deployment, it allows for more sophisticated modeling support features such as version tracking and model reconstruction support (Oppl and Rothschädl 2014).

A combination of both tools is also possible from a methodological perspective. Card-based modeling does not require any dedicated infrastructure and is easy to use. It consequently is used for in in-situ elicitation at the workplace. The interactive surface provides more comprehensive modeling support that is useful for model reflection and revision, which is an integral part of both, single- and double-loop-learning. Technologically, the tools are integrated via a shared representation, which is discussed in the following section.

7.5.2 Representation

All models, together with their creation history, become a part of the DOKB to be accessible during all stages of the knowledge processing and business processing environment in the KLC. Models must not only be stored as graphical representations but rather needed to be represented on a conceptual level to allow for fine-grain referencing and interlinking with other content.

The digital model versions created by interactive modeling surface can be stored in the DOKB without further transformation. The card-based models are only available as images and need to be processed further for deriving a conceptual model representation. A model recognition engine is used for that purpose (Oppl 2015). If offers web-based and mobile gateways to trigger model recognition and interactive revision support (cf. Fig. 7.9). The recognition engine outputs XML-based model representations using the semantics of the modeling language used during model creation. For the showcase, these could either be semantically open concept maps or role-distributed work process models.

The XML-based model representations are then imported in the learning platform, which serves as the DOKB. Here, each piece of content is augmented with metadata about authorship and its relationship to other content elements to allow for an implementation of a transactive memory system, as described earlier. The manipulation of content within the DOKB is discussed in the next section.

Putting the Framework to Operation: Enabling Organizational... 313

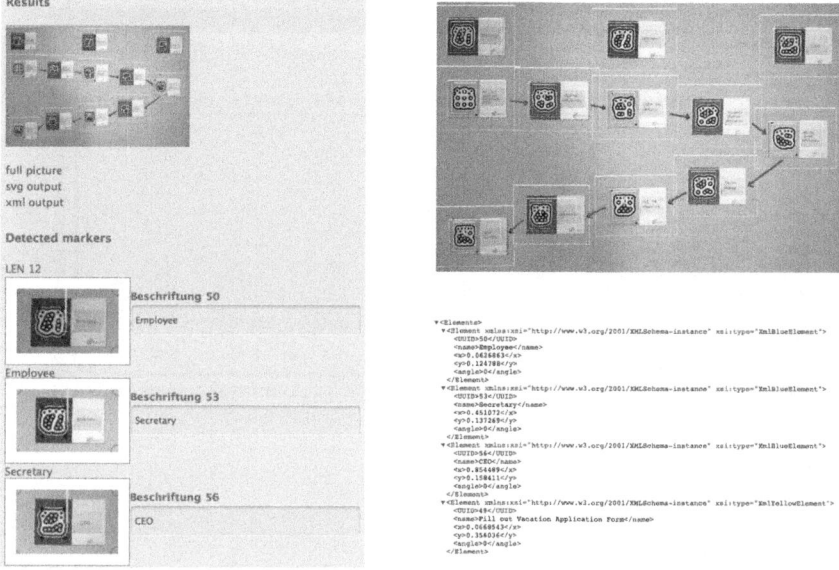

Fig. 7.9 Card-model recognition for conceptual representation: web-interface (left), recognition results (top right), XML-based model representation (bottom right) (released under a Creative Commons Attribution 4.0 International License (CC BY 4.0))

7.5.3 Manipulation

Manipulation of content here refers to embedding it into individually perceived or collectively agreed-upon organizational contexts. This involves searching for, retrieving, and linking relevant content to develop situation-specific perspectives in the DOKB. These perspectives can be shared with others and are available for annotation and discussion. The system implementing these features constitutes the main access path to the DOKB repository. It is built upon the Liferay portal software, which has been adapted to provide the features described earlier.

Figure 7.10 shows sample content from the showcase. The content area (marked "2") shows the individually created process model of an employee's activities in that vacation application process. It has been derived from a card-based model, as described earlier. The navigation area (marked "1") allows accessing the different content types (models,

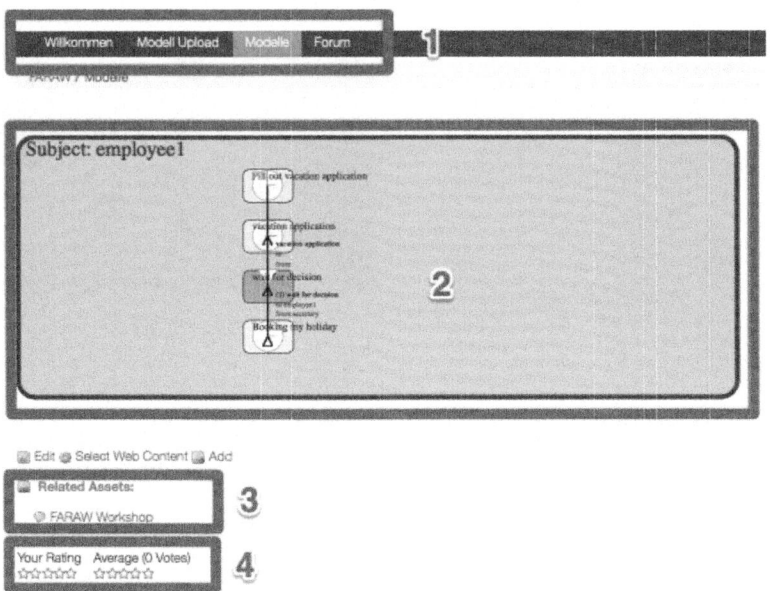

Fig. 7.10 Work process content in the learning environment (released under a Creative Commons Attribution 4.0 International License (CC BY 4.0))

forum-based communication) created for the respective situation. Content cannot only be accessed hierarchically but also contextualized by interlinking different content elements. This is visualized as an example in Fig. 7.10 by a link from the model content to a forum discussion (marked "3"). This link is bidirectional and can also be accessed from the forum, providing immediate context to the discussion. A prototype for a social interaction feature is shown at the bottom of Fig. 7.10 (marked "4"). It allows for rating the usefulness of the provided information.

The linking and rating areas also allow integrating external content providers (such as wikis or document management systems) or social service and communication providers (such as Facebook, Yammer, or Twitter). This allows for integration of platforms already established as means for collaboration and communication in an organization and does not enforce using "yet another" platform. To be able to process information from other applications contributing to the DOBK, an interoperability platform is researched (Weichhart 2014). Technically based on a well-established and simple-to-use REST API approach, the platform provides an enhanced enterprise-service-bus-like environment. This enables not only technical support for the distribution aspect of the DOBK. This approach is in particular suitable for the KLC, as it also conceptualizes the organization as CAS and assumes changes in processes and the different applications and platforms to be the norm, not the exception. This interoperability service also allows for integration of tools for validation of content via simulation. This area of support is discussed in the next section.

7.5.4 Processing

Checking work process models for their feasibility in practical settings, reflecting on them as well as acquiring knowledge about them without prior practical experiences, can be supported by simulation tools. In the KLC, simulation supports the evaluation of knowledge claims and thus feeds back data in the DOKB. The simulation tool is used to execute work process models created with the modeling approaches described earlier. The tool chain described in this paper makes use of an execution engine that can directly process models that are focusing on the collaboration of roles involved in a work process and their activities.

In the showcase, the simulation step is used to assess the adequacy of the vacation application process model with regard to its usefulness in practical settings. The model is executed in a workshop setting, where the involved people check whether the simulated process matches their expectations and covers all potential process variants. Due to the nature of the used modeling approach, which focuses on single cases rather than generic processes, the latter cannot be observed in general. In the showcase, for example, the process steps to be taken in case of a rejection of an application could be missing. Whenever a mismatch or gap in the process model is identified, the model is altered or extended on the fly without restating the simulated process instance. This is technically realized via a workflow engine that allows dynamically changing process models during runtime (i.e., while instances of the process are being executed). In the showcase, this would mean adding the respective activities to handle rejections to the behavior models of each involved role.

The simulation component retrieves its required model information from the repository that is also interfaced by the other components, such as the tabletop modeling tool used for elicitation or the learning platform (cf. Fig. 7.11). This allows to seamlessly switch between these tools even during a running simulation, annotating model variants in the learning

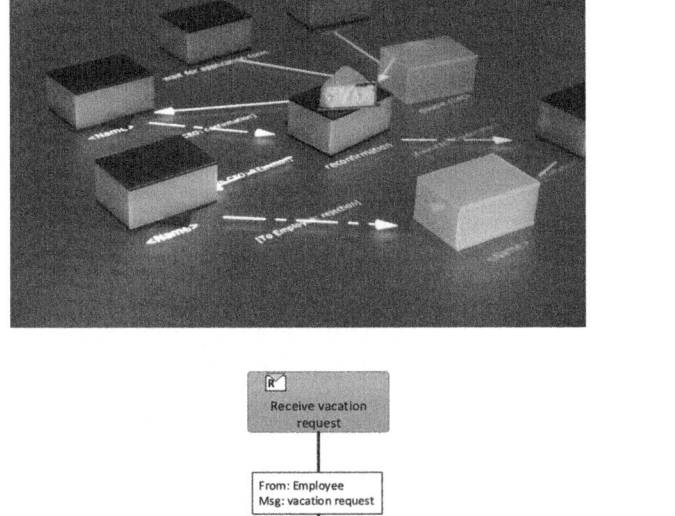

Fig. 7.11 Processing and simultaneous manipulation on an interactive modeling tabletop

platform or performing experimental model changes in the interactive tabletop surface. Consequently, the showcase process could be extended or altered by the original representatives of the roles, using the same tools they used during the initial articulation and elicitation step. Actors thus can perform all modeling steps without the need for expert modelers.

7.6 Conclusions

For effective digital work design, new approaches in involving stakeholders concerning process design are required. New methods and the intermediate usage of technologies could help agile organizations to structure their work on the fly while acknowledging their competencies and keeping their social identity. Concerned parties know best how their work can be improved, and their involvement in adaptation increases their capabilities to keep up with changing business requirements. A respective methodology and a tool chain need to be aligned with individual and collective learning facilities, both on the content and method level. It also requires a living memory to capture content and development processes in an actor-centered way.

In this chapter, we have backed the framework presented in the former chapter with actual technical tools. These tools offer support along the four elementary components of the framework: (i) articulation and elicitation of work knowledge, (ii) knowledge transfer and manipulation, (iii) multifaceted content representation, and (iv) business processes execution for validation and enactment. They thus can be matched to the methods and instruments introduced in the former chapters and thus form an overall coherent picture of how (digital) work design can be supported (by digital means), both methodologically and technically. We pick up this set of mutually interoperable and aligned instruments in the next chapter, where we present a set of case studies that show how they can be deployed (both selectively and in an integrated way, depending on the task at hand) in organizational practice.

References

Chabeli, M. 2010. Concept-Mapping as a Teaching Method to Facilitate Critical Thinking in Nursing Education: A Review of the Literature. *Health SA Gesondheid* 15 (1) (OpenJournals Publishing).

Dann, H.D. 1992. Variation Von Lege-Strukturen Zur Wissensrepräsentation. In *Struktur-Lege-Verfahren Als Dialog-Konsens-Methodik*, Arbeiten Zur Sozialwissenschaftlichen Psychologie, vol. 25, ed. Brigitte Scheele, 2–41. Münster: Aschendorff.

Decker, G., and M. Weske. 2009. Toward Collaborative Business Process Modeling. *Cutter IT Journal* 22 (10): 29.

Eichelberger, Harald, Christian Laner, Wolf Dieter Kohlberg, Edith Stary, and C. Stary. 2008. *Reformpädagogik Goes eLearning: Neue Wege Zur Selbstbestimmung Von Virtuellem Wissenstransfer Und Individualisiertem Wissenserwerb*. Munich: Oldenbourg Verlag.

Engelmann, T., and F.W. Hesse. 2010. How Digital Concept Maps About the Collaborators' Knowledge and Information Influence Computer-Supported Collaborative Problem Solving. *International Journal of Computer-Supported Collaborative Learning* 5 (3): 299–319.

Firestone, J.M., and M.W. McElroy. 2005. Doing Knowledge Management. *The Learning Organisation Journal* 12 (2): 189–212.

Fischer, Frank, and H. Mandl. 2005. Knowledge Convergence in Computer-Supported Collaborative Learning: The Role of External Representation Tools. *The Journal of the Learning Sciences* 14 (3): 405–441. https://doi.org/10.1207/s15327809jls1403_3.

Fishkin, Kenneth P. 2004. A Taxonomy for and Analysis of Tangible Interfaces. *Personal and Ubiquitous Computing* 8 (5): 347–358 (Springer).

Fleischmann, Albert, Werner Schmidt, C. Stary, Stefan Obermeier, and Egon Börger. 2012. *Subject-Oriented Business Process Management*. New York: Springer.

Furtmüller, F.G., and Stefan Oppl. 2007. A Tuple-Space Based Middleware for Collaborative Tangible User Interfaces. In *Enabling Technologies: Infrastructure …*, 400–405. IEEE. https://doi.org/10.1109/WETICE.2006.71.

Fürlinger, S., A. Auinger, and C. Stary. 2004. Interactive Annotations in Web-Based Learning Systems. In *Proceedings of IEEE International Conference on Advanced Learning Technologies, 2004*, 360–365. IEEE. https://doi.org/10.1109/ICALT.2004.1357437.

Goguen, J. 1993. On Notation. Department of Computer Science and Engineering, University of California at San Diego.

Hornecker, E. 2001. Graspable Interfaces as Tool for Cooperative Modelling. *Proceedings of IRIS* 24: 215–228.

———. 2004. Tangible User Interfaces Als Kooperationsunterstützendes Medium. Department of Computing, University of Bremen.

Ishii, Hiroshi, and Brygg Ullmer. 1997. Tangible Bits: Towards Seamless Interfaces between People, Bits and Atoms. In *Proceedings of the SIGCHI Conference on Human Factors in Computing Systems (CHI)*, 234–241. New York: ACM Press. https://doi.org/10.1145/258549.258715.

Jiang, Yingying, Feng Tian, Xiaolong Luke Zhang, Guozhong Dai, and Hongan Wang. 2011. Understanding, Manipulating and Searching Hand-Drawn Concept Maps. *Transactions on Intelligent Systems and Technology (TIST)* 3 (1): 1–21 (ACM Request Permissions). https://doi.org/10.1145/2036264.2036275.

Kaltenbrunner, Martin, and Ross Bencina. 2007. reacTIVision: A Computer-Vision Framework for Table-Based Tangible Interaction. In *TEI '07: Proceedings of the 1st International Conference on Tangible and Embedded Interaction*, 69–74. New York: ACM Press.

Klemmer, S.R., M. Thomsen, E. Phelps-Goodman, R. Lee, and J.A. Landay. 2002. Where Do Web Sites Come From?: Capturing and Interacting with Design History. In *Proceedings of the SIGCHI Conference on Human Factors in Computing Systems*, 1–8. ACM.

Luebbe, Alexander, and Mathias Weske. 2011. Bringing Design Thinking to Business Process Modeling. In *Design Thinking*, 181–195. Berlin and Heidelberg: Springer Berlin Heidelberg. https://doi.org/10.1007/978-3-642-13757-0_11.

Neubauer, Matthias, Stefan Oppl, C. Stary, and Georg Weichhart. 2013. Facilitating Knowledge Transfer in IANES—A Transactive Memory Approach. In *Innovation Through Knowledge Transfer 2012*, Smart Innovation, Systems and Technologies, vol. 18, ed. R. Howlett, B. Gabrys, K. Musial-Gabrys, and J. Roach, 39–50. Berlin and Heidelberg: Springer. https://doi.org/10.1007/978-3-642-34219-6_5.

Neubauer, Matthias, C. Stary, and Stefan Oppl. 2009. Towards Topic Map-Based E-Learning Environments. In *Proceedings of TMRA 2009*, November. Poster at the Fifth International Conference on Topic Maps Research and Applications.

———. 2011. Polymorph Navigation Utilizing Domain-Specific Metadata: Experienced Benefits for E-Learners. In *Proceedings of the 29th European Conference on Cognitive Ergonomics (ECCE 2011)*, 45–52. New York: ACM Press.

Novak, Joseph D. 1995. Concept Mapping to Facilitate Teaching and Learning. *Prospects* 25 (1): 79–86 (Kluwer Academic Publishers). https://doi.org/10.1007/BF02334286.

Novak, Joseph D., and A.J. Canas. 2006. The Theory Underlying Concept Maps and How to Construct Them. Florida Institute for Human and Machine Cognition.

Oppl, Stefan. 2006. Towards Intuitive Work Modeling with a Tangible Collaboration Interface Approach. In *Proceedings of WETICE '06*, June, 400–405. IEEE Press. https://doi.org/10.1109/WETICE.2006.71.

———. 2011. Subject-Oriented Elicitation of Distributed Business-Process Knowledge. In *Proceedings of the 3rd Conference on Subject-Oriented Business Process Modelling (S-BPM ONE 2011)*, Communications in Computer and Information Science, vol. 213, 16–33. Berlin and Heidelberg: Springer. https://doi.org/10.1007/978-3-642-23471-2_2.

———. 2013. Towards Role-Distributed Collaborative Business Process Elicitation. In *Proceedings of the Workshop on Models and Their Role in Collaboration*, ed. Alexander Nolte, M. Prilla, Peter Rittgen, and Stefan Oppl, 33–40. CEUR-WS.

———. 2015. Articulation of Subject-Oriented Business Process Models. In *Proceedings of S-BPM ONE 2015*, 1–11. New York: ACM Press. https://doi.org/10.1145/2723839.2723841.

———. 2017. Supporting the Collaborative Construction of a Shared Understanding About Work with a Guided Conceptual Modeling Technique. *Group Decision and Negotiation* 26 (2): 247–283. https://doi.org/10.1007/s10726-016-9485-7.

———. 2018. Which Concepts Do Inexperienced Modelers Use to Model Work?—An Exploratory Study. In *Proceedings of MKWI 2018*.

Oppl, Stefan, and Christian Stary. 2009. Tabletop Concept Mapping. In *Proceedings of TEI 2009*, 275–282. New York: ACM Press. https://doi.org/10.1145/1517664.1517721.

Oppl, Stefan, and C. Stary. 2011. Towards Informed Metaphor Selection for TUIs. In *Proceedings of the 3rd ACM SIGCHI Symposium on Engineering Interactive Computing Systems (EICS 2011)*, Communications in Computer and Information Science, vol. 213, ed. F. Paternò, 16–33 (Chap. 2). Berlin and Heidelberg: ACM Press. https://doi.org/10.1007/978-3-642-23471-2_2.

Oppl, Stefan, and Christian Stary. 2014. Facilitating Shared Understanding of Work Situations Using a Tangible Tabletop Interface. *Behaviour & Information Technology* 33 (6): 619–635. https://doi.org/10.1080/0144929X.2013.833293.

Oppl, Stefan, and Thomas Rothschädl. 2014. Separation of Concerns in Model Elicitation—Role-Based Actor-Driven Business Process Modeling. In *S-BPM ONE—Setting the Stage for Subject-Oriented Business Process Management*, Communications in Computer and Information Science, vol. 422, ed. Hagen Buchwald, Albert Fleischmann, Detlef Seese, and Christian Stary, 3–20. Berlin and Heidelberg: Springer International Publishing. https://doi.org/10.1007/978-3-319-06191-7_1.

Oppl, Stefan, C.M. Steiner, and D. Albert. 2010. Supporting Self-Regulated Learning with Tabletop Concept Mapping. In *Interdisciplinary Approaches to Technology Enhanced Learning*, 391–410. Münster: Waxmann.

Oppl, Stefan, Christian Stary, and S. Vogl. 2017. Recognition of Paper-Based Conceptual Models Captured Under Uncontrolled Conditions. *IEEE Transactions on Human-Machine-Systems* 47 (2): 206–220. https://doi.org/10.1109/THMS.2016.2611943.

Parkhurst, Helen. 1922. *Education on the Dalton Plan*. Read Books Ltd.

Rentsch, Joan R., Abby L. Mello, and Lisa A. Delise. 2010. Collaboration and Meaning Analysis Process in Intense Problem Solving Teams. *Theoretical Issues in Ergonomics Science* 11 (4): 287–303 (Taylor & Francis).

Resnick, M., F. Martin, R. Berg, R. Borovoy, V. Colella, K. Kramer, and B. Silverman. 1998. Digital Manipulatives: New Toys to Think with. In *Proceedings of the SIGCHI Conference on Human Factors in Computing Systems*, 281–287. New York: ACM Press.

Rothschädl, Thomas. 2012. Ad-Hoc Adaption of Subject-Oriented Business Processes at Runtime to Support Organizational Learning. In *S-BPM ONE*, Lecture Notes in Business Information Processing, vol. 104, 22–32 (Chap. 2). Berlin and Heidelberg: Springer Berlin Heidelberg. https://doi.org/10.1007/978-3-642-29133-3_2.

Sarini, Marcello, and C. Simone. 2002. The Reconciler: Supporting Actors in Meaning Negotiation. In *Proceedings of the Workshop on Meaning Negotiation*.

Shi-Kuo, Chang, E. Hassanein, and Chung-Yuan Hsieh. 1998. A Multimedia Micro-University. *IEEE Multimedia* 5 (3): 60–68. https://doi.org/10.1109/93.713305.

Stary, C. 2007. Intelligibility Catchers for Self-Managed Knowledge Transfer. In *Seventh IEEE International Conference on Advanced Learning Technologies (ICALT 2007)*, 517–521. IEEE.

Steiner, C.M., D. Albert, and J. Heller. 2007. Concept Mapping as a Means to Build E-Learning. In *Advanced Principles of Effective E-Learning*, ed. N.A. Buzzetto-More, 59–111. Santa Rosa, CA: Informing Science Press.

Stoyanova, Neli, and P. Kommers. 2002. Concept Mapping as a Medium of Shared Cognition in Computer-Supported Collaborative Problem Solving. *Journal of Interactive Learning Research 13* (1): 111–133.

Wachholder, Dominik, and Stefan Oppl. 2012. Stakeholder-Driven Collaborative Modeling of Subject-Oriented Business Processes. In *S-BPM ONE—Scientific Research*, ed. C. Stary, 145–162. Berlin and Heidelberg: Springer. https://doi.org/10.1007/978-3-642-29133-3_10.

Weichhart, Georg. 2014. Learning for Sustainable Organisational Interoperability. In *Preprints of the 19th IFAC World Congress*, August. International Federation of Automation and Control, 4280–4285.

Weichhart, Georg, and Chris Stary. 2009. Collaborative Learning in Automotive Ecosystems. In *Proceedings of the 3rd IEEE International Conference on Digital Ecosystems and Technologies*, 235–240. IEEE.

Weichhart, Georg, Christian Stary, and Markus Appel. 2018. The Digital Dalton Plan: Progressive Education as Integral Part of Web-Based Learning Environments. *Knowledge Management & E-Learning: An International Journal* 10 (1): 25–52 (Directory of Open Access Journals).

Zuckerman, Oren, S. Arida, and M. Resnick. 2005. Extending Tangible Interfaces for Education: Digital Montessori-Inspired Manipulatives. In *Proceedings of the SIGCHI Conference on Human Factors in Computing Systems (CHI)*, 859–868. New York: ACM Press.

Open Access This chapter is licensed under the terms of the Creative Commons Attribution 4.0 International License (http://creativecommons.org/licenses/by/4.0/), which permits use, sharing, adaptation, distribution and reproduction in any medium or format, as long as you give appropriate credit to the original author(s) and the source, provide a link to the Creative Commons licence and indicate if changes were made.

The images or other third party material in this chapter are included in the chapter's Creative Commons licence, unless indicated otherwise in a credit line to the material. If material is not included in the chapter's Creative Commons licence and your intended use is not permitted by statutory regulation or exceeds the permitted use, you will need to obtain permission directly from the copyright holder.

8

Case Studies

This chapter demonstrates the use of the proposed framework and the embedded methods. It shows their interplay and describes the impacts that could be observed in real-world cases. This chapter is practice-oriented and could provide students and practitioners with anchors to reflect on the proposed methods and underlying theories from a "doing" point of view.

The first case (Sect. 8.1) demonstrates how meaningful work-model entities can develop in the course of articulation and guide aligned restructuring of work. It stems from a complex setting, namely, planning in clinical health treatment requiring the structured elicitation of contextual knowledge from all stakeholders involved to develop working procedures in time-critical situations. Sharing expert knowledge from doctors, nurses, technicians, administration, and patients was supported by collaborate development of instruments and tools.

The CoMPArE/WP (Collaborative Model Articulation and Elicitation of Work Processes)-case (Sect. 8.2) has its focus on alignment when bridging from intuitive or semi-structured models to techno-centric (formal) models that can be executable for some workflow engine. It demonstrates effective stakeholder participation while eliciting and consolidating elicited work knowledge. The illustrative case shows an application of the respective concepts explained in Chap. 4.

The third case (Sect. 8.3) targets articulation and alignment of educator knowledge, a highly complex task, as it involves domain knowledge, didactic competence, and social skills. However, applying the instruments and the concepts presented in this volume helped generate a working model for digitalizing learning support in a transparent and intelligible way for educators. Thereby, semi-structured elicitation and stepwise refinement to digital support features turned out to be essential facilitators.

The Me2Me2You-case (Sect. 8.4) has its focus on alignment when bridging from intuitive or semi-structured models to techno-centric (formal) models that can be executable for some workflow engine. It demonstrates effective stakeholder participation while eliciting and consolidating elicited work knowledge. The illustrative case represents a proof of the respective concepts explained in Chap. 4.

Our starting point in revisiting the cases is the framework for articulation, alignment, and processing developed in Chaps. 6 and 7. We indicate the involvement of each component by giving visual clues in the framework diagram. Then the case study is described as it evolved in the specific application context. Overarching objective of each presented case was to achieve stakeholder-driven digitalization of work processes, thus transforming existing socio-technical systems to be perceived by actors resilient to socio-technological capabilities rather than disruptive for the respective organization.

8.1 Categorical Knowledge Building Support—A Planning Case

This case concerns the transformation of an expert organization performing critical tasks in healthcare. The frame of reference for digital work design enabling expert participants to gradually develop a model of their planning process in a cooperative way can be mapped to the case, as shown in Fig. 8.1.

This healthcare planning case has been part of an organizational development process of an Austrian healthcare institution. The case targeted

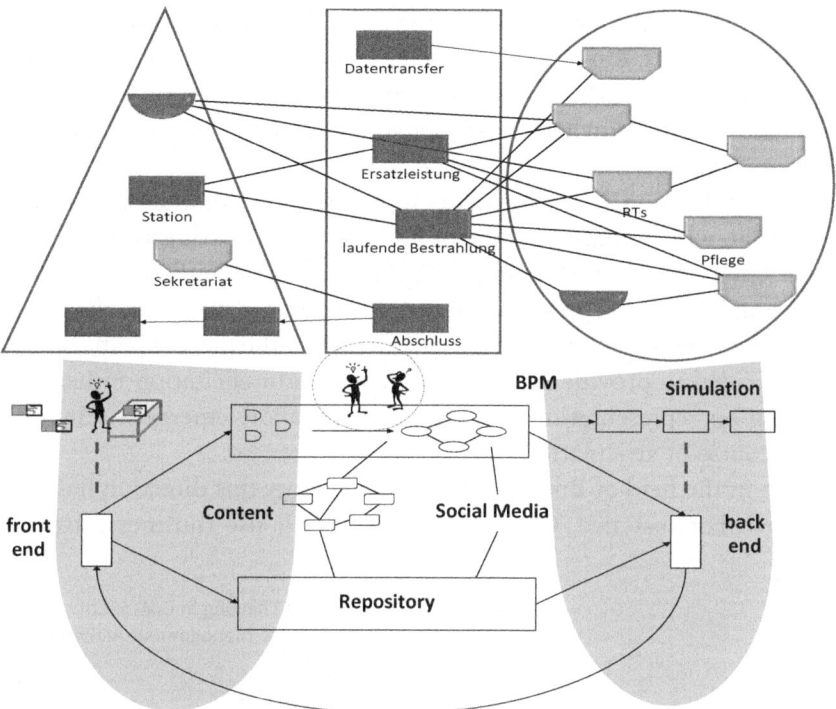

Fig. 8.1 Embodying the planning case into the digital work design framework

qualifying clinical staff on the fly in system thinking and organizational learning, when dealing with a common problem in healthcare—planning (Bardram and Hansen 2010). It affected around 40 experts from different fields and involved 14 of them in developing systemic treatment planning. The organizational learning step concerned aligning human and resource planning. It involved doctors, nurses, technicians, administration, and patients when collaborating in planning.

The facilitator's task was to design and facilitate eliciting mental models of doctors, nurses, and technical and administration staff in order to create a systemic orientation space for changes. That space should enable team learning while reflecting personal mastery (Senge 1990). Content-wise, the organization of planning, in terms of both scope and flow of information, should be re-designed.

Stakeholders also play a crucial role in the organizational change processes. Their reflections and ideas open the opportunity to further develop work systems. Figure 8.2 reflects the cognitive and social involvement of stakeholders and knowledge management activities in the course of organizational change.

Most of the modeling approaches for work knowledge analysis provide a notation, which might be more or less oriented towards execution, such as the UML (Unified Modeling Language) or BPMN (Business Process Modeling Notation) (see www.omg.org). In order to minimize the cognitive burden and bias towards a specific notation, the project management team decided to provide means for articulation or elicitation rather than focusing on representation. Such approaches allow emergent semantics in the course of articulation.

Even in the field of Business Process Modeling, this direction has been followed. For instance, Cohn and Hull (2009) use (business) artifacts

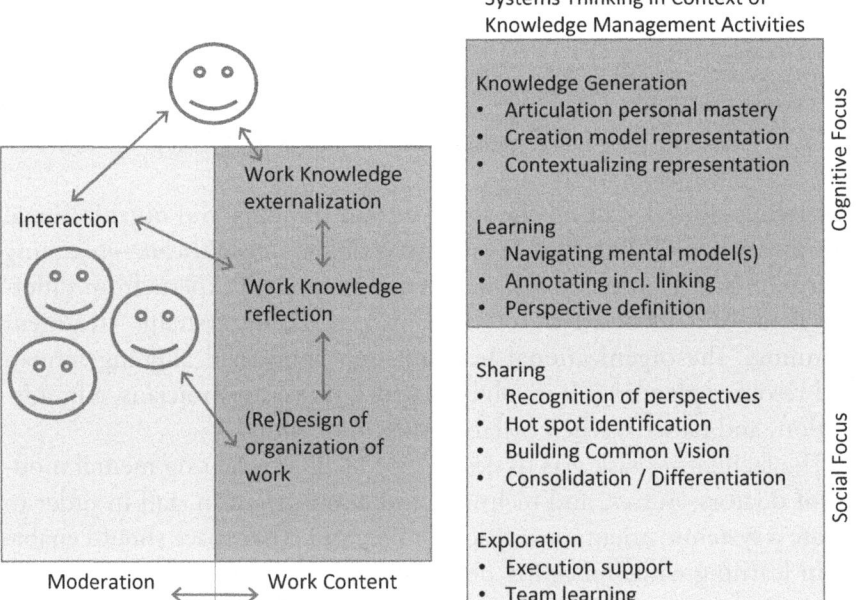

Fig. 8.2 Leveraging stakeholder knowledge for organizational change

combining data and process as basic building blocks of modeling. Artifacts are key business entities (business-relevant objects) evolving when passing through a business's operation. They can be created, modified, and stored. As a result, business operations can be decomposed along various levels of abstraction. Artifacts are typed using both an information model for data about the business objects during their lifetime and a lifecycle model, describing the possible ways and timings that tasks can be invoked on these objects. In this case, the representation "enables strong communication between a business's stakeholders in ways that traditional approaches do not. Experience has shown that once the key artifacts are identified, even at a preliminary level, they become the basis of a stakeholder vocabulary" (ibid.).

From this kind of studies, we could conclude that evolving element and relation categories are of benefit for developing a stakeholder-oriented modeling and analysis approach. In addition, we suspected knowledge that would be conversed from tacit to explicit (Nonaka and Krogh 2009), since we focused on individuals' experiences on a system-critical work situation, namely, a treatment planning procedure in radio oncology, and accurate articulation of these experiences. Such an approach goes beyond accumulating experience, as individuals and groups should figure out what works in which way, and why, and what could be changed in the execution of an organizational task. The facilitation task was driven by forming the group as a team in order to create sustainable models (Hillier and Dunn-Jensen 2013) while avoiding misinterpretations from external stakeholders in the course of articulation (Sandberg 2005).

In work, knowledge articulation teams make a cognitive effort to enhance their understanding of the causal links between actions and outcomes while engaging in collective reflection to gain insight (see also Fig. 8.2). The codification, and thus collective availability of individual work knowledge, are considered key enablers, as they overcome barriers resulting from established relationships and conventions.

Facilitation did not start in the traditional way with predefining an articulation space through a dedicated notation to represent work elements. The facilitation rather targeted the capability of the involved stakeholders to express knowledge using their semiotics according to their individual perception of the functional roles involved in treatment

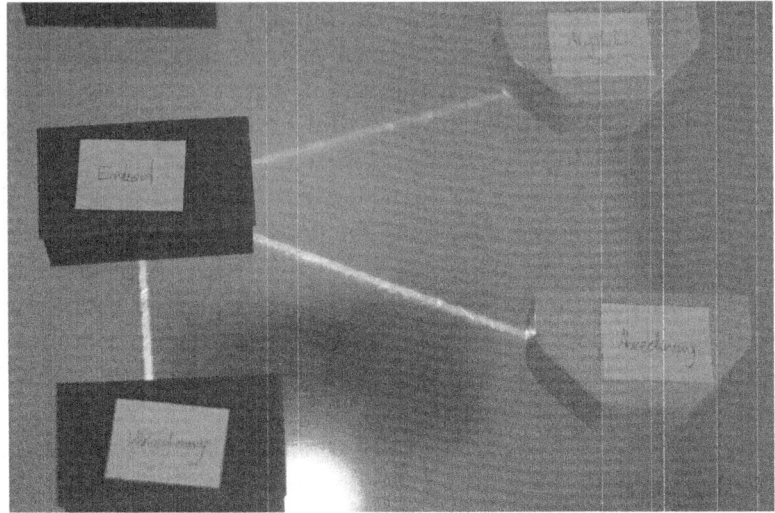

Fig. 8.3 Interactive concept mapping (see also Oppl and Stary 2009, 2011)

planning of the patients. The reflection of the presented meaning has been used to justify the results of the common sessions on the current organization of work tasks, otherwise misinterpretations are likely to occur when post-processing generated knowledge (Sandberg 2005).

In the beginning of the series on change workshops, the involved stakeholders (doctors, nurses, administrators, and device experts) agreed on the goal of the endeavor, namely, maximizing the clinics' treatment service for patients in terms of minimizing planning time. Knowledge codification was performed using concept mapping. It is used by groups to develop a representation of a domain, situation, or procedure (Novak 1995), and to capture content in its systemic context (Trochim 1989). The participants started by drawing or putting nodes (concepts, meaningful items) and relationships on a virtual or paper surface, according to their experiential knowledge—see Fig. 8.3.

8.1.1 Sample Case

Figures 8.4 to 8.7 show the start pattern pictured from the tabletop. It allowed revealing essential relations and language constructs for representing meaningful information from the group (Rentsch et al. 2010). The participants came up with an overview of roles and functional units (concepts) involved in patient treatment planning (red rectangles in Fig. 8.3 referring to doctor roles, e.g., case manager, and experts, e.g., LINAC system specialists). The (blue) half circle (Stereotaxie group) represents a group of people working on a particular topic involving several functional units (out-patient department, LINACs). The relationship the participants set was either 'part-of' ones, such as establishing the Stereotaxie group, or addressing the exchange of patient data ('mutual communication'), the latter being central to coherent and consistent planning. The facilitator asked whether the concept map represented the relevant part of the organization before proceeding. Then, the participants enriched the map with auxiliary and enabling actors/work group, such as the secretary and device management group (yellow hexagons) in Fig. 8.4.

After having created the overview of involved actors and roles, as displayed in Figs. 8.4 and 8.5, the scope (situational context) has been set. Subsequently, the group decided to pick out major actors, such as the out-patient department, and to detail the patient planning-relevant pro-

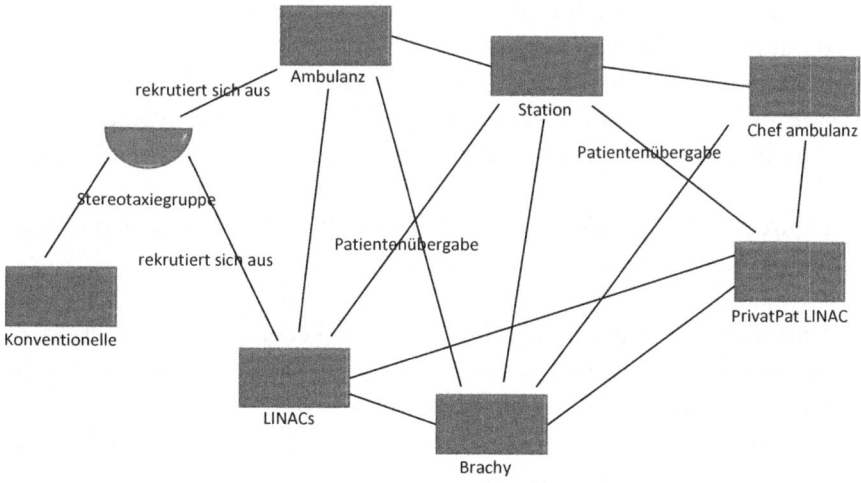

Fig. 8.4 Start map

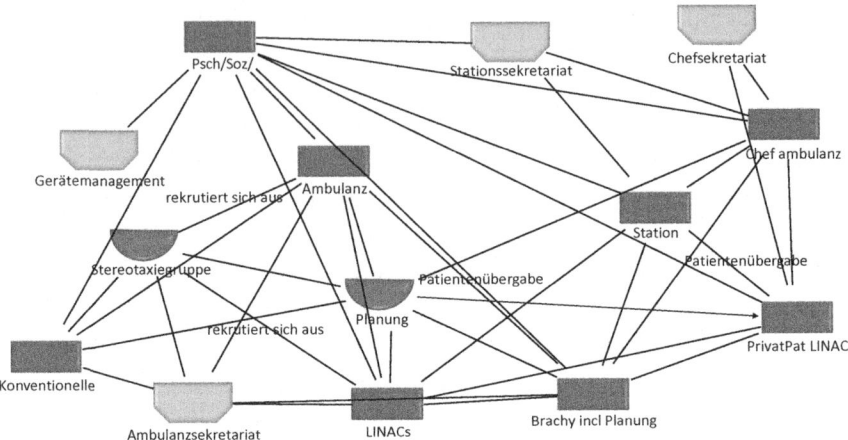

Fig. 8.5 Completing the relevant part of the organization

cess parts. They put the major steps to be followed for patient treatment planning into the middle row (red rectangles) of the map to arrange all other information along actual planning steps.

Figure 8.6 reveals three categories of elements the group needed to detail the patient treatment planning procedure:

- *functions* (red rectangles), that is, process steps according to their temporal order (from top to bottom)
- *decision and operation bodies* (blue half circles) required to proceed
- *roles* (yellow hexagons) representing medical or administration staff handling the process steps either in the sense of front or back office

Figure 8.5 also reveals three categories of relationships required to represent work tasks for further analysis:

- *temporal order* of functions, that is, directed edge between functions
- *hand-over functions*, that is, directed edge between roles, depending on who takes care of the patient in a certain step
- *dedicated assignments of roles* to functions or organizational bodies, that is, directed or non-directed edge

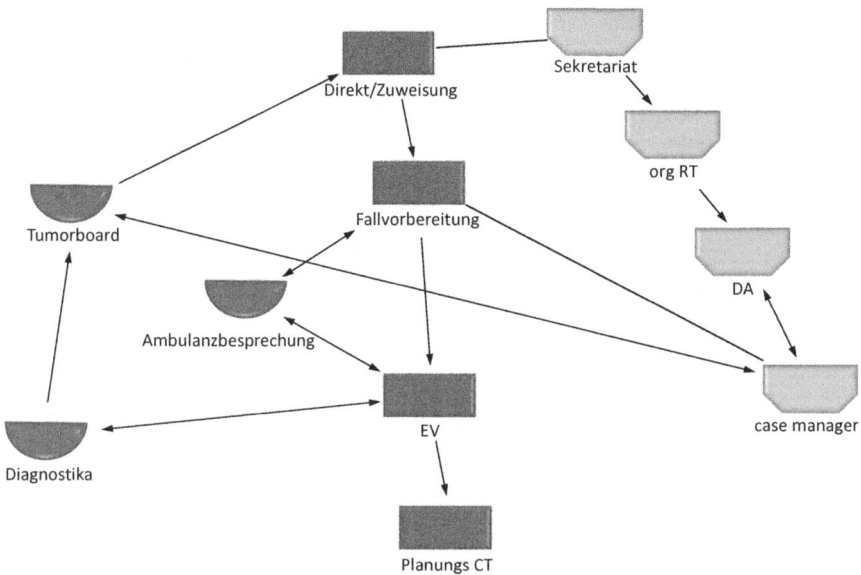

Fig. 8.6 Patient-oriented treatment planning (out-patient department)

Both types of information are required to specify process-relevant settings. Structural elements, such as functions, are the prerequisite to set temporal relationships and need to be put in mutual context with other structural elements, such as roles. Interestingly, the participants specified several flows, namely, the function flow, and the 'responsibility' flow in parallel distinguishing information flow in addition to the function flow.

Figure 8.7 shows how this (re)presentation logic could be kept until the envisioned steps of planning have been articulated. The LINAC team does the fine-grain adjustment of plan data to enable the actual treatment of a patient by the respective technical system.

Figure 8.7 also shows the spatial grouping of notational elements the stakeholders identified when being facilitated to develop their planning process. The middle part was dedicated to the functional core in terms of the objectives, namely, patient orientation throughout treatment planning. The left part reveals the instrument part, both in terms of tools required for planning and treatment, and organizational decision-making, such as the tumor board. The right part denotes the back office roles and

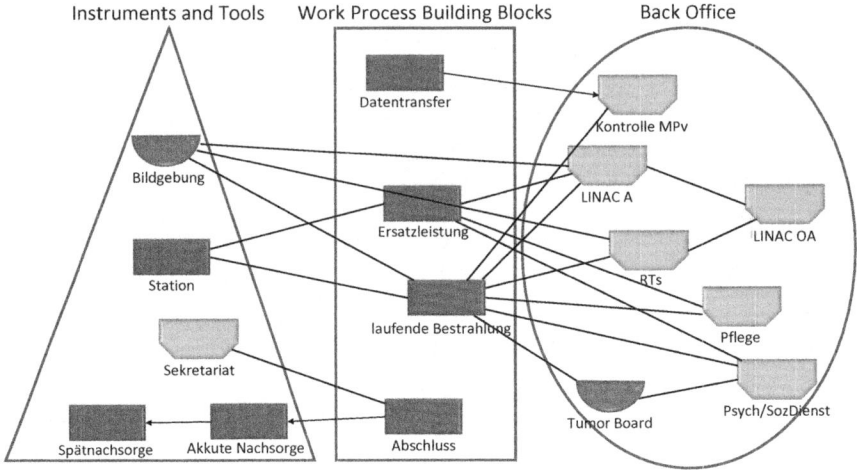

Fig. 8.7 Finalization of treatment planning (LINAC)

exchange patterns that are required to accomplish the core work process tasks shown in the middle part.

8.1.2 Insights

What kind of support could stakeholders need when getting involved actively in transforming work processes? The findings indicate that:

- Eliciting knowledge requires an open format for articulation and collaborative reflection (semantic openness). Hence, predefined notations, such as BPMN (www.bpmn.org), would restrict articulating work knowledge and inputs for change, as the case study reveals, considering functions and actors as integrated concept in the beginning.
- Knowledge codification needs to be accompanied by sharing knowledge to be accessed and reflected by others—representations, such as concepts or business process models, serve as baseline for discussion and discourse.
- Middle-out tops top-down and bottom-up analysis—it reflects social dynamics within the scope of modeling.

- Intertwining the content perspective with social processes helps not only for reflecting a situation 'as-it-is' to come up with ideas 'as-it-could-be', but also setting the context of work procedures in terms of relevant factors for task accomplishment.

Consequently, it seems neither developers nor stakeholder is prepared for effectively participating in developing work (re)designs. Hence, a learning perspective open for content generation and dissemination seems to be appropriate for stakeholder-driven organizational development.

Of crucial importance seems to be the role of the facilitator who should pre-condition the process by clarifying the semantic openness when expressing experiences and ideas for change. Another observation concerns the interface between individual learning and organizational development: Each mental model needs to have its place and space before starting the team learning process.

8.2 CoMPArE/WP Facilitating Project-Based Business Operation

This case reflects on developments towards human-centric modeling of work. The frame of reference enabling process participants to gradually develop a comprehensive model of their business process in a cooperative way can be mapped to this case, as shown in Fig. 8.8.

We provide the illustrative case 'project set up' as it has been performed in the course of validating the approach.

As already mentioned, the design of the CoMPArE/WP method is based on conceptual considerations derived from the aims of intuitive human modeling. Its components are informed by procedures and concepts identified to be supportive in reaching those aims in existing research. The novelty of CoMPArE/WP lies in the combination of those procedures and concepts in order to reach the aims of natural modeling while providing a well-defined bridge towards techno-centric modeling. The goal of validation in this article therefore is to show that the method

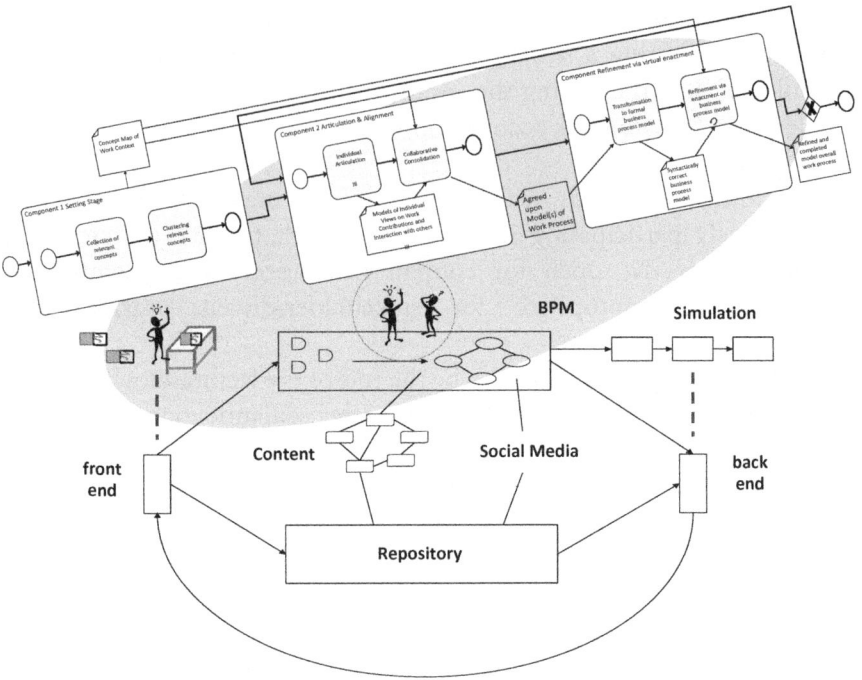

Fig. 8.8 Embodying the CoMPArE approach to the digital work design framework

facilitates natural modeling and at the same time enables participants to produce a techno-centric model of the business process. Consequently, the validation questions can be derived from the design goals, as formulated in Chap. 4:

Q1. Are the modeling participants able to semantically interpret the used notation(s) intuitively in the way specified by the method?
Q2. How do the created models facilitate knowledge sharing and promote negotiation?
Q3. To what extent does the approach enable the modeling language to emerge dynamically based on the situation at hand?
Q4. Do the final modeling results provide the syntactic and semantic quality of techno-centric models and allow for further processing in IT-systems?

These questions imply the existence of an organizational context in which actors can develop different views on a business process, calling for case study research. We thus present in the following an illustrative case study that demonstrates the implementation of the CoMPArE/WP approach in a real-world setting. Methodologically, the validation requires to qualitatively document and analyze both the process and the result of modeling in the different method components with respect to the formulated questions. Consequently, the modeling process of the case study was video-taped and analyzed with respect to the validation questions asynchronously. The modeling results of components 1 and 2 were photographed and transcribed to digital versions for easier assessment. The results of component 3 were exported from the used BPMS. The documented results and observations made in the case are used to discuss how the requirements of natural modeling are met while maintaining the bridge towards a technically interpretable business process model.

8.2.1 Sample Case

The case presented in the following is situated in an organization that undertakes software development projects. At the beginning of every project, the project set-up process is conducted aiming at agreeing upon the project's scope, the relevant stakeholders, the timeframe, and so forth. The project teams always consist of a set of developers, who are led by a team leader. Ongoing communication with the client is ensured by a dedicated contact person (who might also be a developer). In addition, there are mentors who formally do not belong to the team, but are experienced project managers supporting the project teams and acting as backups, in case interventions become necessary.

The aim of the CoMPArE/WP workshop was to investigate the effectiveness of the CoMPArE/WP approach regarding a) the active involvement of process participants in business process design and b) the transition to a comprehensive process model. Representatives of the following roles took part in the workshop: a team leader, a mentor, a contact person, and a client. In addition, a facilitator was involved to guide the process methodologically. One observer was present to document the results and the process of the workshop for later evaluation. The workshop was carried out in two parts. The first three-hour block was dedicated to the first two components of CoMPArE/WP. Based on the outcomes of this first part, a model was built using the CoMPArE/WP language (based on the *who, what, exchange* constructs). This was used for virtual enactment in the second part of the workshop, which lasted two hours.

8.2.1.1 Component 1: Setting the Stage

The four modeling participants implemented the first component of CoMPArE/WP by creating a model that described the relevant concepts in the context of clarifying the scope of a new project. They individually collected concepts each of them considered important, and subsequently consolidated them in a shared model.

The identified concepts were complementary, as the modeling participants focused on different aspects of the business process. Consolidation consequently required effort in making mutually transparent the individually selected foci and explaining their meaning. However, no discussions on the relevancy of certain concepts arose, and all concepts were finally incorporated in the model.

Figure 8.9 shows a conceptualized transcript of the model. On the right, a photo of the workshop's actual card setting is presented. As shown in this photo, the cards bear the visual markers for digital recognition mentioned earlier. Also, a big table constituted the sharing modeling surface, and thus connecting arrows were drawn directly on the cards.

The identified concept classes largely centered on the different involved roles (operative in the project team—*OpRole*; as well as roles that support the process within the organization—*SupRole*; and client-side roles—*ClientRole*) and relevant information items (*InfoItem*) that were backed with sub-items in the case of the project description (visualized at the bottom of Fig. 8.9). In addition, skills required within the project team (*ReqSkill*) as well as the aim of the process (*Aim*) were identified.

The concepts were clustered along two dimensions: the sequence of elements running from top-left to the bottom-right of the model, indi-

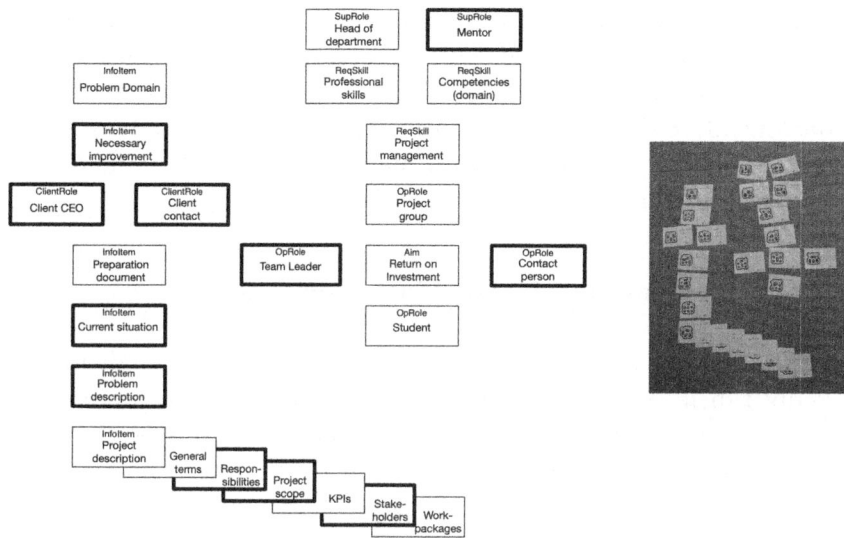

Fig. 8.9 Result of component 1—"Setting the Stage"

cating the fundamental procedure of clarifying the project scope with the customer. It thus can be considered to represent an "external perspective" on the project setup process. The ostensible sequence in the first cluster, however, does not describe a process, as it does not rely on activity-describing concepts, but mixes other, structurally motivated concept classes. The second cluster of concepts can be considered to cover the "internal perspective" on the project setup process and has identified the necessary skills and involved operative and support roles.

The open semantics used in this component enabled both the agreement on relevant conceptual classes (like aims, skills, roles, and information items) and their clustering in terms of perspectives to be considered when thinking about the business process for project setup (internal needs vs. externally visible collaboration and artifacts). The elements marked with bold outlines were directly reused in individual articulation and subsequently were incorporated in the consolidated model version. The remaining elements (drawn with narrow stroke outline) were not incorporated in the following steps but left as contextual information, describing the context of the process.

The outcome of the first modeling step thus clarified the scope of the business process to be reflected upon and outlined its fundamental building blocks. It furthermore validated the selection of the involved roles. Consequently, concepts specified in the first component were reused in later modeling steps indirectly by the modeling participants, who picked them up again during individual articulation.

8.2.1.2 Component 2.1: Individual Articulation

In the second component, the modeling participants individually described their own perceived involvement in the business process and their interaction with others. The individual modeling results are shown in the following. As the connecting arrows were drawn directly on the cards, explicit representations of sources and targets in communication acts have been added in the conceptual transcriptions for easier understandability.

Figure 8.10 (left) shows the model created by a modeling participant representing the client. Content-wise, one notable modeling choice here

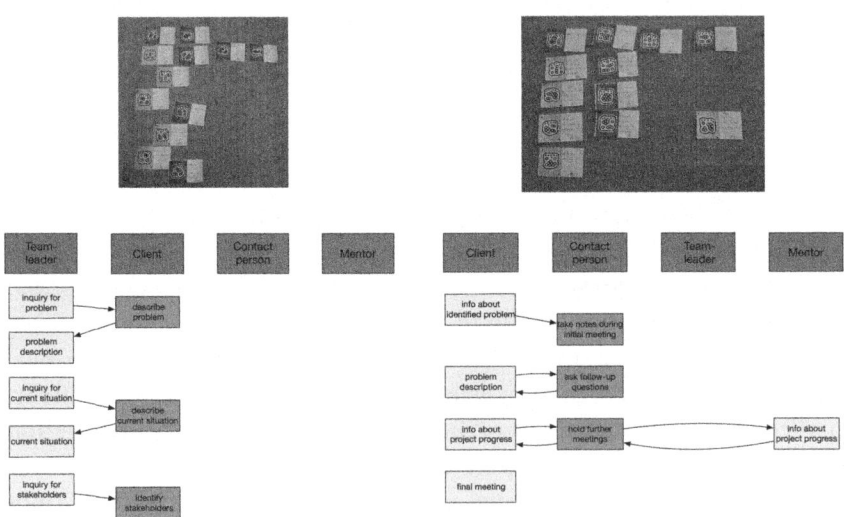

Fig. 8.10 Result of component 2.1—"Individual Articulation" for participants representing "Client" (left) and "Contact Person" (right)

is the strong involvement of the team leader in communication, while at the same time communication with the formally responsible contact person is completely omitted.

The perceived involvement of the contact person is shown in Fig. 8.10 (right). The modeling participant representing the contact person basically described the formally prescribed procedure of acting as the primary contact for the client and involving the mentor during project implementation, after the problem description has been settled upon.

The model incorporates a syntactic deviation from the proposed modeling language as EXCHANGE elements were used to describe mutual communication processes. The proposed syntax defines EXCHANGE elements to always have exactly one source activity and one target activity, representing a unidirectional flow. In terms of natural modeling, however, this is a valid use of the element as it takes a coarser approach to describing exchange of information, which can be refined in later steps when developing towards a model that is useable for workflow execution.

The model shown in Fig. 8.11 (left) represents the mentor's view on the business process. It describes an intervention in the late stage of the scope clarification, where the mentor communicates with a management representative of the client and the operative contact regarding relevant

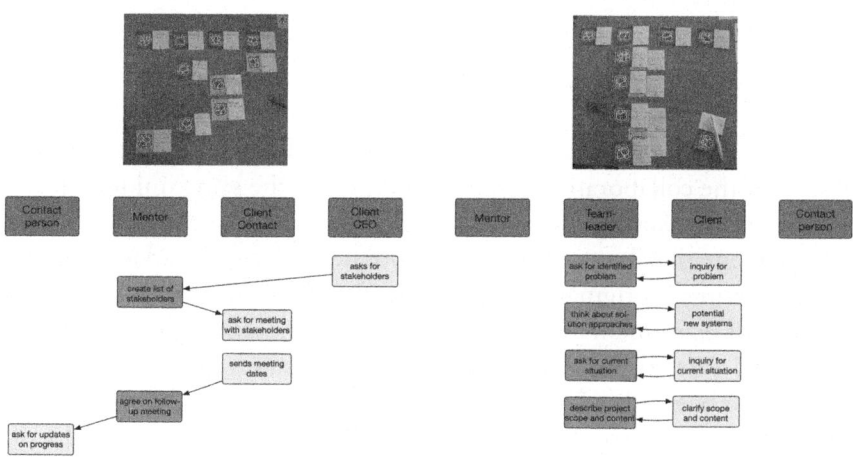

Fig. 8.11 Result of component 2.1—"Individual Articulation" for participants representing "Mentor" (left) and "Team Leader" (right)

stakeholders in the client company and then agrees on a follow-up meeting during the project with the customer contact person in the project team. The mentor was the only modeling participant, who distinguished between different client roles.

The forth individual model shown in Fig. 8.11 (right) represents the team leader's view on the business process. It largely matches the view of the client, in which the main tasks of project setup are shared by the two of them—in contrast to the company-wide guideline, which stated that the contact person should be the sole face to the client. Structurally, the model contains bidirectional exchange attached to single activities, like in the model of the "contact person." Similar to that mentioned earlier, the participant was not able to describe a more detailed interaction process for his perceived tasks and thus—as proposed in principle 3 of natural modeling—dynamically adapted the modeling language to be able to represent his perceptions.

Overall, individual articulation lasted around 30 minutes and was carried out without any communication between the modeling participants. The facilitator intervened methodologically once in clarifying the meaning of EXCHANGE elements for the person representing the customer contact. The other modeling participants did not have any issues with understanding and using the modeling elements according to their description.

8.2.1.3 Component 2.2: Collaborative Consolidation

Figure 8.12 shows the agreed upon card-based model of the business process of the collaborative consolidation, using the same unique identifiers for elements as specified in the individual articulation models. The only element that has not been incorporated in the shared model was "final meeting" (originally contributed by the contact person). This EXCHANGE element was agreed during collaborative modeling to be superficial, as it was beyond the scope of the business process. Some elements have been added, mainly to reflect the activities of the originally underestimated role of the team leader. Added elements are marked with a bold outline in the schematic drawing in Fig. 8.12. In the following, we describe the changes made during consolidation and outline their ratio-

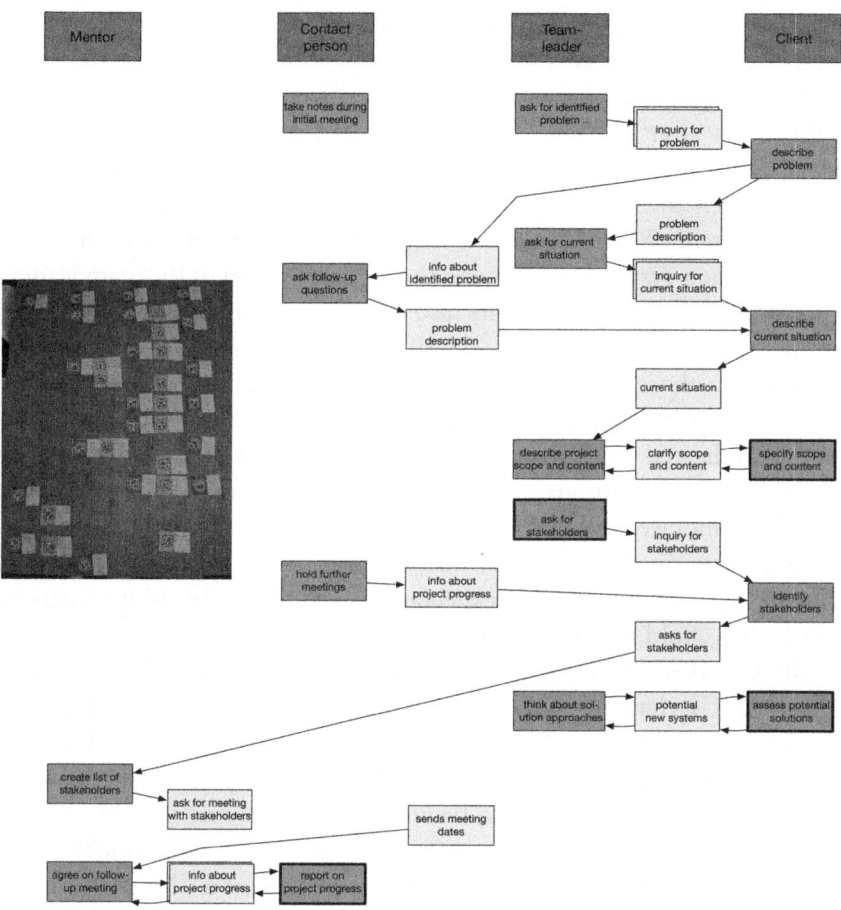

Fig. 8.12 Result of component 2.2—"Collaborative Consolidation"

nale as given by the modeling participants (extracted from workshop recordings).

The consolidated model shows the business process from an overall perspective. In a collaborative effort, the modeling participants reached common ground on the issue of who should be the primary contact to the customer during project setup. The modeling participants followed the argumentation of the client representative, who claimed that it was crucial to involve the team leader in the early phases of a project to create a clear and unbiased image of the client's needs. Consequently, the role of the customer contact was reduced to acting as a supporter for the team leader during the project setup phase and only taking over operative communication after the successful kick-off of the project. The modeling participants also recognized the need for phases of intense communication between the team leader and the client, which is indicated by the double-linked EXCHANGE elements "clarify scope and content" and "potential new systems." Following the argumentation of the team leader, the other modeling participants also refrained from detailing the communication any further and identifying distinct acts of information exchange in those phases. The same holds true for the communication between the mentor and the customer contact at the very bottom at the model (indicated by the matched and merged EXCHANGE elements "info about project progress"). Additions to the model (all elements with a bold outline) were added by the modeling participants representing the affected roles. In all four cases, this was triggered when they were confronted with EXCHANGE expectations of communication partners which they could not meet with existing WHAT-elements.

The CoMPArE/WP methodology should lead to pair-wise matching EXCHANGE elements, one element representing provided EXCHANGE created by the sender and one representing expected EXCHANGE created by the recipient. This matching, however, was done only three times. The lack of further matches can be attributed to the role shift in interaction with the customer, which was not reflected in the individually articulated models of the customer contact person and the mentor. In addition, the EXCHANGE elements "ask for meeting with stakeholders" and "sends meeting dates," originally targeted at the client in the individually articulated model of the mentor, were not matched by the client in the consolidation phase. The representative of the client was not able to describe a WHAT-element that would have been triggered by the received message and would have led to send the response, and thus left those two EXCHANGE elements dangling. This leads to a temporary under-specification of the model, which causes issues that need to be resolved during virtual enactment.

In a final step, the results of component 1 ("Setting the stage") were checked against the outcome of collaborative consolidation. Regarding constructs, the participants were not able to match the concepts describing skills and the aim of the process. These concepts were left aside for later consideration.

As far as content is concerned, the participants discussed the concepts representing roles and information items. They were able to confirm semantic equivalence to WHO and EXCHANGE items, respectively: "Team Leader," "Contact Person," and "Mentor" were directly matched. "Client CEO" and "Client contact" were only used as separate items in the mentor's individual articulation, whereas all other participants only worked with a single "Client" element. During reflection of collaborative consolidation, this issue was addressed again. The participants used a single client element in the consolidated version, as they agreed that distinguishing between the Client CEO and Client contact was not necessary and relevant for the depicted scenario. "Problem description" was directly reused in component 2 by the contact person, "Current situation" was reused by the client. Other InfoItems were identified during reflection on semantical equivalence: "Necessary Improvement" was matched with an element by the contact person, "Responsibilities" and "Project Scope" were covered by elements contributed by the team leader, and "Stakeholders" was subject to modeling in the sequence stating at "ask for stakeholders" in the lower part of the consolidated model. The remaining concepts that were considered to be potentially relevant during component 1 have not been incorporated in the result of component 2. They were still considered relevant for understanding the business process and consequently remained as context information.

8.2.1.4 Component 3: Virtual Enactment

For virtual enactment, the model was transformed to a syntactically correct process model (cf. Fig. 8.13). The source model has some semantic ambiguities that hamper direct enactment, as the BPMN model is semantically underspecified.

The affected elements are EXCHANGE elements of the team leader and the contact person, where the exact point in time of EXCHANGE is not specified. In addition, the EXCHANGE elements of the mentor

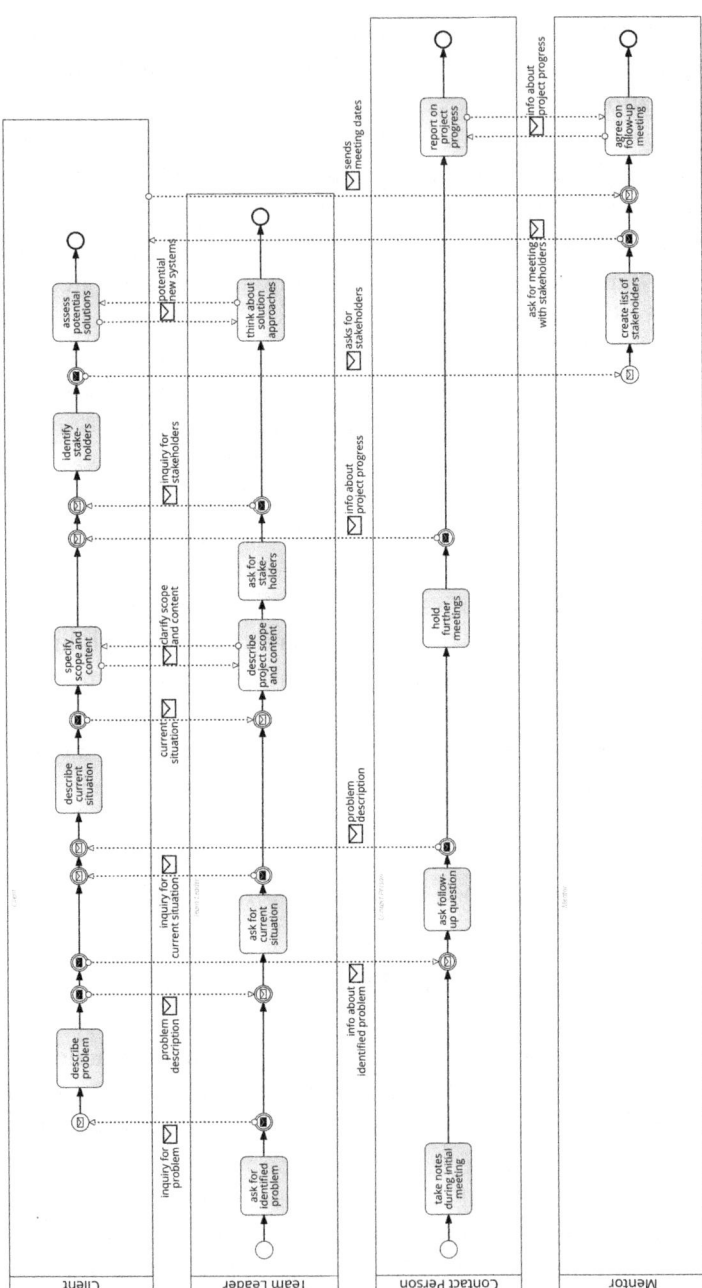

Fig. 8.13 Result of transformation to BPMN

directed at the client are not explicitly considered by the client for receiving and sending, respectively, at all. Consequently, the first group of ambiguities was transformed to mutual message flows connected to the respective activities, whereas the second group of messages was transformed to message flows connected to the targeted pool representing the client. All other exchange elements were mapped to message flows, with corresponding throwing and catching message events.

This model was used for virtual enactment to identify necessary refinements and extensions of the process model. This was done in a second workshop, in which a representative of the team leader role was also involved. An example for refinements through virtual enactment in shown in Fig. 8.14, where the initial refinement step made in the workshop is shown. The original version of the team leader's behavior is shown on the left and the refined description of the behavior is depicted on the right. The elements with task names set in italics have been added during refinement. The refinements in this step do not affect any other pools; thus, no cascading changes were necessary.

In the later phases of virtual enactment, the semantic ambiguities still contained in the model were resolved. For the underspecified EXCHANGE elements, a more detailed description of the communication procedure (to be implemented in future) was created, whereas the dangling EXCHANGE elements of the mentor were removed, reducing the mentor's role to an internal one, only interacting with the client contact person and the team leader. When making these changes, the model gradually evolved from depicting the as-is-process to depicting a to-be-process, envisioning improvements to the collaboration setup via playing

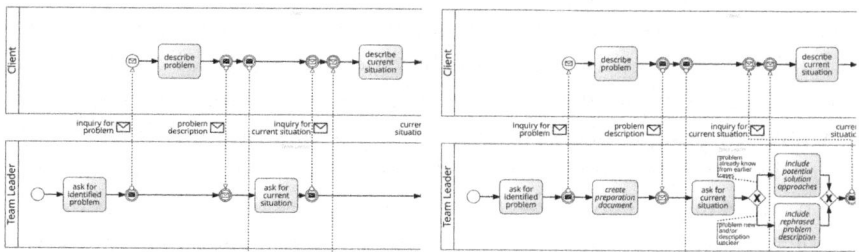

Fig. 8.14 Example of refinement (left: original process; right: refined process)

through the process model. The case study was concluded after this first iteration through the modeling and virtual enactment process.

8.2.2 Observed Effects

The following discussion of the evaluation results is structured along the validation questions formulated earlier.

- *Q1—Intuitiveness of modeling*: Ex-post feedback of the workshop's modeling participants revealed that they enjoyed their engagement in process modeling. They felt that they had generated added value for their understanding of the business process itself and how it is embedded in the process landscape of the organization. In the case study, the incremental rise in the modeling language complexity throughout the phases in particular helped the inexperienced modelers become familiar with reading and understanding models and using them as a form of expression for their own viewpoint, facilitating in this respect the externalization of tacit knowledge. Tangibility of the modeling elements (i.e., their physical presence in the form of cards and the chance to directly manipulate them) seemed to have a positive impact on the "intuitiveness" of the modeling process itself. One participant in the case study, agreed by the others, stated that "not having to master a computer tool before being able to contribute" provided added value over more traditional computer-screen-based means of modeling support.
- *Q2—Facilitation of knowledge sharing and negotiation*: The process of modeling and refining the model through virtual enactment is inherently cooperative in all its components, which have been successfully implemented to this respect in the case. Alignment of concepts and constructs in particular has been facilitated in the second component, which, by design, focuses on uncovering ambiguities and different perceptions and facilitates the development of a shared understanding. The fundamental content-wise revision of the business process during collaborative consolidation in contrast to the individually created model parts is an indicator that knowledge was not only successfully

shared among the modeling participants but also has been actively co-constructed via negotiation processes. This observation is confirmed by results of further studies of a variant of this component reported on in Oppl (2016).

- *Q3—Emergence of modeling language semantics*: During concept mapping applied in component 1, the used language constructs emerged fully dynamically during modeling. In component 2, the set of language constructs was more restricted, but still left room to adapt to the situation at hand due to their abstract nature. The modeling elements used in component 2 (WHO, WHAT, EXCHANGE) were intuitively used correctly (i.e., according to their prescribed semantics). A drawback of the reduced set of modeling elements, however, became apparent during collaborative consolidation. The lack of a structured approach to specify the content of EXCHANGE elements led to "vague" definitions (Herrmann and Loser 1999) that neither reflected nor facilitated the achievement of agreement on the transferred information or artifacts. This, however, could be compensated for during virtual enactment, when the resulting "vague" message flows were refined with scaffolds provided by the facilitator.

As it can be seen in the case description, nearly half of the concepts identified in component 1 were reused in component 2 as a foundation for individual articulation and for collaboratively reflecting on the outcome of consolidation. The benefit of open semantics as used in component 1 is that it makes visible how to reconcile fundamentally diverging viewpoints on the scope of the process and the vocabulary used to describe it. Both issues were hardly present in the case study, so that the added value of component 1 was to confirm the already shared understanding of what the project setup process was about and to produce an artifact that later could be used for reflection of the process modeling results.

- *Q4—Evolution of techno-centric models*: The model resulting from performing component 2 semantically depicted a single scenario of the complete process and was syntactically compatible to BPMN. The transformation process led to a model that already met the aim of producing a syntactically correct business process model. This model was then used for semantic refinement through virtual enactment in component 3. Only at this point, a semantically fully refined modeling language (BPMN) was used for representing the process. During virtual enactment, the participants, however, were not directly confronted with the BPMN model representation, but performed refinement by describing their additional or altered process steps in the BPMS. The process of refinement, however, was perceived to be cumbersome due to the lack of appropriate tool support in the prototype. Participants had difficulties to appropriately describe their additional process steps appropriately, in particular when additional message exchange was required. Picking up sent messages on the receiving side was confusing for the participants, as the user interface did not appropriately guide them to resolve such temporary process inconsistencies. Although these situations could be resolved by the facilitator, they require further research and development.

Summary: According to our overall experience acquired through the case study, the method has succeeded in implementing the principles of natural modeling and has achieved to actively involve process participants in modeling, leading at the same time to the production of a BPMN model, which can act as the basis for further techno-centric processing. The case study, however, also illustrated challenges in the design process, in particular at the gateways between the methodological components. The role of a facilitator still appears to be of high importance for guiding through the articulation and consolidation process. The major challenge here seems to be prompting participants in a way that facilitates description of their work so that the semantics of BPMN elements the model is transformed to later on is accommodated. This has not been fully successful in the described case, which caused higher effort during transformation to BPMN. Facilitator's guidance appears also to be required for applying correctly the modeling guidelines. It is notable that participants failed to correctly refine the labels of the EXCHANGE elements, after their transformation to BPMN message flows, for use in component 3. In component 2, they partially used verbs instead of nouns that are normally used to indicate exchanged messages in BPMN and were not aware of the need to change that until the facilitator intervened.

8.2.3 Insights

The approach aims at actively involving participants in business process modeling to adjust elicitation and modeling steps of work processes. Active involvement of process participants creates several challenges, as the latter are not expected to have modeling skills, and thus require facilitation for elicitation and formulation of the models in a way that allows for technical processing of the results. The CoMPArE/WP approach meets successfully this goal by operationalizing the principles of natural modeling while, at the same time, providing a transition to a representation of a business process that can be enacted by a BPMS.

As also revealed by the case study, the gateways between the methodological components constitute the major challenge in the application of the approach. CoMPArE/WP has tackled this issue by introducing a simple intuitive modeling language (consisting of the fundamental process concepts *who*, *what*, and *exchange*) that bridges the gap between the human-oriented card-based model of the first components, which uses open semantics and the techno-centric process model created in the last component.

The approach enables participants to gradually develop structured business process models and does not confront them with the complexity of fully elaborated process models. While the transparency of the complexity of the developed model has been a design goal, it can be, at the same time, considered the most fundamental disadvantage of the approach, as it prevents to develop an in-depth understanding of the resulting process model by the modeling participants. Furthermore, the elicitation strategy of the methodology is focused on the individual perceptions of the business process contributed by the participants and does not consider potentially divergent process views of other stakeholders, which are not directly involved in the modeling process.

8.3 Articulating and Aligning Digital Learning Support Features

This case reveals the benefits of eliciting, encoding, and different perspectives on information elements relevant for human-centered work design. The case ranges from articulating educational designs and tagging didactic content to purposeful navigation and traceable digital learning spaces, featuring concept maps as overarching representation scheme. By understanding such application development as a learning process itself, representation techniques need to enforce systemic understanding (Christian Stary et al. 2015). The frame of reference for digital work design enabling educators to elicit and align didactic concepts with learning support for collaborate classroom design can be mapped to the case, as shown in Fig. 8.15.

Since this case requires some insights into the domain of digital learning support, we will briefly provide the rationale for the addressed work practices and knowledge representation concept. Details on this case can be found in in the works by Auinger et al. (2007), Neubauer et al. (2011), Oppl and Stary (2009, 2011), Christian Stary (2016), Christian Stary et al. (2015), Weichhart (2014b), and Weichhart and Stary (2014).

Although digital learning support has been investigated and developed for quite a while, the quest for goal setting in technology-supported education and digital learning support is still valid (Feldstein 2014). With respect to the effectiveness of pedagogical models, one of the commonly agreed cornerstones of learning support developments, a shift in design thinking seems to be required; quoting George Siemens (from Feldstein 2014):

> The connectionist view that learning is a network creation process significantly impacts how we design and develop learning within corporations and educational institutions. When the act of learning is seen as a function under the control of the learner, designers need to shift the focus to fostering the ideal ecology to permit learning to occur. By recognizing learning as a messy, nebulous, informal, chaotic process, we need to rethink how we design our instruction. Instruction is currently largely housed in courses and other artificial constructs of information organization and presentation. Leaving this theory behind and moving toward a networked model requires that we place less emphasis on our tasks of presenting information,

Case Studies 353

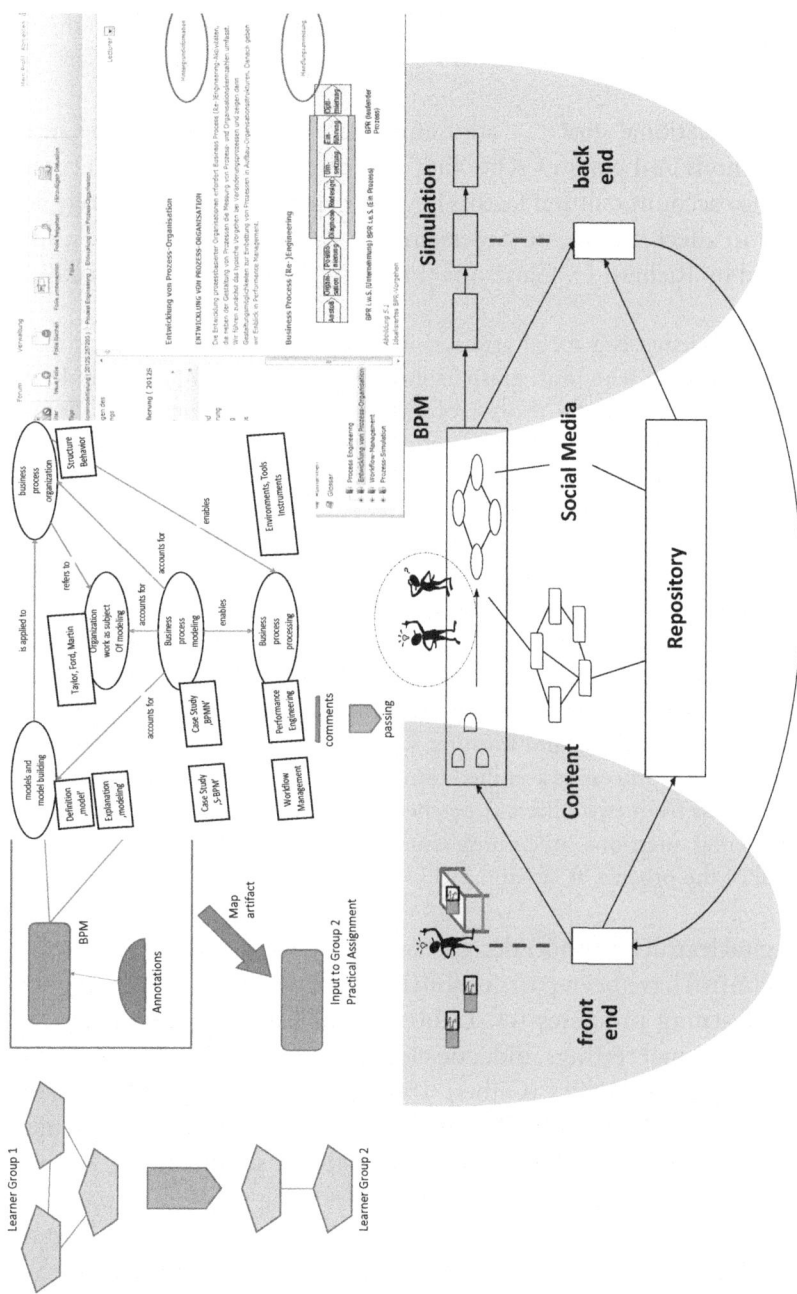

Fig. 8.15 Embodying the educator case to the digital work design framework

and more emphasis on building the learner's ability to navigate the information (i.e. connectivism). (Siemens 2005)

Such "educational goals ... are framed in direct contrast to the traditional methods and goals of schooling" (Feldstein 2014, p. 4). They need to take into account cultural factors beyond cognition and technology and are likely to affect the role of understanding of teachers and learners, such as induced by Richard D. Garrison's (unifying) transactional perspective:

> While knowledge is a social artefact, in an educational context, it is the individual learner who must grasp its meaning or offer an improved understanding. The purposeful process of facilitating an outcome that is both socially and personally worthwhile goes to the heart of the teaching and learning transaction. This transaction is common to all educational experiences, including digital learning support.
>
> Hence, an educational experience has a dual purpose. The first is to construct meaning (reconstruction of experience) from a personal perspective. The second is to refine and confirm this understanding collaboratively within a community of learners. At first glance, this dual purpose would seem to reflect, respectively, the distinct perspectives of the teacher and student. However, closer consideration of the transaction reveals the inseparability of the teaching and learning roles and the importance of viewing the educational process as a unified transaction. We are simply viewing the same process from two different perspectives. These two perspectives raise fundamental questions concerning issues of responsibility for learning and control of the process. (Garrison 2011, p. 62)

In digital learning support designs, reflection of educators and increased learner control have been parts of shifting from teacher-controlled to self-directed learning processes (cf. Dabbagh and Kitsantas 2012). Since it affects educational settings, didactic elements increasingly get questioned by principles of mathetics (Gilbert 1962; Scott 1968). When educators share the responsibility of the learning process with learners, the preparation of the environment becomes essential for self-managed learning (Eichelberger et al. 2008). It is for digital learning support of particular importance to get learners interested in being exposed to various learning modes (termed polyvalent by Leclercq et al. 1977), exploiting a variety of methods and resources on provided content elements (Duckworth 2006).

As such, digital learning support designs require not only transparent acquiring and representing how content is prepared for learning, but also revising interaction facilities and information structures, for example, recognizing the social character of transfer processes (Derouin et al. 2005).

Concept maps (Novak and Canas 2006) are widely used as effective and valid means to elicit, represent, and share knowledge (Moon et al. 2011). Albeit being traditionally utilized in educational settings (Markham et al. 1994; Novak 1995; Kinchin 2000), they have been introduced to organizational learning (Peris-Ortiz et al. 2014; Kolb and Shepherd 1997), as they allow

- making 'thinking visible' in a socially accepted way (Collins et al. 1991)
- embodying cognitive and social learning experience (Roth and Roychoudhury 1993; Roth and Roychoudhury 1992)

Their fundamental structure and handling is kept simple and can easily be conveyed to different stakeholders. As such, they qualify for engaging the various stakeholders in learning processes and knowledge management activities, including experts (Coffey et al. 2002). The ease of use while ensuring a high degree of expressiveness due to their diagrammatic nature lays ground for user-/usage-centered design. The various stakeholders, in particular curriculum designers, educational content providers, authors, tutors, facilitators, and learners, need to interact within and across their peer group when aiming to put to practice the interactionist and connectionist stance addressed earlier. A coherent use of concept maps should bring digital learning support developments closer to achieve Dewey's objective that, finally, there can be no difference between an educator and a learner's understanding, in particular, in democratic educational institutions (Dewey 2013).

In the course of learning and interaction, the complex cognitive and social fabric develops dynamically, requiring stakeholders, on the one hand, to stay tuned to their role and its adjunct perspective(s)—for example, educators being domain expert and knowledge transfer designers—while, on the other hand, meeting contextual objectives at the same time—for example, formal (institutional) qualification requirements and sense-making skill development for individual learners. To that respect, concept maps allow not only encoding different types of relevant information but also elaborating different perspectives on information elements (Kinchin and Alias 2005). By exchanging perspectives (Boland and Tenkasi 1995), they allow stakeholders' reflection (McAleese 1998), concerning the meaning of conveyed content and features for interaction in the case of digital learning support developments (Hughes and Hay 2001).

The successful use of concept maps as tools for orientation, such as in navigation in digital learning support systems (Hwang et al. 2011), in addition to content organization, recommends their use when increasingly focusing on learner-centered designs besides presenting information (in the sense of Siemens 2005). Since concept maps allow for both, non-intrusive and non-disruptive user- and usage-centered design of learning environments should become possible.

Finally, the more self-organized the process of (re-)constructing knowledge can be organized, the better problem-solving capabilities can be developed by learners (Hwang et al. 2014). Although from these empirical findings it can be concluded that integrating concept mapping into digital learning support environments helps learners acquire knowledge in a more effective way, a recent study reveals "it remains an open issue to find a suitable way of integrating concept maps into the learning process without introducing too much extra cognitive load" (Hwang et al. 2014, p. 77). The connectionist view on learning (Siemens 2005) together with intertwining roles according to the interactionist approach as proposed by Garrison (2011) could help to minimize cognitive load along learning processes.

Consequently, we have tested concept maps for eliciting mental models of educators (instructors, content providers, etc.), including their domain and didactic understanding for a certain education task (Kinchin et al. 2008), for example, in terms of subject-specific learning paths. Subsequently, we offer learners to use representations of such kind as a means of orientation for navigation and individual learning path development (as part of content individualization). Implementing this concept should increase problem-solving capacity without burdening learning with existing domain and educational structures.

We introduce informed learning design along the following structure:

(i) Articulation support for intentional education
(ii) Semantic navigation
(iii) User-/usage-centered design spaces

Articulating educational design and using it for navigation lay ground for structuring design spaces (iii), as they link features of learning environments to domain structures and didactic models. They contain all required information for contextual design due to their systemic representation, enabled by concept maps. All conceptual findings have been tested in the field, allowing to present concrete data and to instantiate methodological or technological concepts in each section. All sample cases refer to learner-centered didactics and/or the same application domain, namely, Business Process Management (BPM). BPM is applied in practice across disciplines, in particular economics, organization, and information and communication technology (Weske 2010). Moreover, coherent design in higher education, as proposed by Kinchin (2014), requires rethinking learning in terms of processes—Business Process Management captures these essentials from an organizational and technology perspective.

8.3.1 Articulation Support of Intentional Education

In this section, concept mapping for eliciting educator knowledge is discussed. Being part of various acquisition approaches when designing learning environments, concept mapping allows identifying several categories of relevant knowledge (Novak 1995; Trochim 1989):

- Domain structures
- Didactic patterns, including envisioned learning paths
- Context of learning processes, such as situations of use

Knowledge articulation is primarily a (meta-)cognitive effort to reflect on inputs to actions, such as educational resources and causal links between actions and outcomes, triggering learning activities through engaging resources (cf. Strauss (1988) referring to explicit Articulation Work). Concept maps, in particular when scaffolding (meta-)cognitive processes as hierarchy, cluster, or chain (O'donnell et al. 2002), codify knowledge—a necessary precondition to enable others accessing and using externalized or generated knowledge (Swan et al. 2010). Such documentations serve well as focal point for further processing, for example, curriculum design (Toral et al. 2007); however, they require to justify elicited knowledge (cf. Sandberg 2005).

In the following, we apply concept mapping for educational knowledge generation, i.e. for identifying and documenting concepts or nodes in their mutual context. Once a topic or question is provided (Novak 1995; Novak and Canas 2006), the setting can be designed differently for effective utilization. We start with the open format by giving a certain topic, such as the design of a course. Such a scenario fits well for educators starting to reflect on their experiences and skills from a perspective of their choice, such as domain, institutional, or didactic perspective (Kinchin et al. 2008). It also meets the objective when 'an empty sheet' approach is required to open up for novel ideas. As Lee and Nelson (2005) revealed, generative concept maps could outperform prefabricated ones.

We proceed with elicitation procedures via structured interviews that turned out to set the stage for designing digital learning support application in a comprehensive but focused way. It fits well to concept mapping, as concept maps facilitate analyzing existing learning resources, such as textbooks, in a structured way. Explicit content structures, finally, allow designing learning support systems including the didactic arrangement of content and its context, such as social interaction features.

8.3.1.1 'Open' or Non-directed Elicitation and Reflection

This type of concept mapping starts with an objective, which the participants need to agree upon. It may concern either an individual topic or a group task. Typically, the trigger to elicit and document educational knowledge and resources for educational design is the (re-)development of a course, or the occurrence of an educational challenge. The involved stakeholders start constructing a concept map by identifying nodes (concepts, meaningful items) and relationships on a virtual or paper surface, articulating their experiential knowledge. A variety of media for interaction can be provided, in particular paper, GUI-based applications, such as the Cmap tools (Novak and Canas 2006; Canas et al. 2004), and tabletop approaches, such as Comprehand (Oppl and Stary 2014, 2011)—see Fig. 8.16 and the introduction of the system in Chap. 7.

Interactive tabletop mapping in that context targets at tangible information spaces. Correspondingly, concepts/nodes as physical representations can be put on a tabletop surface and linked by pushing two nodes against each other. Nodes and links may be provided with text that is then displayed on the tabletop. The 3D-elements also allow 3D-nodes to be opened, in order to put in other artifacts.

The example given in Fig. 8.17 stems from the preparation phase of the International Summer School on Subject-Driven Role-Guided Externalization of Organizational Models (Erasmus Intensive Programme

Fig. 8.16 Tabletop concept mapping

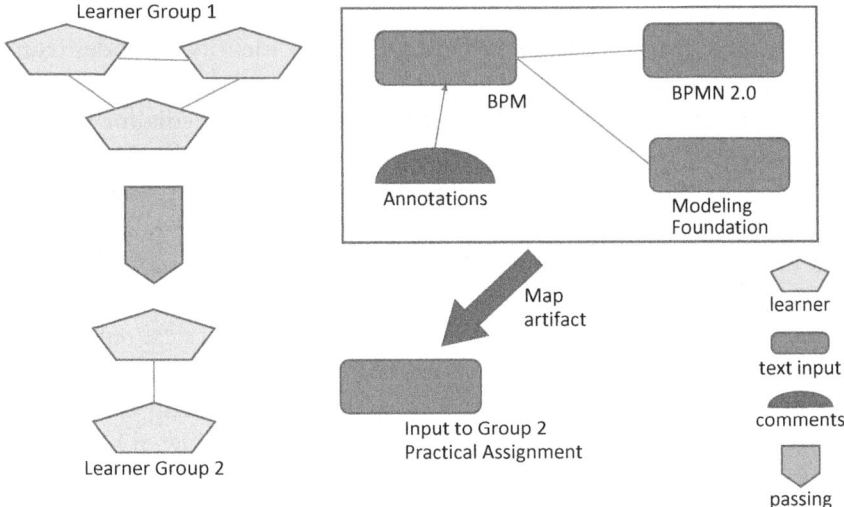

Fig. 8.17 Tabletop concept mapping for articulating educational design—sample patterns

sponsored by the Lifelong Learning Programme of the European Commission). The figure shows some educational design principles for an introductory lecture on Business Process Management (BPM). Since the summer school is intended for students from different European countries and curricula (economics, organizational studies, computer science, business computing, information systems), the crucial task is to align their understanding with respect to major concepts of the field and their nature.

The tabletop map reveals on the left side a chain (sequence) of two learning steps involving different learner groups. In a first step, Learner Group 1 (upper-left part of the figure) receives a bundle of information on BPM, composed of modeling foundations and the language standard on the Business Process Modeling Notation (BPMN) 2.0 (upper-right part). Learner Group 1 is asked to annotate the BPM lifecycle that can be found in 'Modeling foundations' with examples according to their own experiences and background (Haslhofer et al. 2014). These annotations, together with the other resources, are passed on to Learner Group 2 to accomplish a practical BPM modeling task, namely, Process Analysis for a service industry. The container function of the tabletop system (Map artifact in Fig. 8.17) has been used to put the map denoted by the rectangle into the red tabletop element (by using the artifact marker for a Snapshot) shown at the lower part (Input to Group 2—Practical Assignment).

In contrast to paper-based concept mapping, the process of mapping may be recorded according to the needs of the users. Hence, the process of elaborating a structure may be traced and variants may be developed starting from any recorded state. In the presented system, due to the import into a GUI-based editor, each map/snapshot may be processed further and be manipulated. For the tabletop system, an export to the Cmap tool format (Canas et al. 2004, cmap.ihmc.us) has been implemented, in order to allow processing the maps with a widely used GUI tool set. For procedural chains, such as shown on the left side in Fig. 8.17, an export has been developed to a business process suite.

8.3.1.2 Setting Up Didactic Requirements

Benefits for education design can be created from reflecting and exploring didactic approaches, again using concept mapping. In this section, we exemplify such an endeavor for progressive education, a learner-centered approach oriented towards self-organization and constructivism (cf. Eichelberger et al. 2008; Weichhart 2012, 2014a; Weichhart and Stary 2014). Such comparative analyses for educational design follow a four-step procedure:

1. Specifying the universe of discourse, such as identifying didactic approaches relevant for progressive education.
2. Detailing each constituent, collecting and structuring according to the information available, for example, procedures, assumptions, empirical findings.
3. Cross-checking according to capabilities, for example, degree of self-organization, effort of preparation.
4. Consolidating for further action, in particular requirements for digital learning support.

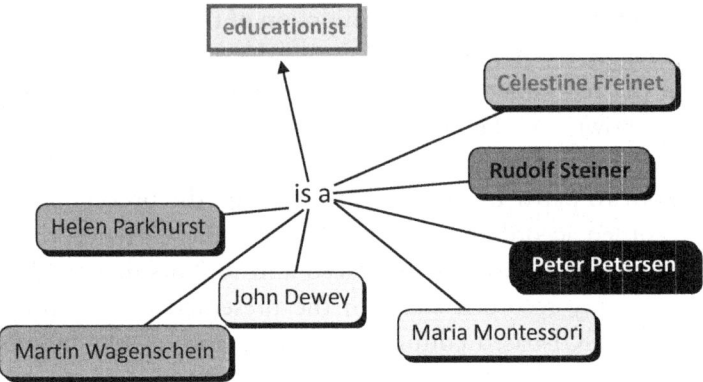

Fig. 8.18 Approaches to progressive education, according to Weichhart and Stary (2014)

Figure 8.18 exemplifies step 1 for progressive education, naming all analyzed educationists, and thus scoping the universe of analysis. Color codes are introduced, facilitating traceability when cross-checking findings.

In step 2, each approach is detailed according to the source of information in the sample case documented findings. Dewey (1928) (Fig. 8.19) puts emphasis on educating children using democratic principles, and educating them to acquire experimental, self-organized learning capabilities, thus allowing them to contribute actively to societal developments. Parkhurst (1922) (Fig. 8.20) appreciated Montessori and Dewey. She developed the role of the teacher further, namely, towards guiding learners rather than controlling them. The developed pedagogy is centered on two instruments, which allow the provision of guidance and progress monitoring. Assignments provide scaffolds instead of details of how to solve a task. The progress of the students along these scaffolds is monitored, using process graphs. Learning incorporates group work and cooperation.

In step 3, cross-check according to educational tasks is performed. Hereby, parts of the aforementioned concept maps on the individual pedagogical approaches are put into a single map, thus providing an aggregated view on progressive education. In order to be able to identify the source of information of each concept and link, they are colored differently, as indicated in Fig. 8.21. Concepts that are represented by a rectangular shape represent the core concept of the particular map.

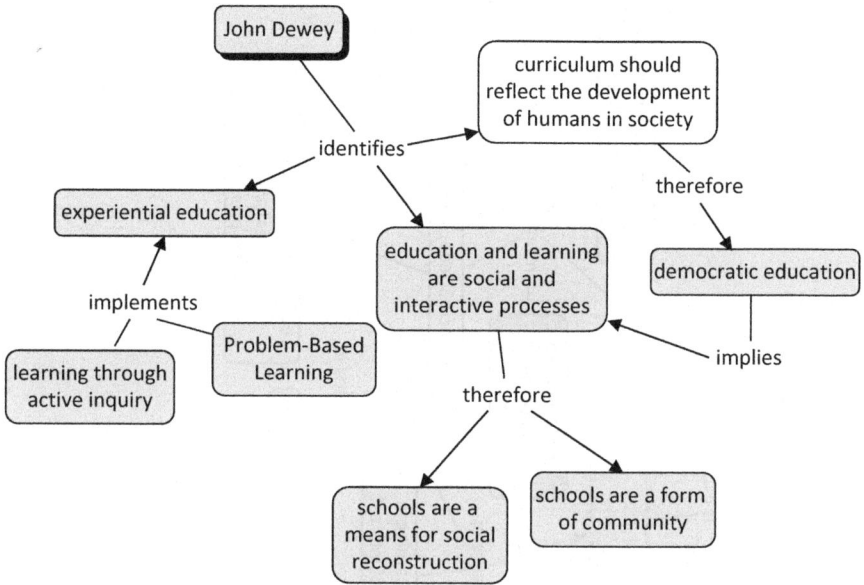

Fig. 8.19 John Dewey's approach, according to Weichhart and Stary (2014)

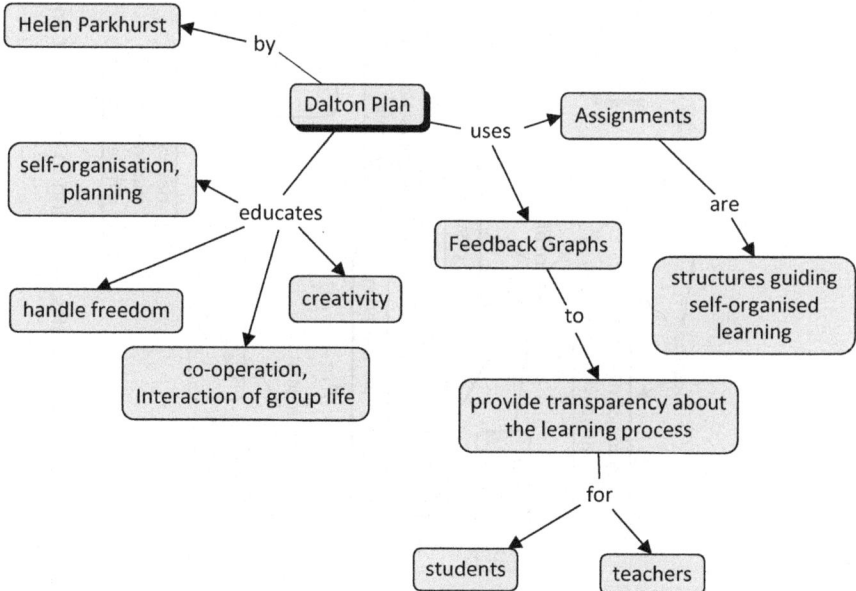

Fig. 8.20 Helen Parkhurst's approach, according to Weichhart and Stary (2014)

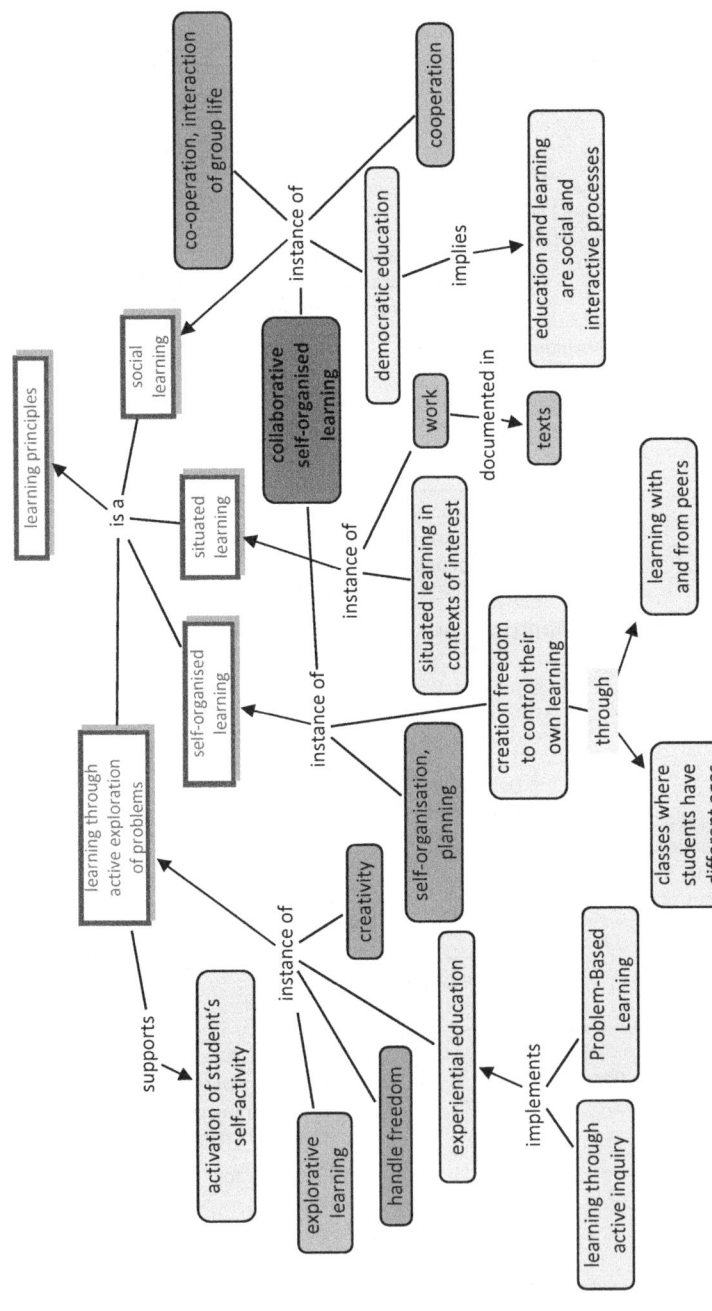

Fig. 8.21 Learning principles, according to Weichhart and Stary (2014)

The following map shows learning principles facilitating learner-centered capacity building. The analyzed approaches require learners to take care about the freedom to select or develop their individual problem-solving capability in a self-responsible manner. The requested active exploring of problems promotes analytical thinking, creativity, practical abilities, and social capabilities for problem solving, since learning should also occur in groups.

Finally, in step 5, requirements for educational design in digital learning support environments may be derived from the map in Fig. 8.21. The concept map in Fig. 8.22 conceptualizes a learning environment providing learning facilities according to the aforementioned principles, by showing enablers to achieve major objectives of progressive education.

8.3.2 Developing Digital Learning Support Baselines (Course and Content Models)

Although Rye and Rubba (1998) could not demonstrate essential benefits for generating knowledge, incorporating concept maps into interviewing, their work laid ground to structure narratives according to concepts, and thus apply concept mapping in the context of collecting educators' experiences for further engineering (Middleton et al. 2008). The presented content engineering process has been developed and evaluated in the projects ELIE (E-Learning in Engineering) (Auinger et al. 2007) and mobiLearn (Zaharieva and Klas 2004; Ferscha et al. 2004). It has been enriched with concept mapping, not only facilitating note taking through providing a structure according to the interview, but also encoding domain structures that can be annotated with additional information. Of particular interest are domain-specific refinements and educational metadata.

The approach comprises five main steps: preparation, preliminary document analysis, structured interview, extended document analysis and mapping of didactics, and the actual content authoring and delivery to a digital learning support system (Fig. 8.23). The core process steps aim to identify domain-didactic items based on relevant learning items and interview findings from domain experts, and to specify didactically enriched learning content.

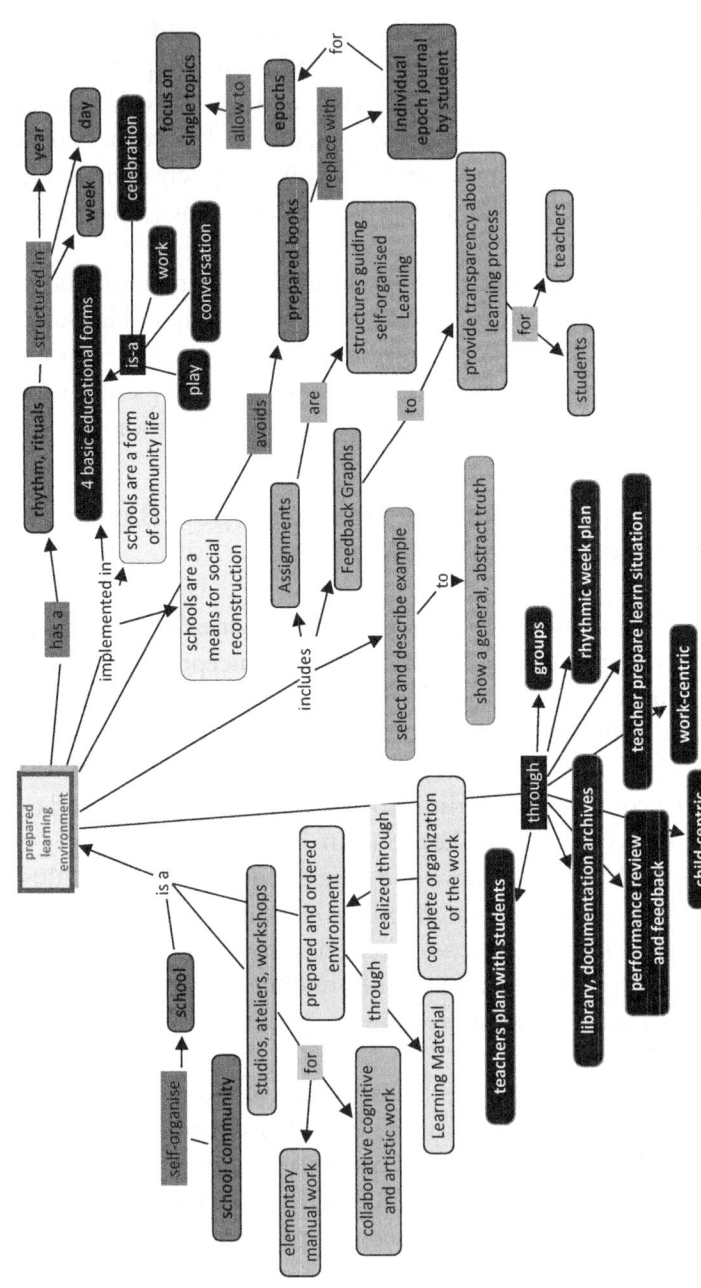

Fig. 8.22 Progressive learning environment requirements, according to Weichhart and Stary (2014)

Case Studies 367

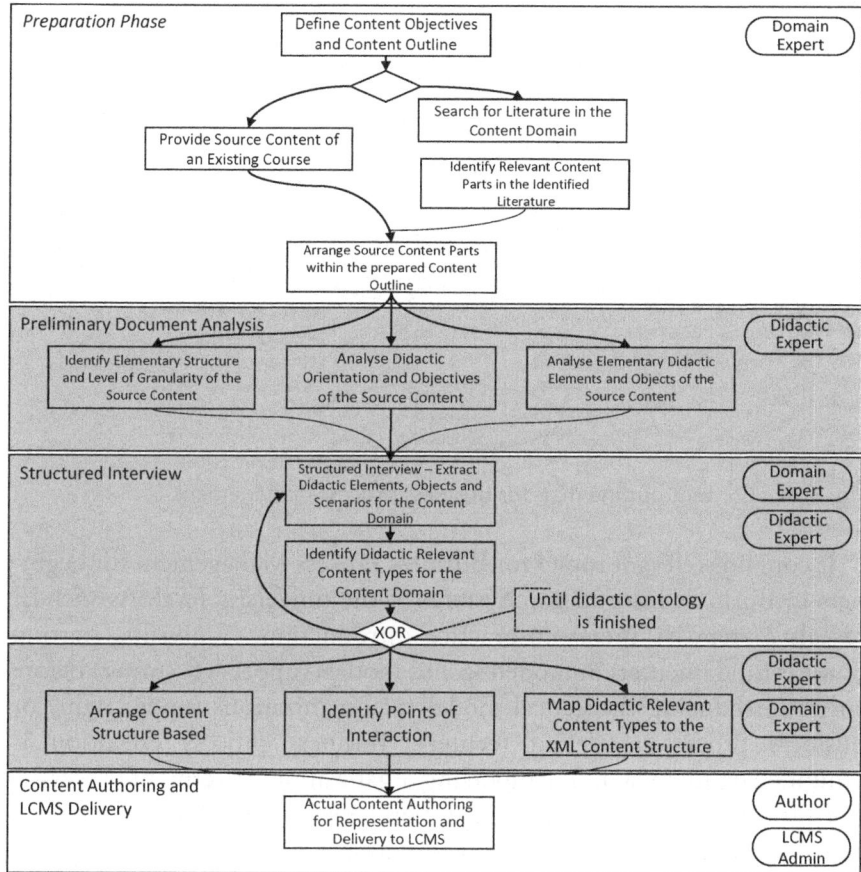

Fig. 8.23 Process map for digital learning support content engineering according to Auinger et al. (2007)

In the course of the preparation phase, resources for content development have to be identified, mainly by educators who are also domain experts. A content outline map, including building blocks of a course, such as learning goals, target learner group, basic structure, depth, and granularity of content, is specified. According to that structure, resources can be structured and analyzed. A set of resources forming an educational baseline serves as input for the didactic enrichment (tagging) process. Figure 8.24 shows an outline map (step 1).

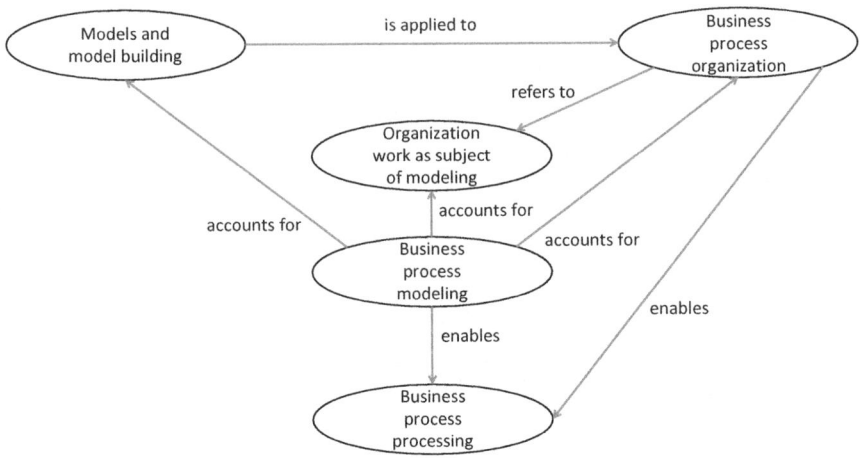

Fig. 8.24 Content outline map for business process management

It contains relevant topics for Business Process Management for beginners in Business Information Systems at the university level. As such, it reveals a stepwise theory-to-practice introduction. A possible starting point is fundamentals in modeling and models (upper-left corner) before either introducing theoretical models of organizations (upper-right) or business process modeling (center). Business process execution is grounded on understanding modeling organizations in terms of processes.

Figure 8.25 shows the annotated map (step 2). The elementary structure displayed in Fig. 8.24 allows annotating:

- Refinements of the fundamental structure, such as detailing business process execution in terms of performance engineering and workflow management (lower-left part of Fig. 8.25)
- Essential aspects, such as 'structure' and 'behavior' for understanding 'business process organization'
- The assignment of elementary didactic tags along refinements, such as 'case study', 'definition', 'explanation'
- Information on didactic orientation according to objectives of a course, such as assigning theory- or practice-laden didactic terms to topics, for example, 'tool' to 'business process processing'

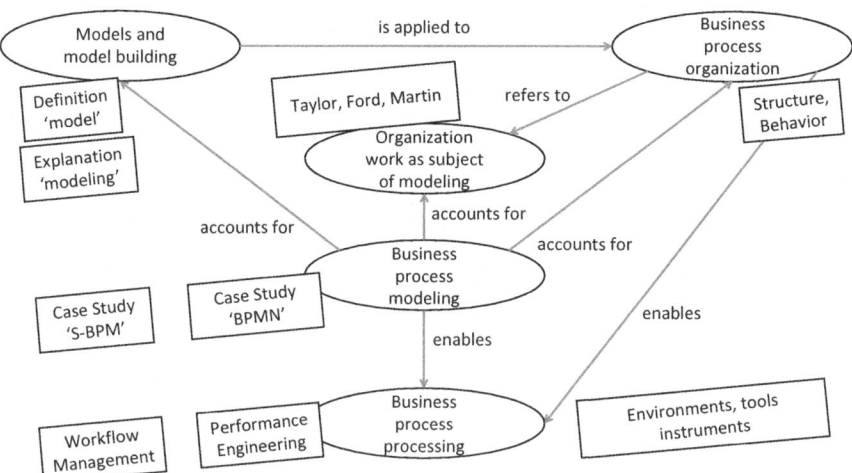

Fig. 8.25 Annotated structure map

In the course of the preliminary document analysis, source content chunks and documents are scanned to identify the level of granularity, content for orientation and navigation, and elementary didactic elements. The level of granularity of resources can be quite different: presentation slides, textbook elements, animations, and apps. In the concept map, annotations are used to identify relevant content items. Depending on the intended use of the content, different levels of detail may be useful. Finally, elementary didactic elements, such as definition (e.g., case study), can already be identified. A concept map structuring all sources of relevant input also contains the rationale why this element should be included, relationships between the documents, and metadata, such as modality of information (video, text, etc.). Hence, the final map contains all relevant associations (links) including navigation and navigational guidance. It forms the guide for the structured interview to validate the findings so far.

The structured interview with the educators concerns the following issues (Auinger et al. 2007), supported by a structured mind map (see Fig. 8.26) to condense all provided inputs:

Fig. 8.26 Structure map for interviewing and result presentation

1. *Organizational context.* Organizational issues include content profiles, learner profiles, and the organizational learning environment:

   - Number of educators and learners
   - Current didactic quality of resources, including metadata of different kinds
   - Structure and procedure of educating and facilitating learning
   - Criteria most important for facilitating learning processes, ranging from quality and adaptability of content to learner satisfaction and innovation
   - Target group(s) in terms of background, motivation, literacy, learning style, professional orientation (technical, business, and the like)
   - Guiding principles of learning processes: (i) to make new knowledge accessible, (ii) to practice and deepen linking knowledge, (iii) to link existing knowledge, and (iv) to embed knowledge in a global context
   - Type of education in terms of learning (self-directed/instructor-driven, project/assignment-driven, etc.) processes

2. *Individual positioning.* This section should clarify the individual approach of educators with respect to supporting learning processes:

   - Time spent with learners (either face-to-face or in virtual settings)
   - Fundamental individual didactic principles and preferences, for example, less is more
   - Potential of (re-)designing learning resources

3. *Learner/learning support.* It comprises

   - activities of educator along

     – preparation phase, for example, selection of content elements, establishment of specialized didactics, learner consultancy
     – implementation of the course, for example, classroom teaching, feedback sessions, quality checks
     – assessment
     – evaluation

   - improvement of learning resources, didactic approach, and tools (based on evaluation results)
   - didactically motivated content elements utilized for learning (codification such as text, pictures, multimedia, drawings; content types such as examples, cases, definitions, directions; interactive elements)
   - structure of learning resources: linear/sequencing, linked/hyper medial, hierarchical, hybrid;
   - completeness of learning resources with respect to didactic design
   - organization of learning support, including feedback to learners
   - grading and examination

4. *Communication.* Social interaction and skills of the interviewed educator refer to

   - frequency of contact with other stakeholders (educators, learners, etc.)

- particularities of interaction, such as Hot Potatoes, organizational issues, tools, taking lead

5. *Technical support.* It addresses

- categories of ICT-tools, such as content management, social media
- technical interface issues when linking of two or more tools is required for learning support
- meta-cognitive (learning-to-learn) support tools
- learner profiling, identity management, and integrity/security issues

The structured interview should clarify individual, organizational, and technical aspects of the learning support process. In the core part of the interview, didactically motivated elements such as didactic content types, and interactive elements are identified by the interview partner.

In the next phase, the didactic elements and structures are mapped to the (XML-)content structure. In case content has been already tagged, as some text books are generated according to metadata or didactic ontologies (Meder 2006; Schluep et al. 2006; C.-M. Chen 2009), these data can also be generated automatically (Tseng et al. 2007) or semi-automatically (Leake 2006; Larrañaga et al. 2008).

Since the early days of digital learning support, the need for encoding didactic quality into content has been demanded (Schulmeister 2017). Content elements should not only contain but also visualize metadata, such as definition, for orientation and selection. Figure 8.28 shows such an approach (Auinger et al. 2007). Learning units are part of modules courses are composed of. They contain content blocks with various domain- and education-relevant tags assigned to content elements. These elements can be text, graphics, video, or audio information.

Table 8.1 shows part of a typical didactically enriched structure developed for a course on Business Process and Communication Modeling at the University of Linz, Department of Business Information Systems. The course is given as an introduction to BPM to students in the Business Information Systems curriculum in the first year of the corresponding bachelor degree program. Modules and Learning Units can also be shared with other courses (Initiative 2004), either in Computer Science or

Table 8.1 Example of tagging a BPM content structure

Module	Learning unit	Block	Didactic tag
Process engineering	Development of process organizations	Business process re-engineering	Background information
		Design	Case study
		Performance engineering	Explanation
		Implementation	Example
	Workflow management	Ontology	Explanation
		...	
	Process simulation	Objectives	Content
		...	

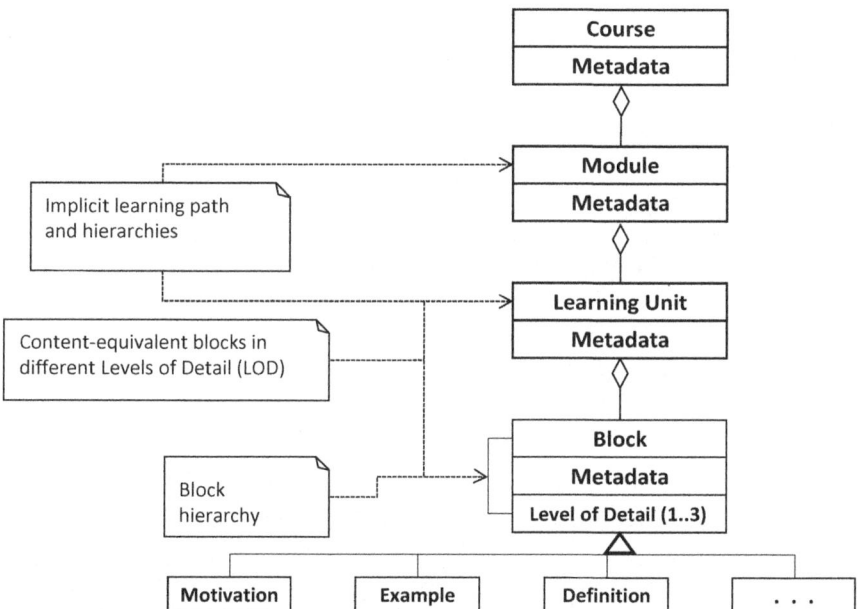

Fig. 8.27 Educational metadata structure

Business Information Systems, such as Communications Engineering. In those cases, the assignment of metadata (block types in Fig. 8.27) needs to be reconsidered (Leidig 2001), as, for example, some definition in computer science may need to be re-categorized as explanation in Business Information Systems due to its explanatory character when focusing on application of computer science theories and concepts.

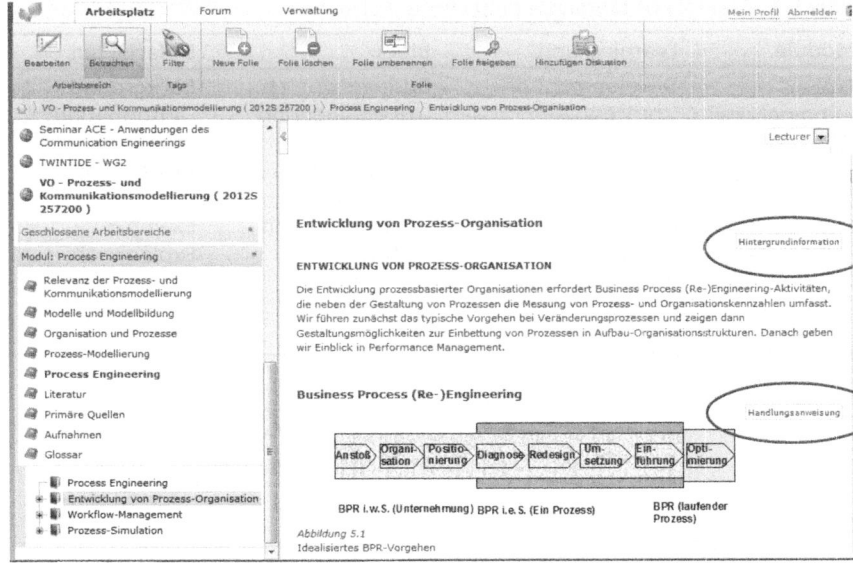

Fig. 8.28 Tagged BPM content—'background information' and 'practical guideline' on the development of process-based organizations (released under a Creative Commons Attribution 4.0 International License (CC BY 4.0))

Tagging follows the structure of the content outline map shown Fig. 8.25, leading to the following modules (see also Fig. 8.28—navigation area on the left side of the screenshot):

- Introduction, providing the relevance of the field
- Models and modeling, giving some background on abstraction and representation
- Organizations and processes, introducing the nature of business processes and their history in organization science
- Process modeling, detailing functional, object-, und subject-oriented approaches to business process modeling, with practical guidelines on how to construct models in the respective paradigm
- Process engineering, providing fundamentals of performance engineering, architecture designs, and workflow management, in order to implement business process models by ICT systems

In Table 8.1, one of the modules, 'Process Engineering', is detailed with respect to some of its learning units (Development of Business Organizations, Workflow Management, Process Simulation), and its content elements (blocks), and their tags for the first learning unit.

In addition to tags, distinguishing various levels of detail has turned out to be useful for targeted content delivery. Using several LODs (levels of detail), content developers can structure learning resources on three different levels of granularity. A common instantiation of that concept is to provide slides for classroom presentation on LOD1, text book elements for reading and self-studies on LOD2, and additional information or further resources (links, files, videos, and the like) for exam preparation and in-depth studies on LOD3.

For learners, tags are visualized when content elements are displayed. The content area in the center of the screen (see Fig. 8.30) corresponds to the work space of stakeholders. Navigation is provided initially as tree view on the left side of the screen. It supports nesting of content elements, in order to facilitate structured access to content elements, such as displayed for Process Engineering on the bottom-left of the screen in Fig. 8.28.

Explicit tags also allow filtering according to learning styles, for example, selection of all examples of a learning unit, in case a learner is more practically oriented when acquiring knowledge. Given the proper functionality (see third entry from left 'Filter' in the toolbar beyond the navigation space), the LSS (Learning Support System) displays only those parts in the navigation and content area that contain the selected tags. Hence, both domain structures and didactic expertise contribute to semantic richness of the provided BPM content.

In Fig. 8.28, on the left side of the screen, a tree view for navigating the nested content is shown, whereas in the center, the selected content is displayed, in this case 'Development of process organizations' being part of the module 'Process Engineering'. The tags are 'background information' (Hintergrundinformation) and 'practical guideline' (see marked areas on the right side of the screen) concerning some text to motivate developing process-based organizations, and a practical guideline on the development of process organizations revealing BPM phases that should be followed in the course of development.

The latter case reveals the intention of tagging, expressing the context of how the content is addressed and could be used. For learners who should find orientation of how to set up BPM projects and participate in BPM lifecycle activities, the tag 'practical guideline' indicates this educational intention. In case learners are focusing more on becoming acquainted with frameworks, such as when comparing lifecycles from various BPM approaches, they could be supported effectively using a tag like 'operational frame of reference' or 'value chain'.

8.3.3 Semantic Navigation

Navigation makes up most of the user's experience (Smolnik and Erdmann 2003). Consequently, navigation features should facilitate the access to domain- or user-relevant information including content and its manipulation features. When using those features, users should build up and maintain a coherent mental representation of the traversed environment, the so-called cognitive map. Such a representation serves as a baseline for learners and facilitators when interacting with a learning support system (Rovine and Weisman 1989). However, for content-rich applications, there is no consensus on (re)presenting content and manipulation features in a user-centered way (Godwin et al. 2008).

The learner support presented so far (see previous section) featured the dynamic selection of metadata, such as 'explanation', which allows learners navigate through content and experience it individually. Its design is led by domain concepts which can be created by mining techniques from documents (N.-S. Chen et al. 2008) and could be utilized for adapting to learner needs, such as planning individual learning paths (C.-M. Chen 2009). Tseng et al. (2007) constructed concept maps for achieving adaptive learning. Hereby, they automatically created predefined concept map of course descriptions (ibid.) that could be adapted to individualize learning paths. They can help educators and learners to locate and assign learning resources according to recognized learning goals. However, intentional elements need to be visualized and accessible interactively (Sumner et al. 2005).

In the following, we report on the concept map-based tool developed by Neubauer et al. (2011) that allows encoding of intentional information dynamically, such as learning objectives, domain, and didactic metadata. Using the learning support system shown in Fig. 8.30, they had found that the deep hierarchy levels had been time-consuming for learners with respect to navigation, and thus were hindering learning processes. They developed an associative navigation design, enriched with educational and domain-specific metadata. It allows individual exploration of content and is displayed as concept map. Learners select learning their paths according to the prepared links and may navigate beyond hierarchies (as encoded in the tree view), and across domains or courses.

Figure 8.29 depicts a concept map for the learning unit on 'Enterprise Architecting' being part of 'Process Engineering'. Educational metadata (motivation, definition, etc.—see also Fig. 8.29) semantically describe links to information resources. Hence, the associative navigation provides learners additional structural navigation information that shapes their learning paths.

Individualization support considering the associative navigation is similar to the navigation concept introduced in Chap. 5. It is enabled through features like annotating a concept map and its elements, editing such as adding individual concepts, and filtering links to information resources according to didactic content types, content modality, or user profiles and preferences. Compared to the hierarchic approach, the

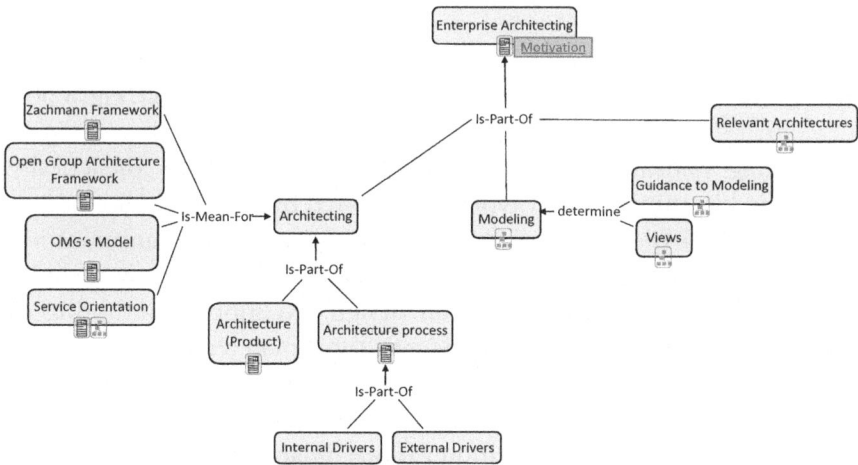

Fig. 8.29 Didactically enriched concept map navigation

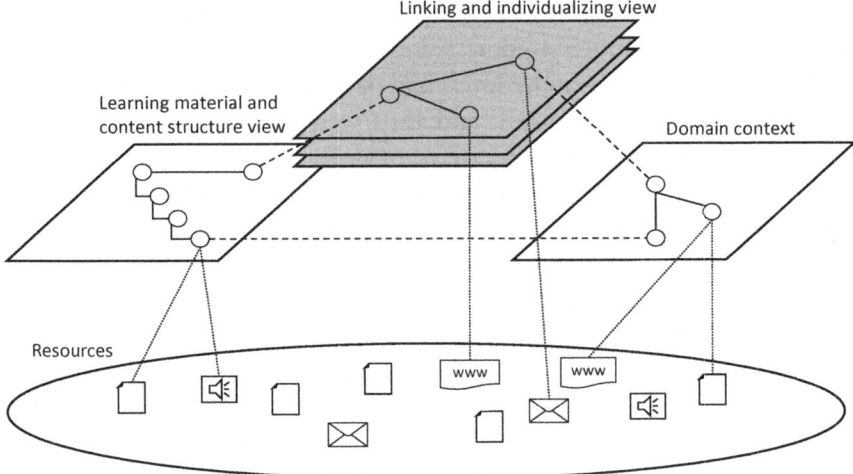

Fig. 8.30 Relationships between main views according to Neubauer et al. (2011)

concept map approach also enables annotation referring to concepts, relations, and links to resources.

In order to support both approaches, a polymorph representation scheme has been developed based on the ISO standard of Topic Maps (Neubauer et al. 2011). For the implementation of the dual navigation approach, the differently organized information structures have been recognized applying an intertwined view concept. The following view types match the different approaches: learning content (structure) view, linking and individualizing view, domain context, and access context (cf. Fig. 8.30):

- *Learning content (structure) view.* This view contains didactically enriched learning content typically authored by educators. It serves to present the basic structure of learning resources and communication features. Regarding the given navigation designs, this view includes parts of the hierarchic navigation design. To support authoring of learning resources in this view, didactic topic map templates are useful (Schmiech 2006). Such templates aim to ensure consistent authoring, and finally consistent navigation. Furthermore, didactic topic map templates take into consideration various didactic attempts and singularities of knowledge.

- *Linking and individualizing view.* The aim of this view is to allow users embed arbitrary content in their individual learning process or in collaborative learning processes, and thus supporting knowledge transfer. Within this view (individual), semantic relationships between arbitrary content elements are represented, such as relationships between learning content and communication items, learning content and domain concepts, learning content/domain concepts and additional information in the web. Nevertheless, content elements (such as block, communication item, domain concept) provide the focus and serve as anchor to represent associated information. Further aspects of individualization such as annotations, metadata, or comments are also represented within this view. Thus, linking and individualizing views allow recording the knowledge construction process of learners (Fürlinger et al. 2004). Moreover, through allowing relationships between arbitrary content elements, new navigation paths can be offered in contrast to hierarchy-driven navigation paths. Since linking and individualizing views record the knowledge construction process of learners, for learning in teams, sharing and merging facilities for views are necessary to support collaboration among learners. Topic Maps provide an integrative concept to that respect. For efficient migration, Published Subject Identifiers (PSI) are recommended (Sasha Rudan and Rudan 2008).
- *Domain context view.* Within this view, concept of a given knowledge domain and respective associations are represented. Additionally, this view includes domain-overlapping relationships. Besides concepts and associations, relationships between concepts and information resources are depicted within the domain context. Information resources can either be arbitrary content elements of the learning resources or other information resources, such as external web pages. In order to allow individualizing the description of a given domain, individual views can also be represented upon domain contexts.
- The *access context view.* This supports adapting navigation and presentation of content according to different user preferences, devices, or learning situations. It allows adaptive navigation experience for learners, for example, by retrieving content in different levels of detail (e.g., bullet points—LoD1, text—LoD2, additional information—LoD3) and different modalities (e.g., text, audio, video).

The integration of the aforementioned views provides a holistic perspective on learning content embedded in individual, didactic, communication, and domain context. Considering navigation in such a multifaceted environment, content elements provide a focal point of learning processes. Content elements represent anchors for switching between different views (e.g., domain context, learning resources, and content structure view) or for combining different views.

Finally, reconsidering Topic Maps for the representation of the given views, it is necessary to distinguish the representation of structure (topics+associations) and the representation of content (occurrences). Structure focuses on navigation and supports retrieval of content, while occurrences represent the link to information resources (content). Different statement types support filtering of navigation paths (cf. association types) as well as content types (cf. occurrence types). For instance, occurrence types allow representing various modalities (e.g., audio, video) for a topic, and hereby selecting content according to the desired modality.

Annotating learning content (using hierarchic navigation) with a concrete domain concept allows switching between hierarchic navigation and concept map-based navigation. Besides switching between different navigation designs, the topic map representation approach allows (cf. Fig. 8.31):

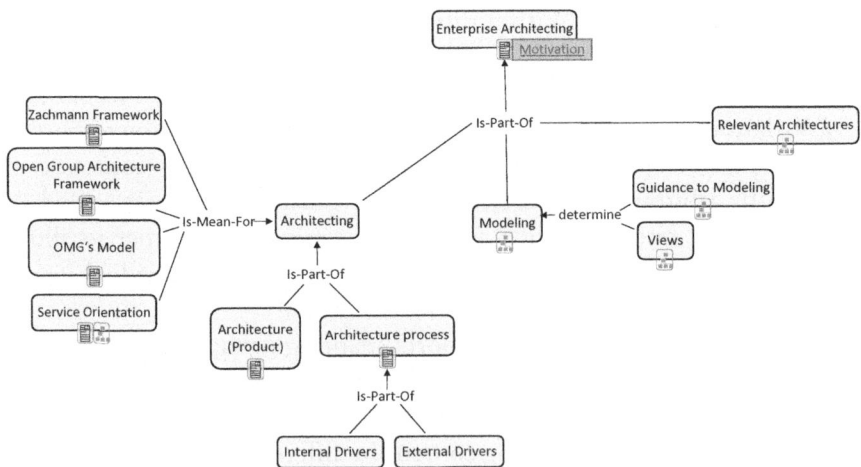

Fig. 8.31 Linking hierarchical and associative navigation design

- Flexible embodiment of didactic information into the navigation design
- Domain-specific adaptation and navigation
- Reusing of content elements in different contexts (e.g., traditional tree view, concept map)
- Filtering content according to didactic type, modality, and granularity
- Filtering navigation paths (associations) within the concept map navigation
- Individualizing of learning resources, for example, linking blocks or concepts with communication items in order to represent context-sensitive discussions

Such an implementation enables a highly flexible learning support system, as it can be adapted to user preferences and navigation styles promoting individual learning experiences. Learners who have been using the associative navigation design mentioned that it helped them to get an overview regarding the content of the lecture and to identify relationships between content elements. However, they indicated to have used the concept map in addition to the provided text book for the lecture, and not as primary source when learning. The content types displayed in the associative navigation have been experienced to support learner navigation. The depicted relationships between concepts as part of associative navigation have been intelligible to most of the learners.

In this way, the empirical findings confirmed some expected benefits, and affirmed that both navigation designs used by learners complement each other (Neubauer et al. 2011). While associative navigation design seems to be used by learners primarily to get an overview of a domain and to recapture associations between the domain-specific concepts and content, hierarchic (tree) navigation seems to be preferred by "top-down learners," working with content primarily in a linear way.

8.3.4 Alignment in User-/Usage-Oriented Design Spaces

From the findings elaborated earlier, in particular for semantically enriched navigation design, various design dimensions to provide meaning of learning content have become evident—see also Fig. 8.32:

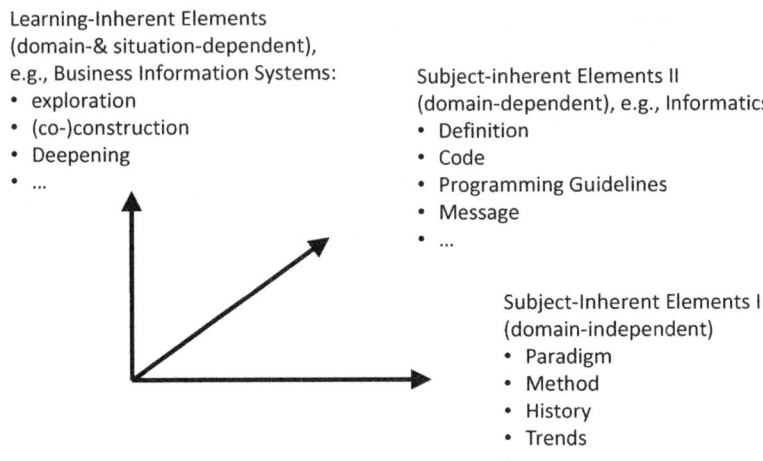

Fig. 8.32 Categories of design elements

- *Subject-inherent and domain-independent elements*: They can be found in most of the educational subjects, as they constitute disciplines. Among these elements are origin, concept, and paradigm. In BPM, typical origins are organizational development or software engineering, concepts are modeling elements to represent business processes, and paradigms are communication-orientation and functional specification.
- *Subject-inherent and domain-dependent elements*: These elements are typical for certain domains, and allow differentiating domains, such as software project management and BPM. In BPM, typical instances for domain-dependent elements are business process models, analysis methods, and lifecycle. They concern fundamental elements to understand the field.
- *Learning-inherent elements* that are domain- and situation-dependent: This category refers to elements directly influencing the style of presentation, location, and reception of resources as well as learner behavior (Farmer and Hughes 2005). For instance, in progressive education, self-regulated learning, exploration, and informed problem solving are of eminent importance. The domain-dependence is given by looking whether the technical domain, such as BPM, allows such an approach.

The same holds for the situation, as the format of lectures influences learner behavior. A course providing project assignments is likely to allow self-organized problem solving in contrast to focused method training.

When it comes to implementing didactic settings, the underlying services are of importance (Hung 2012). More particular, a variety of tools supports digital learning today and are part of respective environments. Besides traditional content management, Web 2.0 technologies, such as blogs, wikis, chat rooms, and video streaming, are widely used (ibid.). Few of them aim to create an integrated learning support system (Alario-Hoyos et al. 2013). Hence, a mapping from didactic requirements to services allows for traceability of the development process. Hereby, a middle design layer (see Fig. 8.35) as a focal point in terms of feature bundles turned out to be useful.

Once the underlying education scheme is considered to be a starting point for learning, design (Zardas 2008) features need to be derived from pedagogic elements in terms of technological functionality in the course of development. Concept maps also help to structure and guide this process. In Fig. 8.33, the top layer consisting of domain and didactic struc-

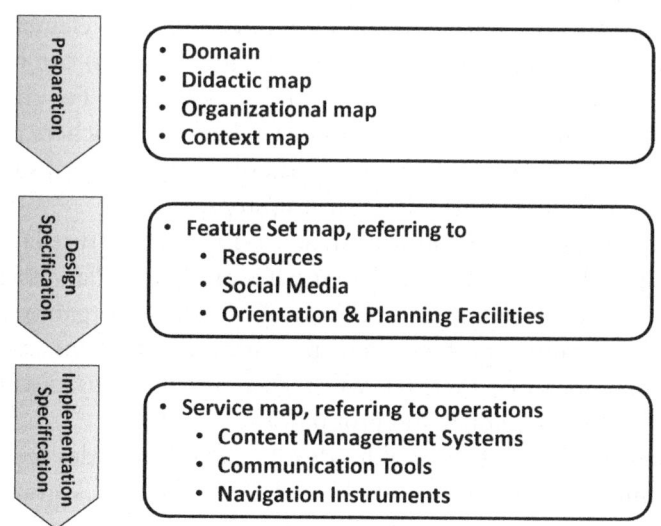

Fig. 8.33 A layered approach to a user-/usage-centered learning design space

tures is related to feature bundles located in the middle layer that allow identifying classes of systems for implementation and refining them in terms of their specific features or services (cf. www.archimate.org).

Figure 8.34 exemplifies the principle of this design mechanism based on input presented in the previous sections. For the sake of intelligibility, the link structure of the map is only sketched between the top and the middle layer. The middle layer exemplifies typical 'design cornerstones', such as a feature bundle for content management, integration social media into content management, and supportive transfer structures. Each set of features is detailed in terms of tools or tool sets on the bottom layer.

For instance, following the progressive education approach, the Dalton Plan, as introduced by Parkhurst (1922), has been implemented (Weichhart 2012). The Dalton Plan primarily uses assignments and feedback graphs in conjunction with bulletin boards and conferences. An implementation in a learning support system requires a prepared environment, as shown in the top-right of the figure. From the middle design space layer, Content & Communication and Transfer Structures are addressed in line with recent findings with respect to effective digital learning support processes (Dabbagh and Kitsantas 2012).

According to the concept mapping guidelines, each element of the upper layer (encoding the didactic and domain concepts) can be related to one or more elements of the upper and middle layer. For the Dalton Plan implementation, a link needs to be set between 'teachers plan with students' (upper layer) and 'transfer structures', as the Dalton plan is based on a work plan structuring learning steps.

Using the Dalton Plan editor (systems and specific feature layer in the design map of Fig. 8.34), the different parts of Dalton Plan assignments and their relationships can be specified. Assignments organize learning processes by detailing problems and providing descriptions, namely, in terms of documentation (Written Work) and cognitive activities (Memory Work) involving individual and group tasks.

The Dalton Plan facility enables deadlines and provides feedback to learner achievements (see Figs. 8.35 and 8.36). Feedback graphs allow transparent progress reports. Meetings and the so-called conferences are

Case Studies 385

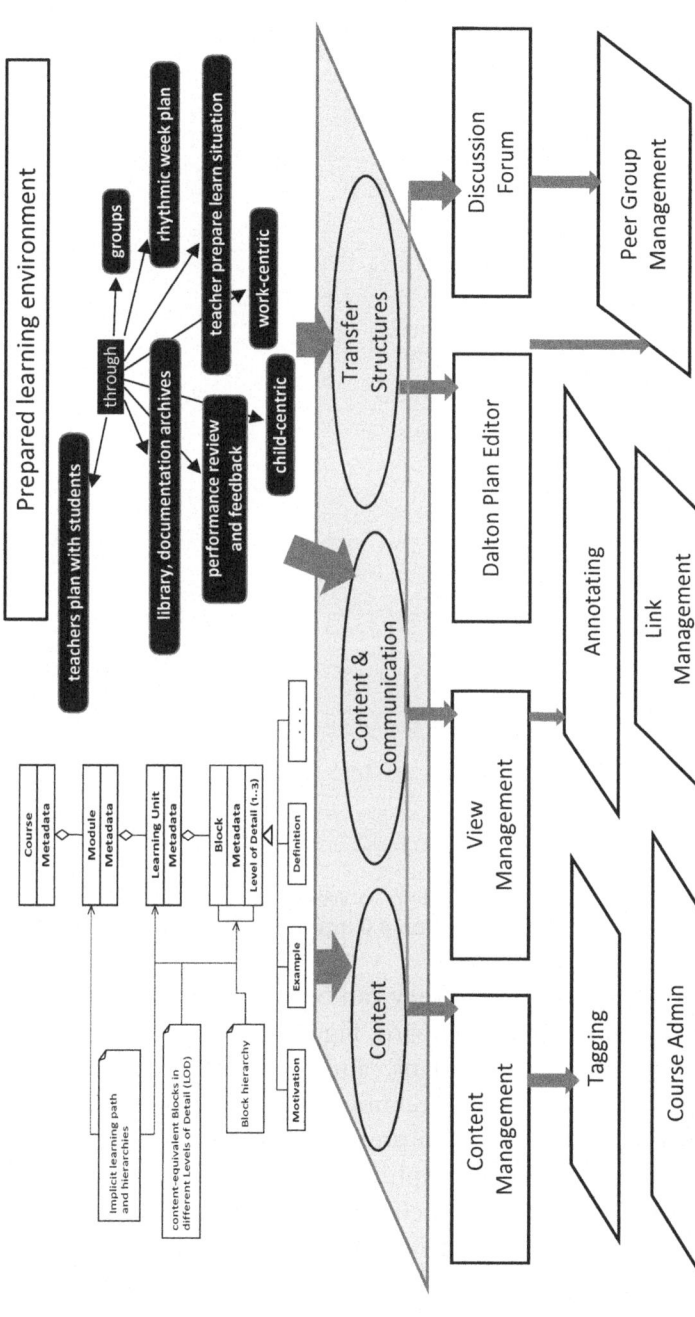

Fig. 8.34 Schematic instance of design map according to Weichhart and Stary (2014)

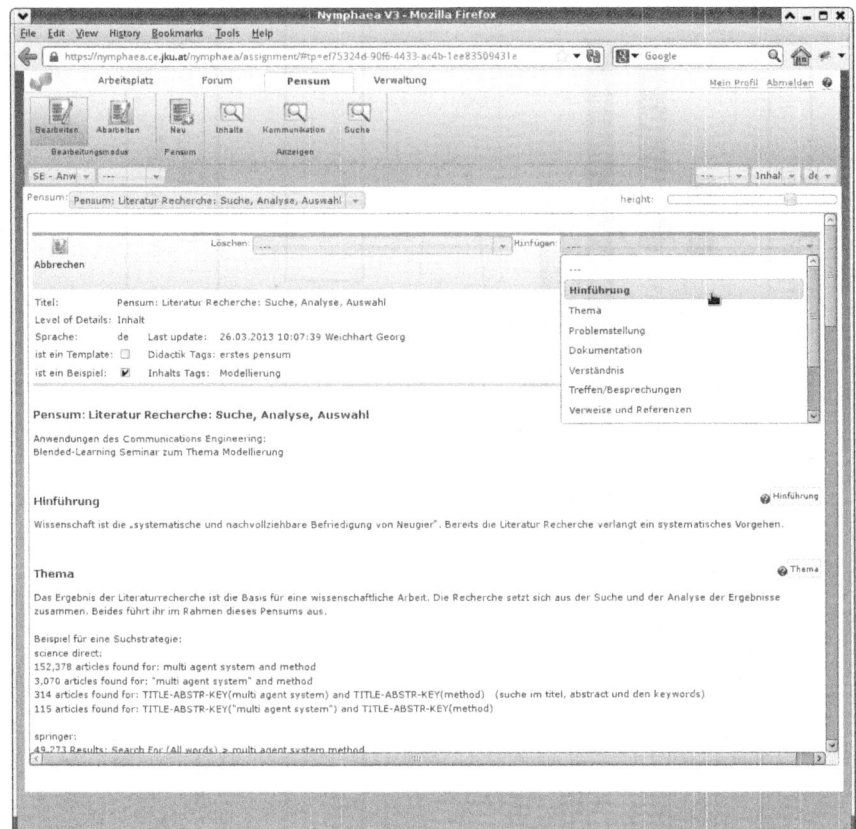

Fig. 8.35 Dalton Plan editor according to Weichhart and Stary (2014) (released under a Creative Commons Attribution 4.0 International License (CC BY 4.0))

also part of the Dalton Plan. They can be scheduled on a regular basis or announced on the bulletin board. Figure 8.35 shows the assignment editor for specifying work plans, and feedback graphs (Fig. 8.36) implemented using a web 2.0 technology stack (Tiropanis et al. 2012) in the Learning Support System presented earlier. Each learner can be (re)presented by a feedback graph once working on a specific assignment. For each assignment, all currently involved learners can be displayed according to their state of affairs, both in terms of self- and educator assessment.

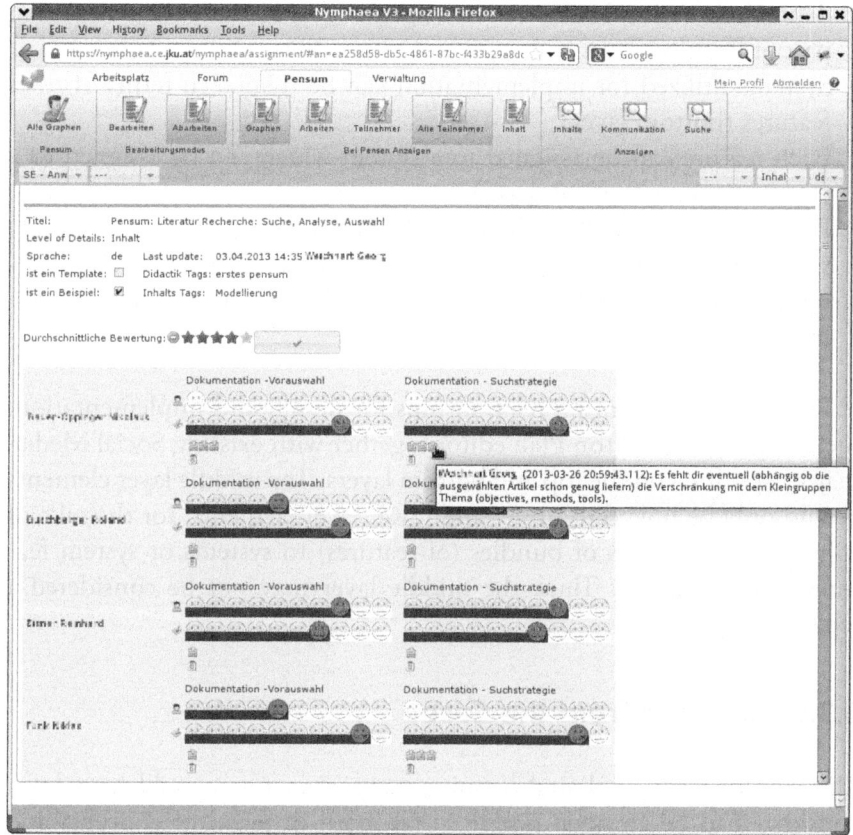

Fig. 8.36 Feedback graphs according to Weichhart and Stary (2014) (released under a Creative Commons Attribution 4.0 International License (CC BY 4.0))

- In general, the introduced design space approach for user-/usage-centered learning designs bridges the gap between educational requirements and technical system features by a middle layer that serves top-down and bottom-up specifications.
- Educational inputs can be refined to requirements, in terms of domain, didactic, or situational structures (top layer).
- For each of these maps from the top layer, one or more points of reference in terms of bundles (of features) in the middle layer can be

defined, for example, content management for didactic elements being part of learning units.
- Systems utilized for implementation can be refined in terms of their features (bottom layer).
- Each feature can be assigned to a system which can be assigned to a class of systems (bottom layer).
- Each set of features (middle layer) is implemented through (a set of) systems (bottom layer), and vice versa, and each class of systems, system, or feature can be assigned to a bundle of features on the middle layer.

Finally, all neighboring relationships for design and implementation, such as using the Dalton Plan editor together with existing Social Media, may be specified on the top and bottom layers. The middle layer elements should only be linked to upper and lower layer elements, for the sake of coherent assignments of bundles (of features) to systems or system features (bottom layer). Thus, the middle layer may not be considered a separate map.

8.3.5 Insights from the Case

The development of digital learning support systems could have been considered to be an open puzzle so far, both in terms of concepts and instruments, in particular when putting progressive didactic concepts to practice. In this case, we utilized concept maps as overarching scheme and representational glue to support articulation and alignment, once relevant items have been identified. In this way, we could capture educational intentions, meaningful content, and learning process specifications.

When intertwining emotional, social, cognitive, and technological issues, means of orientation and documentation become essential, not only for those who are carrier of these processes, but also for those who initiate and facilitate these processes, namely, educators, content providers, and developers. A living design memory has to keep information in a topic-specific and context-sensitive way, in order to organize knowledge for sharing digital learning support expertise and for providing learning process support.

By Articulation Work on educator knowledge and education-relevant mappings for learner-centered design, we could up a work-relevant alignment and design spaces. It allowed proceeding with content production and navigation design based on intentional and meaningful design elements. Metadata are key to implementing design maps with semantic technologies which can be captured in a layered design space according to generic feature classes. Educational metadata stemming from domain didactics can be effectively used for content and navigation structuring. Concept map-based navigation design, complementary to nested tree structures, can be created using topic maps and support learners along individualized learning processes. Hence, the primacy of didactic design together with dynamic adaptation forms the base for user- and usage-centered interaction, and thus work design. The underlying technologies, such as intelligent content management and social media, need to become part of an integrated system, in order to provide effective stakeholder support.

8.4 Subject-Oriented Organizational Management

In this chapter, we exemplify how subject-orientated digitization works given the communication-oriented perspective on work (knowledge). The presented Me2Me2You technique is based on capturing business operations in terms of pragmatic qualities including role awareness, task accomplishment, and interaction with other stakeholder roles, as reported in a study by Christian Stary (2018). The starting point includes meaningful entities for the articulating stakeholder with respect to each of these aspects. Based on experiential data, a reference procedure can be proposed. It could help articulating behavior in critical situations and for regular or routine tasks.

The frame of reference enabling process participants to gradually develop a comprehensive model of their business process in a subject-

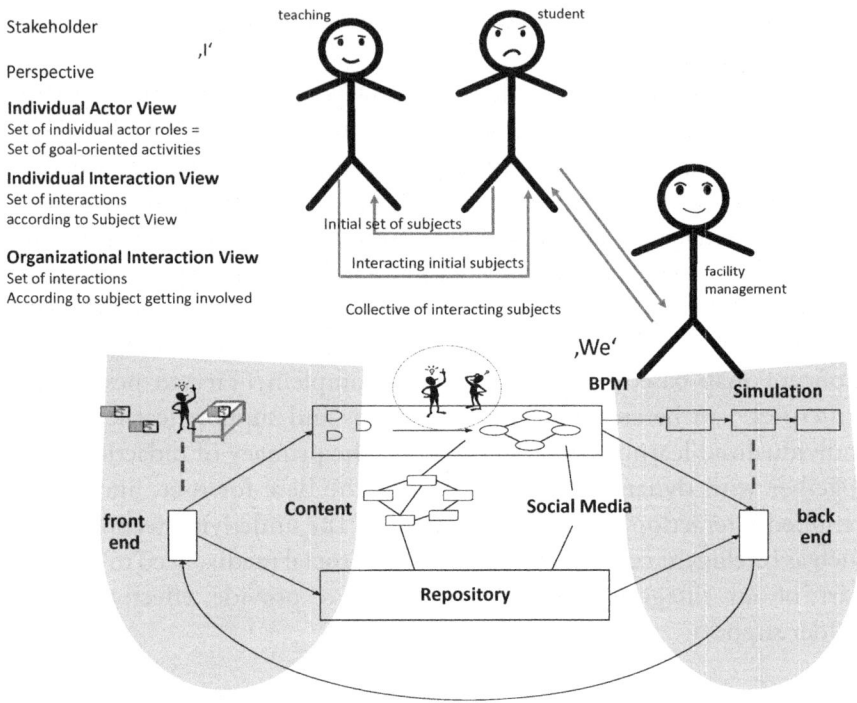

Fig. 8.37 Embodying the organizational management case to the digital work design framework

oriented way can be mapped to the overall framework, as shown in Fig. 8.37. It reveals that the articulation and execution parts are affected.

Since the case explores novel ways of designing organizations, and thus digital work, we provide some relevant background information for this study.

8.4.1 Organizational Management

In organizational management, meaningful behavior has already been recognized as a highly individual construct. As Shchedrovitsky (in Khristenko et al. 2014) in his analysis on the engineering nature of organization, leadership, and management of work has pointed out, it requires understanding the semantics of a situation (Shchedrovitsky 2014, p. 42ff):

What is 'meaning'? It is a tricky question. Really, there isn't any meaning. Meaning is a phantom. But here's the trick. I can say a sentence, like 'The clock has fallen off the wall' in two situations with two completely different meanings: 'The *clock* fell' and 'The clock *fell*.' The change of accent corresponds to two fundamentally different situations. Imagine this: when I am lecturing, I have got used to the fact that there is a clock here on the wall. At some point, I turn, I see an empty space, and someone in the audience says, 'The clock fell off the wall.' They might simply have said 'it fell' because, in this instance, the word 'clock' carries no new information. I look at the clock, I have got used to it and everyone in the lecture hall has got used to it. We look at that place and someone says 'it fell off the wall', and that phrase provides new information.

But now imagine a different situation. I am giving a lecture and all of a sudden there is a crash behind me. What has made it? I am told, 'The clock fell off the wall.' The situation is entirely different because what is new in this instance is the message about the clock. I heard something fall—that is a given—and I am told that it is the clock that fell. We pin this down in terms of 'subject' and 'predicate' in their functional relationships: in the first case, the clock is the subject, and in the second case the subject is the falling. We carry out syntactical analysis and highlight a difference between the two oppositions 'noun–adjective' and 'subject–predicate'. The distinction between subject and predicate is this: when we have a text, the subject is what we are talking about and the predicate is the characteristic that we ascribe to it. So when I hear any text, I understand it through an analysis: I work out what is the subject. Why do I work it out? I relate it to the situation.

The subject might be an action. In an algorithm I always treat actions as items, to which characteristics are ascribed. So I am always doing a particular sort of work: I parse the text syntactically, identify its syntactical organization, its predicate structure, and map this onto the situation. This is a process of scanning, of relating the text to the situation. When you understand my text now, you carry out this complex relational work. You are constantly identifying what is being talked about and what I am saying about it. This is the standard work that goes on automatically, you understand what is being said to the extent that you can find these objects and relate the text to them.

These paragraphs reveal several insights that are not only relevant when one perceives a specific situation at hand, but also when aiming to represent or modeling it. Providing information, that is, giving meaning to perceived data, needs to be considered a context-dependent process itself. Simply by focusing on a specific part of a sentence, like shown earlier for 'The clock has fallen off the wall', different meanings can be conveyed, and thus different situations and adjacent work practices could be revealed. Shchedrovitsky considers ascribing meaning to a situation as relational work. It requires an active entity identifying elements of concern (perceived) information can be assigned to.

After rephrasing subject-oriented representations, a model of eliciting and structuring perceptual knowledge of stakeholders in a certain situation is proposed based on exemplifying stakeholder articulations. In these samples, several persons were asked to describe how they make meaning when 'The clock has fallen off the wall' in a classroom situation. The articulation model contains several perspectives helping to structure individually perceived situational information for further operation. Each perspective can be enriched with another one, leading to a cascade of perspectives, finally allowing to create subject-oriented process models.

8.4.2 Subjects As Carrier of Work Behavior

We follow the aforementioned example. When learning facilitators in a classroom are asked to describe how they react when 'The clock has fallen off the wall', they could identify several carriers of behavior, that is, subjects. Figure 8.38 shows a set of possible subjects, Clock, Facility Management, and Clock Producer that could be considered of relevance

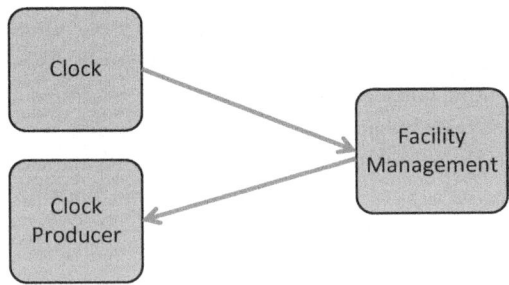

Fig. 8.38 Sample universe of discourse for 'The clock has fallen off the wall'

for 'The clock has fallen off the wall.' The directed links denote the interaction pattern for message exchange.

According to subject-oriented modeling (Fleischmann et al. 2012), any setting or situation can be structured as a set of individual actors or behavior elements. They can be humans or technological artifacts and are encoded in subject diagrams according to their communication with each other. When subject needs to communicate directly with another subject, as required in case of maintenance, a subject-behavior diagram also encodes this link. It is executed during runtime (after technical implementation).

On the modeling layer, the corresponding activity is a request sent to another subject. The sending subject waits until it receives an answer. Then, it processes the received answer—see Fig. 8.39 for that pattern. The rectangles denote the messages which the subjects exchange.

Figure 8.40 shows a Subject Interaction Diagram (SID). SIDs provide a global view of a situation, comprising the subjects involved and the messages they exchange. The SID contains a maintenance support process in Fig. 8.40. It comprises several actors (subjects) involved in communication: Facility Management coordinating all maintenance activities, a Clock Producer taking care of providing a working clock, and the Clock providing scheduling support in classroom management. They exchange messages in case of operational problems, as shown along the links between the subjects (rectangles).

Subject Behavior Diagrams (SBDs) provide a local view of the process from the perspective of individual actors (subjects). They include sequences

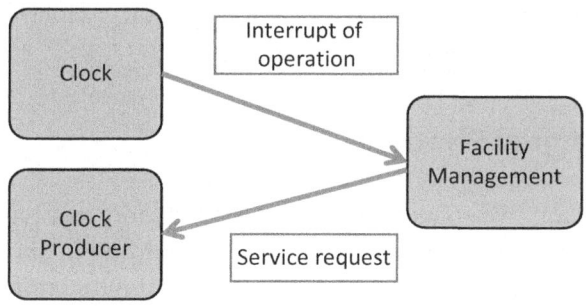

Fig. 8.39 Sample interaction pattern for 'The clock has fallen off the wall'

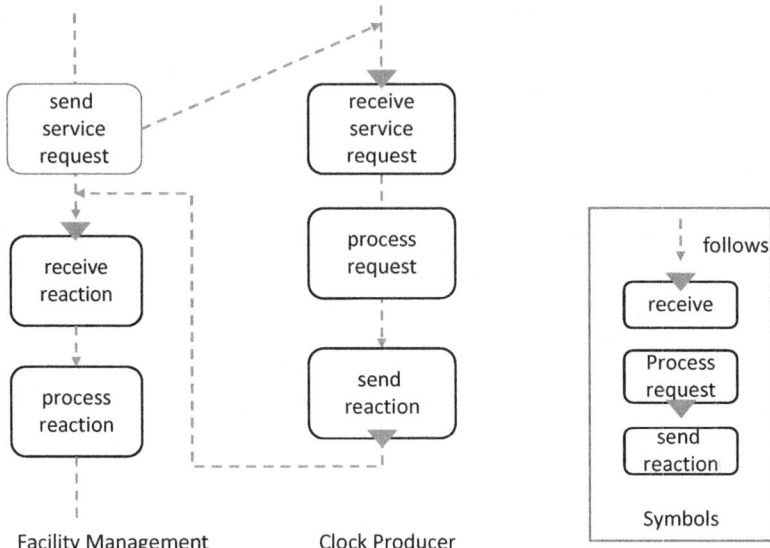

Fig. 8.40 Sample Behavior Synchronization of 2 SBDs

of states representing local actions and communicative actions including sending messages and receiving messages. Arrows represent state transitions, with labels indicating the outcome of the preceding state (see Fig. 8.40). The part shown in the figure represents a service request to the Clock Producer subject from the Facility Management subject.

Given these capabilities, subject-oriented representations can be utilized for articulation due to their (i) a simple communication protocol (using SIDs for an overview) and thus (ii) standardized behavior structures (enabled by send-receive pairs between SBDs), which (iii) scale in terms of complexity and scope.

8.4.3 Essential Principles

In the following, we introduce relevant articulation and representation principles from organizational management according to Shchedrovitsky (2014). We start with the identification of meaningful entities, proceed with interactions of identified entities, and complete the set of basic principles with the alignment of interactions recognizing systemic operations.

8.4.3.1 Identifying Meaningful Entities

When stakeholders perceive situations, they start with spotting relevant elements according to their current perspective:

> Now imagine the following device. I project a ray of light from my consciousness as I compare things—first, second, third thing—all the time extracting information and drawing it to myself. And there is a little paint brush with black paint attached to this ray and every time I send out the ray the brush leaves a mark. When I jump to something else the brush leaves a mark again; when I go back it makes another mark. In this way the brush leaves a kind of grid behind it. Then we look at the grid and we say that it is *meaning*. So, meaning is a particular structural representation—a sort of freezeframe—of the process of understanding. We can look at this another way, by asking a trick question: does movement have parts or not? I make a movement what parts can there be in it? And, generally, how can you stop it and capture it temporally? You cannot do any such thing because in order to obtain parts, you have to cut it up. But my movement isn't capable of being cut up!
>
> But see what we actually do. Here is a movement. For example, something falls. It leaves a trail. Now we begin to slice this trail into sections, we get parts of the trail and we transfer it to the movement.
>
> So, the movement obtains parts secondarily, by transfer onto it of the parts of its trail. Otherwise, we cannot work with movements in thought. In order to cut them up, transform them, or do something else with them, we have to stop them—to represent some 'frozen' part of the movement structurally. This is how we work with any process—whether of understanding, work or something else. We divide it into stages and phases, but in order to do this we have to find and register the traces (the trail) of this process. (Shchedrovitsky 2014, p. 43)

The 'trail' may range from realizing the trigger event for the clock's falling down to watching how the broken glass spreads over the floor in the classroom. Evaluating this trail allows to scope the entire scene in terms of all relevant elements involved, for example, the holder went off the wall, the clock fell down, and the clock fell apart when touching the floor. Hence, meaning could be action-triggered which, in turn, is relevant for the stakeholders in the room. Assuming that nobody got hurt through the event, for the students in the room it may be an event of low complexity, as they do not have to care about the time and are able to watch their steps when avoiding stepping on the clock's broken parts. For the learning facilitator, it is a major event, as he/she needs to take care about the time and the safety of the students.

As we can see, each stakeholder constructs meaning using some role-specific view. It may require immediate action or reaction to an event. The learning facilitator may take action through interrupting the process of teaching and switching to the role of caretaker of classroom safety, warning the student to be careful when leaving the classroom. From the facilitator's perspective, in a second step, the time problem needs to be addressed, assuming classes are structured along time slots. The facilitator needs to interact with somebody from the class or facility management to ensure correct timing, in case he/she relies on an external source of information with respect to time. Finally, the facility management needs to be addressed for taking care of all the damage. From a representational perspective, several entities are involved to make meaning out of a situation:

- The event—being an action itself (falling off the wall ending another operation, namely, the time ticking), or 'sliced', a set of small actions or events
- The role—student, learning facilitator, caretaker, facility management
- Actions and interactions, such as teaching and warning the students
- Concerned objects, the clock and the classroom

Each of these elements is constitutional to subject-oriented representations. Subjects denote roles and encapsulate behavior in terms of doing, sending, and receiving messages. Finally, the concerned objects are addressed in or passed through messages exchanged between subjects.

8.4.3.2 Conveying Meaning to Others

Situations trigger not only certain behavior, but also need to be documented and transferred to others, for example, to guide further behavior.

> We ought to speak in such a way that those listening cannot fail to understand. How they understand is a very complex question. We all understand

through the prism of our own peculiarities. And very often understanding is richer than what the speaker or writer of the text intended. The text always contains much that the speaker, the author of the text, did not personally put into it. This is due, first of all, to the fact that the author uses the tools of language. It is fair to say that language is always smarter than us, because all the experience of humankind is stored and accumulated in it. Language is the principal battery for storing experience. Second, the person who understands carries their own situation with them and always understands in the light of that situation, and often sees something more or something else in the text than its author. (Shchedrovitsky 2014, p. 44)

It could happen that communication is not documented through documentation, and very likely reduced to technical behavior. Subject orientation goes beyond that—it enforces to think in terms of communication and interaction of stakeholders or systems, as behavior specifications cannot exist without interaction. For instance, the teacher subject (i.e., a role) activates the caretaker, who, in turn, activates the facility management.

8.4.3.3 Aligning People

In order to run an organization, it may not be sufficient to develop a chain of interactions from a single perspective. For instance, administration, technically not involved into the clock falling off the wall, needs to be activated to ensure whether the classroom can be utilized by students for the next class.

> Everything starts with engineers who master the principles. They do not discover what was already in nature, but create a structure, something fundamentally new something that was not there in nature. They collect the elements and create—by assembling, joining together, 'bootstrapping'—completely new things not made by nature, and in doing this they are supported by creative—bold, 'crazy'—thought. All this is bound together in a unity, which does not follow the laws of nature, discovered by science: there was nothing to 'discover' until an engineer created something.
>
> The work of organizers, leaders and managers has the character of engineering work: it is structural and technical. Organizers, leaders or managers must always be one step ahead; they have to come up with something new.

> *Technical knowledge.* Suppose that you have to lead or manage people. You must determine their future actions, make a decision concerning their actions. As a result, you have a goal in advance, and you consider this person as a means or tool to achieve this goal. This is how things always are if you are an organizer, leader or manager. But people might resist, 'break loose', or act in some unforeseen way. You say one thing to them, and they—perhaps they are creative individuals—do something else. And you do not know whether you need to regulate their manner of execution or if you only need to set the goal. In short, each time you need to have knowledge about the individuals and their actions, but this knowledge must be oriented from the very outset to your goals. You have to achieve a certain goal through these people. And so, your knowledge answers the question: how can you achieve your goal through these people, and adjust their actions and your relations with them as a function of your goals? Such knowledge is what we call technical knowledge. (Shchedrovitsky 2014, p. 7f)

Shchedrovitsky, in the aforementioned statement, indicates that the matter of including or recognizing perspective can be a matter of goal setting, and in this way, scoping responsibilities.

> Technical knowledge gives us the answer to a question about an object, its mechanism and its action. However, this knowledge does not have a general nature: it is specifically geared to the achievement by us of our goals. It shows how adequate the object is for achieving these goals, and what we must do with it, how we must act on it in order to achieve our goals.
>
> Technical knowledge is very complex. It is actually much harder than scientific knowledge. And the work of an engineer is actually much more difficult than the work of a scientist. The work of a practical worker is even more complex. … Technical knowledge is not just a matter of goals, it is also about your means of influence. You are not interested in the object in itself, but in the achievement of the goal using your existing tools and methods of action. And you see this object in this context. … Necessary and sufficient information is needed. You need to have adequate knowledge. (Shchedrovitsky 2014, p. 8ff)

According to Shchedrovitsky (2014, p. 11), a stakeholder needs to pursue a specific goal and to know whom to involve in which way for further operation. As we will see in the following, the goal can help identifying intentional actor performing self-contained tasks according to the perception of a situation. In addition, the means of organizing work could be subject-oriented business process which needs to be probed by applying the model.

8.4.4 Structuring Articulation

In this section, the insights of Shchedrovitsky presented earlier are used for developing a cascaded model of perspectives. It is introduced in Subsect. 8.4.4.1 before a report on a field test is detailed. In this test, interviews were conducted with five stakeholders. Their perception of a situation when a clock has fallen off the wall in a classroom has been captured and structured. The interviews reveal some empirical evidence on its plausibility, also in terms of utilizing subject-oriented modes for representing operational work activities for goal-oriented actors (represented by subjects).

8.4.4.1 Cascading Perspectives

The model takes into account the structured findings revealing that perspectives on the situation trigger

- *technical entities* encapsulating behavior by focusing on activities needed to be performed to achieve an objective or implement an intention (usually referring to some task), and thereby establishing some functional role
- *communication acts* identifying which entity needs to be interacted with
- the *mutually adjustment of encapsulated behavior specifications*, as it plays a crucial role not only in acting as a collective in a specific situation but also in completing work processes or reaching intended goals

Accordingly, the model contains several perspectives helping to structure individually perceived situational information for further operation. Once started with an individual perspective, stakeholders can enrich its result with another one, and so on, thus leading to a cascade of perspectives. Since this cascade contains behavior encapsulations and interactions, it finally allows developers to create subject-oriented process models.

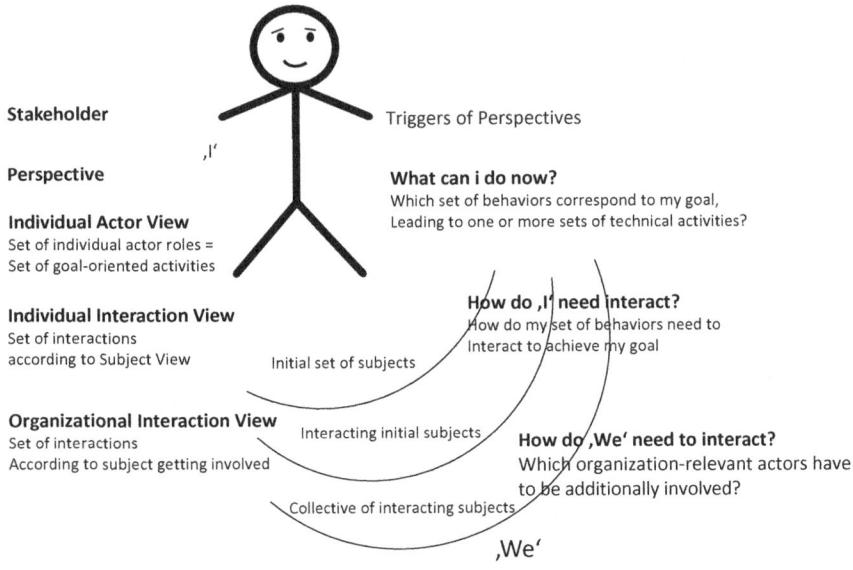

Fig. 8.41 Cascading perspectives

Figure 8.41 shows the model serving as frame of reference of building organizational capacity based on individually perceived situations. It instantiates Shchedrovitsky's approach in terms of structuring behavior in a goal-oriented way. The left part shows the cascade of perspectives that finally captures the evidence of a specific stakeholder when perceiving and reflecting on a situation:

- *Perspective 1—Individual Actor View*: This perspective captures a set of individual roles in which this stakeholder can act and think about in a specific situation. For instance, assuming the clock has fallen off the wall in a classroom having a teacher and students, the teaching role of the teacher addresses all duties related to classroom teaching, whereas the safety-responsibility role of the teacher concerns the physical safety of students in the classroom. Since humans are intentional beings, we can assume that each stakeholder has at least one role or objective to (inter)act that constitutes an actor view. This role or a set of roles corresponds to the individual (task) profile of a person. Each role refers to a specific behavior that has a driver, namely, an intention. For instance, the driver of the teaching role is increasing the level of competence of students, whereas the driver of the safety-responsibility role is ensuring the safety of all the students in the classroom. Since each role has an intention, each stakeholder can pursue a set of specific goals in a situation, depending on the set of roles.

- *Perspective 2—Individual Interaction View*: This perspective looks on the same situation but builds upon the results from taking perspective 1 and the identified roles. It keeps the considered role/objective/intention at the center of interest, but additionally captures a set of individual interactions based on that previously defined intentional behavior set(s). Hence, the set of interactions also depends on the roles in which this stakeholder can act and think about in a specific situation. For instance, we assume the stakeholder identifies the role of the teacher (addressing all duties related to classroom teaching) and the safety-responsibility role (ensuring the physical safety of students in the classroom). Then, from this perspective, the stakeholder needs to think about interactions between these two roles. In case the teacher interrupts the class due to the clock's falling off the wall, the safety-responsibility role takes over to ensure the safety of the students in the room. It may lead to ending the class, if the teacher cannot guarantee the safety of the students in this situation, as perceived by this stakeholder. In case the safety-responsibility role does identify safety risks, the safety-responsibility role informs the teaching role to continue teaching. In each case, the stakeholder can provide and specify a set of interactions, for sending and receiving information on a certain topic, involving relevant objects, such as safety measures.
- *Perspective 3—Organizational Interaction View*: This perspective analogously builds upon existing results, this time from taking the previously described perspectives 1 and 2. They already include roles and interactions, but both from an individual perspective. This perspective captures a set of roles this stakeholder perceives to be relevant for a specific situation in addition to the ones he/she can act him/herself, for example, taking a community or network perspective. It concerns a set of roles the stakeholder having perspective 1 and 2 cannot take or has no privilege to take. For instance, assuming the clock has fallen off the wall in a classroom with a teacher and students, and has been damag-

ing some interior, neither the teaching role nor the safety-responsible role is sufficient to continue with giving a lecture in this classroom, like from perspective 1, another individual actor view is driven by an intention. In the sample case, the goal could be to keep the classes running that are assigned to this room. Then, the interior needs to be restored, which brings in facility management. Its specific behavior needs to be coupled to the safety-responsible role, in order to accomplish the respective tasks. Finally, there may be several perspectives related to the 'We', for example, evolving from an internal community of practice to formal department, networks, regions, and global connections.

Since each perspective builds upon a previous one, a cascade of perspectives evolves in the course of specifying work- and process-relevant information. The middle part of Fig. 8.41 reveals the evolving complexity according to refined and networked behavior specifications. The generation of actors and their interaction relations are based on a set of questions that trigger the definition of subjects and their interactions.

Initial set of subjects: The Individual Actor View leads to a set of intentional actor roles that allow stakeholders to perform goal-oriented activities. The stakeholder at hand identifies the initial set of behavior abstractions (subjects) by dealing with the question 'What can I do now?' This question targets those behavior abstractions that a stakeholder can name, once a goal to be achieved in this situation becomes evident. For instance, in case the clock falls off the wall of the classroom, the ultimate goal of a teacher is to ensure the students' safety before proceeding with the lecture. In order to achieve that goal, the stakeholder can perform a set of technical activities.

Interacting initial subjects: The Individual Interaction View leads to a set of intentional actor roles that synchronize their behavior. The stakeholder at hand identifies all those interactions between the initial set of behavior abstractions (subjects) by dealing with the question 'How do 'I' interact?' when having identified more than one role for handlings a specific situation. For instance, in case the clock falls off the wall of the classroom, the safety-responsible role interrupts the teacher to ensure the students' safety before signaling him/her to proceed with the lecture.

Hence, the interactions are defined in order to achieve the stakeholder goal determined upfront.

Collective of interacting subjects: The Organizational Interaction View leads to a set of intentional actor roles and synchronization of their behavior beyond the stakeholder at hand. This time, he/she needs to answer the question 'How do "We" need to interact?' when embedding further actor roles for handling a specific situation. For instance, in case the clock falls off the wall of the classroom, the safety-responsible role informs facility management, in case he/she cannot ensure the students' safety. Every interaction with facility management needs to be defined in order to achieve the upfront determined stakeholder goal.

Figure 8.42 exemplifies the cascaded perspective. In this case, the stakeholder has identified 'teaching' and 'safety responsible' as role representatives for perspectives 1 and 2 which need to interact sensitive to the safety of the students. For the repair of the clock and classroom restoring, this stakeholder activates facility management through respective interactions.

The 'We' perspective can be extended to bring in additional stakeholders, such as authorities managing school infrastructures, that are con-

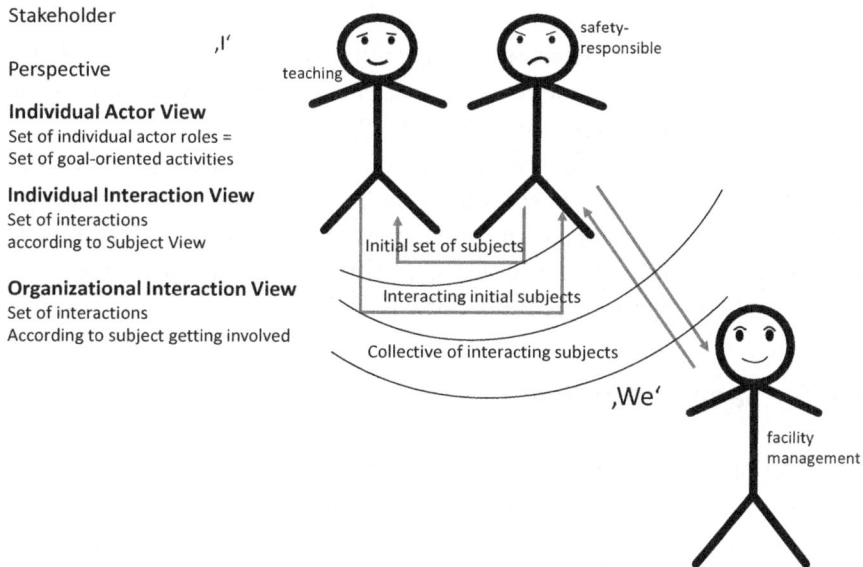

Fig. 8.42 Sample diagrammatic representation

tacted in case needed, for example, by facility management to improve the interior. Hence, the number of cascaded perspectives depends on the intention and the goal of the stakeholder, and results in a systemic view. On the one hand, the schema allows focusing on a perceived part of a situation, while on the other hand extending perspectives limiting contextual or systemic thinking by enabling interaction links to actor roles valid from other perspectives.

Both elements are essential, as they allow handling complex situations or events without reducing the complexity itself, but rather offering a multipartite structure. This structure facilitates handling complexity, namely

- by starting with familiar, since ego-centric behavior encapsulations (roles), and then
- stepwise enriching this set of roles by
- sets of interactions between ego-centric behavior encapsulations
- including non-familiar behavior encapsulations (roles), and
- coupling them through sets of interaction to all other behavior encapsulations

Hence, without predetermining the number of perspectives and the number of modeling elements (behavior encapsulations, interactions), a stakeholder is encouraged to express his/her perception of a situation based on interacting behavior elements. These elements represent subjects allowing stakeholders to detail pragmatic information in terms of role-specific (internal) behavior. The latter is represented in SBDs. Given the interaction between the subjects, a SID, and thus a stakeholder, can create a coherent pragmatic model of a situation.

8.4.5 Sample Applications

This section contains a report on several field tests. They have been performed to validate the approach. The model has been probed with five persons, aged between 39 and 67, three of them females, three of them instructors or teachers, the others a service provider or consultant, but both with teaching experience. Three of the persons had leadership and organizational management experience. The guide aims to reveal whether the perspectives can be cascaded as proposed by the scheme presented in the previous subsection. The interview guide contains the following items:

Consider a setting in a classroom and you are teaching a couple of students. Suddenly, you recognize the clock has fallen off the wall.

- What is your first concern?
- Which role(s) can you identify when you consider yourself acting in this situation?
- What is your (set of) intention(s) allowing to encapsulate your behavior by the time of the event?

  – What does that mean in terms of interaction and communication?

Briefly indicate direction and exchange of information or goods for each of the identified roles representing intentional activities.

- What are your further concerns?

  – Which role(s) can you take by yourself in addition to the previously identified ones?
  – What does the inclusion of these role(s) mean in terms of interactions and communication?

Briefly indicate direction and exchange of information or good for each of the additional identified roles.

- Who else do you think should you also involve in the situation and address due to the event?

  – Which further role(s) do you consider relevant to meet your objectives in that situation and should become part of handling the event?
  – What does the inclusion of these role(s) mean in terms of interactions and communication with your (existing) ones?

Briefly indicate direction and exchange of information or good for each of the external roles.

The interviews lasted about 15 minutes each. They included laddering, in case some context appeared to be relevant for fully grasping some of the answers. For instance, the interview with a teacher, who also has extensive experience in managing schools, has led to the following insights—the collected information is structured according to the items of the interview guide:

Considering the situation where the clock has fallen off the wall.

First concern of person A:

- *Role(s)*: Role being responsible for safety—since the clock has fallen off the wall, I need to interrupt teaching and deal with the new situation immediately.
- *Interaction and communication*: Look at students whether somebody is in danger. In case there is danger, I need to help.

Further concerns of person A: Ego-centric role(s): none

Further concerns external to own role of person A:

- *Role(s)*: Role being responsible for facility management I would need to inform about the event and whether additional action needs to be taken.
- *Interaction and communication*: Look at the damage and student situation—inform facility management accordingly, for example, to address cleaning staff, to order a new clock, to adjust schedule.

The acquired knowledge can be conveyed, as depicted in Fig. 8.43. Person A has taken the three perspectives as guided by the interview items and intended by the scheme.

Figure 8.43 also shows how we could enrich the cascaded representation to specify role behavior in terms of subject-oriented models. The short description person A has provided indicates a set of subjects—teaching,

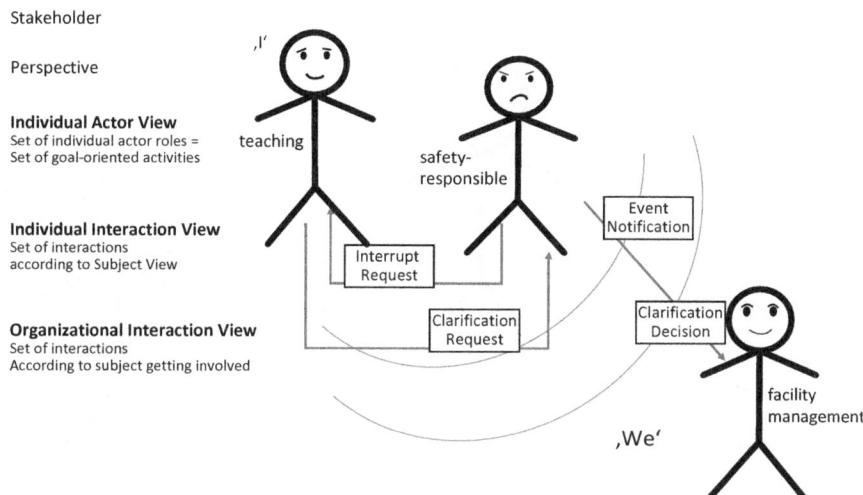

Fig. 8.43 Sample of elicited knowledge and sample of subject-oriented representation

safety-responsible, facility management—relevant for handling that situation. Person A was able to refine the interaction and communication relationships between the subjects and assign the remaining activities to one of the roles she had identified. The refinements allow creating SIDs, as indicated in Fig. 8.43, by the message exchanged between the actors. The assignments allow generating SBDs and capturing sequences of activities.

In contrast to person A, person B, being a consultant, is teaching only occasionally. He identified a single actor for handling the situation. When being asked for the initial concern, it turns out he manages the situation by delegation—a student will be assigned the task to handle the unforeseen event. Person B perceives the situation to be responsible for teaching exclusively, which excludes any other responsible action in case of disturbance. Figure 8.44 shows the cascade involving 'teaching' and 'student' and the interaction representing the task delegation.

Person C considers involving responsible actors to be essential. We could term that approach another form of 'management-by-delegation', but have to acknowledge that not only a student will be involved but rather a decision-making process is instantiated by activating the head of school. Figure 8.45 shows the resulting SID, in which the subject "teach-

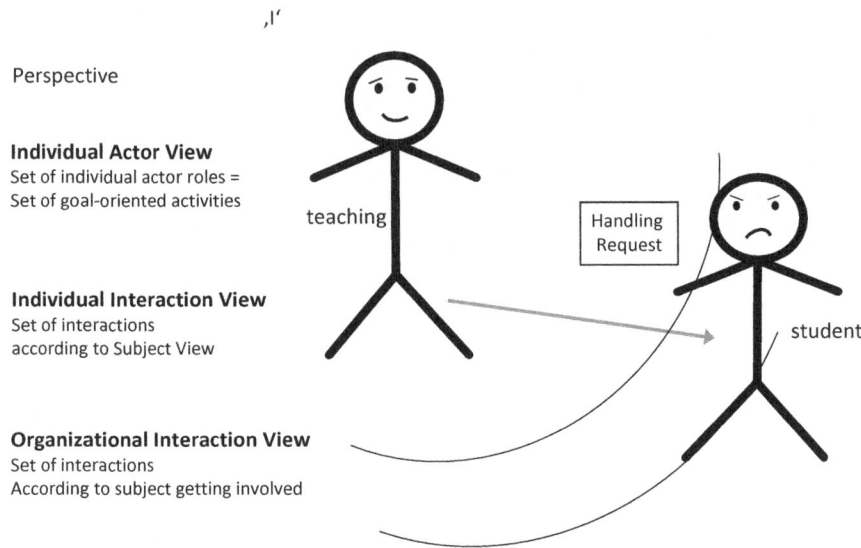

Fig. 8.44 Person B's 'management-by-delegation'

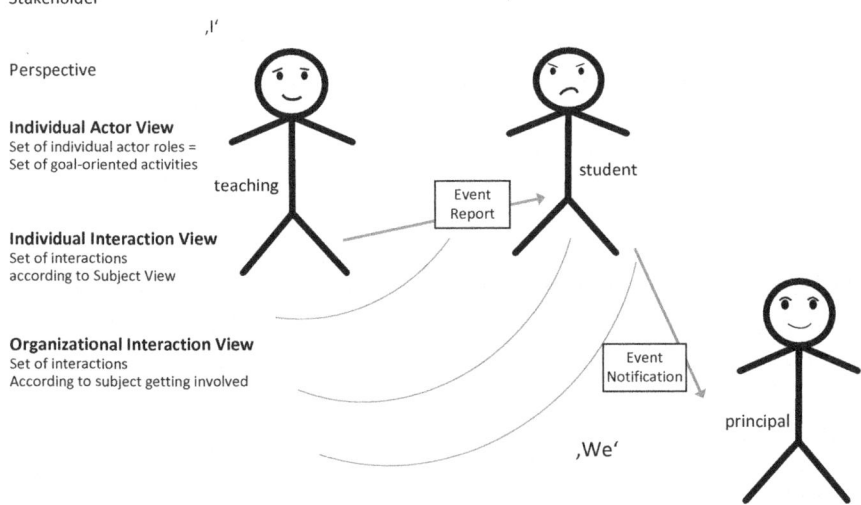

Fig. 8.45 Person C—getting responsible actors involved

ing" provides the event report, which becomes part of the event notification to the principal of the student.

These small examples indicate how situations or events can be captured by individual stakeholders, giving them the freedom to cascade as many perspectives they consider relevant according to their perception and knowledge. The last case could be valid for all persons not trained as school teachers who have to inform responsible actors about unforeseen events immediately. It could become part of a behavior guide of the organization for handling unforeseen events to be studied by external teachers.

8.4.6 Insights from the Case

This case explored an orthogonal concept based on cascaded actor behavior for capturing stakeholder pragmatic perceptions of situations. We started out with Shchedrovitsky's work on the engineering nature of managing today's enterprises, concluding that addressing the pragmatic qualities of business operations allows stakeholders articulating work knowledge. Cascading is based on technical entities identified by intentional objectives and interaction of identified entities. It starts with familiar, behavior encapsulations (roles), and proceeds with enriching this set of roles by sets of interactions between individual behavior encapsulations. The latter include non-familiar behavior encapsulations (roles), finally leading to complete business operations from a stakeholder perspective.

Stakeholders can be encouraged to express their perception of a situation based on interacting behavior elements. These elements represent subjects as known from subject orientation, allowing stakeholders to detail pragmatic information in terms of role-specific (internal) behavior. Given the interaction between the subjects, a stakeholder can create a coherent pragmatic model of a situation. The models are designed to probe representations for operation. For instance, once the Facility Management subject is instantiated, it has to be decided (i) whether a human or a digital device (organizational implementation), and (ii) which actual device, is assigned to the subject, acting as technical subject carrier (technical implementation). Typical subjects are devices and their process-specific services, including smart phones, tablets, laptops, healthcare devices, and so forth. Subjects can also be role carriers controlling or executing tasks. Both types of instantiations can be supported by subject-oriented runtime engines. For an overview, see Krenn et al. (2017). These engines provide services linked to some ICT infrastructure.

Once the runtime engine is tightly coupled to model representations, ad-hoc and domain-specific requirements can be met dynamically. The situation-sensitive formation of systems and their behavior architecture need to be validated before being executed without further transformation. Hence, stakeholders can adapt model representations and proceed to implementation according to their articulation needs.

References

Alario-Hoyos, Carlos, Miguel L. Bote-Lorenzo, Eduardo GóMez-SáNchez, Juan I. Asensio-PéRez, Guillermo Vega-Gorgojo, and Adolfo Ruiz-Calleja. 2013. GLUE!: An Architecture for the Integration of External Tools in Virtual Learning Environments. *Computers & Education* 60 (1): 122–137 (Elsevier).

Auinger, Andreas, Franz Auinger, C. Derndorfer, J. Hallewell, and Christian Stary. 2007. Content Production for E-Learning in Engineering. *International Journal of Emerging Technologies in Learning (iJET)* 2 (2): 1–8 (International Association of Online Engineering).

Bardram, J.E., and T.R. Hansen. 2010. Why the Plan Doesn't Hold: A Study of Situated Planning, Articulation and Coordination Work in a Surgical Ward. In *Proceedings of the 2010 ACM Conference on Computer Supported Cooperative Work*, 331–340. ACM.

Boland, Richard J., Jr., and Ramkrishnan V. Tenkasi. 1995. Perspective Making and Perspective Taking in Communities of Knowing. *Organization Science* 6 (4): 350–372 (Informs).

Canas, A.J., G. Hill, R. Carff, N. Suri, J. Lott, T. Eskridge, G. Gómez, M. Arroyo, and R. Carvajal. 2004. CmapTools: A Knowledge Modeling and Sharing Environment. In *Concept Maps: Theory, Methodology, Technology, Proceedings of the 1st International Conference on Concept Mapping*. Pamplona, Spain: Universidad Pública De Navarra.

Chen, Chih-Ming. 2009. Ontology-Based Concept Map for Planning a Personalised Learning Path. *British Journal of Educational Technology* 40 (6): 1028–1058 (Wiley Online Library).

Chen, Nian-Shing, Chun-Wang Wei, Hong-Jhe Chen, others. 2008. Mining E-Learning Domain Concept Map from Academic Articles. *Computers & Education* 50 (3): 1009–1021 (Elsevier).

Coffey, John W., Robert R. Hoffman, Alberto J. Cañas, and Kenneth M. Ford. 2002. A Concept Map-Based Knowledge Modeling Approach to Expert Knowledge Sharing. In *Proceedings of IKS*, 212–217.

Cohn, David, and Richard Hull. 2009. Business Artifacts: A Data-Centric Approach to Modeling Business Operations and Processes. *IEEE Data Engineering Bulletin* 32 (3): 3–9.

Collins, Allan, John Seely Brown, and Ann Holum. 1991. Cognitive Apprenticeship: Making Thinking Visible. *American Educator* 15 (3): 6–11.

Dabbagh, Nada, and Anastasia Kitsantas. 2012. Personal Learning Environments, Social Media, and Self-Regulated Learning: A Natural Formula for Connecting Formal and Informal Learning. *The Internet and Higher Education* 15 (1): 3–8 (Elsevier).

Derouin, Renée E., Barbara A. Fritzsche, and Eduardo Salas. 2005. E-Learning in Organizations. *Journal of Management* 31 (6): 920–940 (Thousand Oaks, CA: Sage Publications).

Dewey, John. 1928. *Progressive Education and the Science of Education*. Washington, DC: Progressive Education Association.

———. 2013. *The School and Society and the Child and the Curriculum*. University of Chicago Press.

Duckworth, Eleanor. 2006. *The Having of Wonderful Ideas and Other Essays on Teaching and Learning*. New York: Teachers College Press.

Eichelberger, Harald, Christian Laner, Wolf Dieter Kohlberg, Edith Stary, and C. Stary. 2008. *Reformpädagogik Goes eLearning: Neue Wege zur Selbstbestimmung von virtuellem Wissenstransfer und individualisiertem Wissenserwerb*. Munich: Oldenbourg Verlag.

Farmer, R.A., and B. Hughes. 2005. A Situated Learning Perspective on Learning Object Design. In *Fifth IEEE International Conference on Advanced Learning Technologies (ICALT'05)*, 72–74. IEEE.

Feldstein, Michael. 2014. The MOOC and the Genre Moment: MOOCs and Technology to Advance Learning and Learning Research (Ubiquity Symposium). *Ubiquity* 2014 (Sep.): 2 (ACM).

Ferscha, Alois, Clemens Holzmann, and Stefan Oppl. 2004. Team Awareness in Personalised Learning Environments. In *Proceedings of MLEARN 2004*, Bracciano, Italy, 67–72.

Fleischmann, Albert, Werner Schmidt, C. Stary, Stefan Obermeier, and Egon Börger. 2012. *Subject-Oriented Business Process Management*. New York: Springer.

Fürlinger, S., A. Auinger, and C. Stary. 2004. Interactive Annotations in Web-Based Learning Systems. In *Proceedings of IEEE International Conference on Advanced Learning Technologies, 2004*, 360–365. IEEE. https://doi.org/10.1109/ICALT.2004.1357437.

Garrison, D. Randy. 2011. *E-Learning in the 21st Century: A Framework for Research and Practice*. New York: Routledge.

Gilbert, Thomas F. 1962. Mathetics II: The Design of Teaching Exercises. *Journal of Mathetics* 1 (2): 7–56.

Godwin, Stephen, Patrick McAndrew, and Andreia Santos. 2008. Behind the Scenes with OpenLearn: The Challenges of Researching the Provision of Open Educational Resources. *Electronic Journal of E-Learning* 6 (2): 139–148.

Haslhofer, Bernhard, Robert Sanderson, Rainer Simon, and Herbert Van de Sompel. 2014. Open Annotations on Multimedia Web Resources. *Multimedia Tools and Applications* 70 (2): 847–867 (Springer).

Herrmann, Thomas, and K.U. Loser. 1999. Vagueness in Models of Socio-Technical Systems. *Behaviour & Information Technology* 18 (5): 313–323 (Taylor and Francis Ltd).

Hillier, Janet, and Linda M. Dunn-Jensen. 2013. Groups Meet… Teams Improve: Building Teams That Learn. *Journal of Management Education* 37 (5): 704–733 (Los Angeles, CA: Sage Publications).

Hughes, Gwyneth, and David Hay. 2001. Use of Concept Mapping to Integrate the Different Perspectives of Designers and Other Stakeholders in the Development of E-Learning Materials. *British Journal of Educational Technology* 32 (5): 557–569 (Wiley Online Library).

Hung, Jui-long. 2012. Trends of E-Learning Research from 2000 to 2008: Use of Text Mining and Bibliometrics. *British Journal of Educational Technology* 43 (1): 5–16 (Wiley Online Library).

Hwang, Gwo-Jen, Fan-Ray Kuo, Nian-Shing Chen, and Hsueh-Ju Ho. 2014. Effects of an Integrated Concept Mapping and Web-Based Problem-Solving Approach on Students' Learning Achievements, Perceptions and Cognitive Loads. *Computers & Education* 71: 77–86 (Elsevier).

Hwang, Gwo-Jen, Yen-Ru Shi, and Hui-Chun Chu. 2011. A Concept Map Approach to Developing Collaborative Mindtools for Context-Aware

Ubiquitous Learning. *British Journal of Educational Technology* 42 (5): 778–789 (Wiley Online Library).

Initiative, Advanced Distributed Learning. 2004. Sharable Content Object Reference Model (SCORM) 2004. http://www.adlnet.gov/pages/default.aspx.

Khristenko, Viktor Borisovich, Andrei G. Reus, and Aleksandr Prokof'evich Zinchenko. 2014. *Methodological School of Management*. A&C Black.

Kinchin, Ian M. 2000. Using Concept Maps to Reveal Understanding: A Two-Tier Analysis. *School Science Review* 81 (296): 41–46 (ERIC).

———. 2014. Concept Mapping as a Learning Tool in Higher Education: A Critical Analysis of Recent Reviews. *The Journal of Continuing Higher Education* 62 (1): 39–49 (Taylor & Francis).

Kinchin, I.M., and M. Alias. 2005. Exploiting Variations in Concept Map Morphology as a Lesson-Planning Tool for Trainee Teachers in Higher Education. *Journal of in-Service Education* 31 (3): 569–592 (Taylor & Francis).

Kinchin, Ian M., Lyndon B. Cabot, and David B. Hay. 2008. Using Concept Mapping to Locate the Tacit Dimension of Clinical Expertise: Towards a Theoretical Framework to Support Critical Reflection on Teaching. *Learning in Health and Social Care* 7 (2): 93–104 (Wiley Online Library).

Kolb, Darl G., and Deborah M. Shepherd. 1997. Concept Mapping Organizational Cultures. *Journal of Management Inquiry* 6 (4): 282–295 (Thousand Oaks, CA: Sage Publications).

Krenn, F., C. Stary, and D. Wachholder. 2017. Stakeholder-Centered Process Implementation: Assessing S-BPM Tool Support. In *Proceedings of the 9th Conference on Subject-Oriented Business Process Management*, 2. ACM.

Larrañaga, M., J.A. Elorriaga, and A. Arruarte. 2008. Semi-Automatic Generation of Didactic Resources from Existing Documents. In *International Conference on Intelligent Tutoring Systems*, 728–730. Berlin, Heidelberg: Springer.

Leake, A.V.D. 2006. Jump-Starting Concept Map Construction with Knowledge Extracted from Documents. In *Proceedings of the Second International Conference on Concept Mapping (CMC 2006)*.

Leclercq, Dieudonné, Université de Liège Laboratoire de pédagogie expérimentale, Jean Donnay, and Robert de Bal. 1977. *Construire Un Cours Programmé*. F. Nathan; Ed. Labor.

Lee, Youngmin, and David W. Nelson. 2005. Viewing or Visualising—Which Concept Map Strategy Works Best on Problem-Solving Performance?. *British Journal of Educational Technology* 36 (2): 193–203 (Wiley Online Library).

Leidig, Torsten. 2001. L 3—Towards an Open Learning Environment. *Journal on Educational Resources in Computing (JERIC)* 1 (1es): 7 (ACM).

Markham, Kimberly M., Joel J. Mintzes, and M. Gail Jones. 1994. The Concept Map as a Research and Evaluation Tool: Further Evidence of Validity. *Journal of Research in Science Teaching* 31 (1): 91–101 (Wiley Online Library).

McAleese, Ray. 1998. The Knowledge Arena as an Extension to the Concept Map: Reflection in Action. *Interactive Learning Environments* 6 (3): 251–272 (Taylor & Francis).

Meder, Norbert. 2006. *Web-Didaktik: Eine neue Didaktik web-basierten, vernetzten Lernens*. Bielefeld: Bertelsmann.

Middleton, James, Stephen Gorard, Chris Taylor, and Brenda Bannan-Ritland. 2008. The 'Compleat' Design Experiment: From Soup to Nuts. In *Handbook of Design Research Methods in Education: Innovations in Science, Technology, Engineering, and Mathematics Learning and Teaching*, 21–46. Citeseer.

Moon, Brian, Robert R. Hoffman, Joseph Novak, and Alberto Canas. 2011. *Applied Concept Mapping: Capturing, Analyzing, and Organizing Knowledge*. Boca Raton, FL: CRC Press.

Neubauer, Matthias, C. Stary, and Stefan Oppl. 2011. Polymorph Navigation Utilizing Domain-Specific Metadata: Experienced Benefits for E-Learners. In *Proceedings of the 29th European Conference on Cognitive Ergonomics (ECCE 2011)*, 45–52. New York: ACM Press.

Nonaka, Ikujiro, and Georg Von Krogh. 2009. Perspective—Tacit Knowledge and Knowledge Conversion: Controversy and Advancement in Organizational Knowledge Creation Theory. *Organization Science* 20 (3): 635–652 (Informs).

Novak, Joseph D. 1995. Concept Mapping to Facilitate Teaching and Learning. *Prospects* 25 (1): 79–86 (Kluwer Academic Publishers). https://doi.org/10.1007/BF02334286.

Novak, Joseph D., and A.J. Canas. 2006. The Theory Underlying Concept Maps and How to Construct Them. Florida Institute for Human and Machine Cognition.

O'donnell, Angela M., Donald F. Dansereau, and Richard H. Hall. 2002. Knowledge Maps as Scaffolds for Cognitive Processing. *Educational Psychology Review* 14 (1): 71–86 (Springer).

Oppl, Stefan. 2016. Articulation of Work Process Models for Organizational Alignment and Informed Information System Design. *Information & Management* 53 (5): 591–608. https://doi.org/10.1016/j.im.2016.01.004.

Oppl, Stefan, and C. Stary. 2011. Effects of a Tabletop Interface on the Co-Construction of Concept Maps. In *Proceedings of the 13th IFIP TC13 Conference on Human-Computer Interaction (INTERACT 2011)*, 443–460. Berlin and Heidelberg: Springer. https://doi.org/10.1007/978-3-642-23765-2_31.

Oppl, Stefan, and Christian Stary. 2009. Tabletop Concept Mapping. In *Proceedings of TEI 2009*, 275–282. New York: ACM Press. https://doi.org/10.1145/1517664.1517721.

———. 2014. Facilitating Shared Understanding of Work Situations Using a Tangible Tabletop Interface. *Behaviour & Information Technology* 33 (6): 619–635. https://doi.org/10.1080/0144929X.2013.833293.

Parkhurst, Helen. 1922. *Education on the Dalton Plan*. Read Books Ltd.

Peris-Ortiz, Marta, Diana Benito-Osorio, and Carlos Rueda-Armengot. 2014. Applying Concept Mapping: A New Learning Strategy in Business Organisation Courses. In *Innovation and Teaching Technologies*, 41–49. Cham: Springer.

Rentsch, Joan R., Abby L. Mello, and Lisa A. Delise. 2010. Collaboration and Meaning Analysis Process in Intense Problem Solving Teams. *Theoretical Issues in Ergonomics Science* 11 (4): 287–303 (Taylor & Francis).

Roth, Wolff-Michael, and Anita Roychoudhury. 1992. The Social Construction of Scientific Concepts or the Concept Map as Device and Tool Thinking in High Conscription for Social School Science. *Science Education* 76 (5): 531–557 (Wiley Online Library).

———. 1993. The Concept Map as a Tool for the Collaborative Construction of Knowledge: A Microanalysis of High School Physics Students. *Journal of Research in Science Teaching* 30 (5): 503–534 (Wiley Online Library).

Rovine, Michael J., and Gerald D. Weisman. 1989. Sketch-Map Variables as Predictors of Way-Finding Performance. *Journal of Environmental Psychology* 9 (3): 217–232 (Elsevier).

Rudan, Sasha, and Sinisha Rudan. 2008. SocioTM—Relevancies, Collaboration, and Socio-Knowledge in Topic Maps. *Tmra* 2008: 285.

Rye, James A., and Peter A. Rubba. 1998. An Exploration of the Concept Map as an Interview Tool to Facilitate the Externalization of Students' Understandings About Global Atmospheric Change. *Journal of Research in Science Teaching: The Official Journal of the National Association for Research in Science Teaching* 35 (5): 521–546 (Wiley Online Library).

Sandberg, Jörgen. 2005. How Do We Justify Knowledge Produced Within Interpretive Approaches?. *Organizational Research Methods* 8 (1): 41–68 (Thousand Oaks, CA: Sage Publications).

Schluep, Samuel, Marco Bettoni, and Sissel Guttormsen Schär. 2006. Modularization and Structured Markup for Learning Content in an Academic Environment. *International Journal on E-Learning* 5 (1): 35–44 (Association for the Advancement of Computing in Education (AACE)).

Schmiech, M. 2006. *Didaktische Ontologien zur Organisation digitaler Objekte in der Arbeit von Lehrkräften*. Flensburg: University of Flensburg.

Schulmeister, Rolf. 2017. *Lernplattformen für das Virtuelle Lernen: Evaluation und Didaktik*. Walter de Gruyter GmbH & Co KG.

Scott, R.O. 1968. Mathetic and Progressive Chain Strategies for Instructional Sequencing. Doctoral dissertation, University of Michigan.

Senge, P.M. 1990. *The Fifth Discipline: The Art and Practice of the Learning Organization*. New York: Doubleday/Currency.

Shchedrovitsky, G.P. 2014. Part I Selected Works: A Guide to the Methodology of Organisation, Leadership and Management. In *Methodological School of Management*. A&C Black.

Siemens, George. 2005. Connectivism: Learning as Network-Creation. ElearnSpace. Accessed July 12, 2008. Http://Www.Elearnspace.Org/Articles/Networks.Htm. [WebCite Cache].

Smolnik, S., and I. Erdmann. 2003. Visual Navigation of Distributed Knowledge Structures in Groupware-Based Organizational Memories. *Business Process Management Journal* 9 (3): 261–280 (Emerald Group Publishing Limited).

Stary, Christian. 2016. Open Organizational Learning: Stakeholder Knowledge for Process Development. *Knowledge Management & E-Learning* 8 (1): 86 (The University of Hong Kong, Faculty of Education).

———. 2018. 'The Clock Has Fallen Off the Wall'—Emergence of BPM-Relevant Knowledge Based on Cascading Stakeholder Perspectives. In *Workshop—Proceedings of S-BPM ONE 2018*, 2074.

Stary, Christian, Matthias Neubauer, Stefan Oppl, and Georg Weichhart. 2015. Stakeholder-Centered Ontologies for Educational Designs. *Ontology of Designing* 5 (2): 149–178.

Strauss, A. 1988. The Articulation of Project Work: An Organizational Process. *The Sociological Quarterly* 29 (2): 163–178.

Sumner, Tamara, Faisal Ahmad, Sonal Bhushan, Qianyi Gu, Francis Molina, Stedman Willard, Michael Wright, Lynne Davis, and Greg Janée. 2005. Linking Learning Goals and Educational Resources Through Interactive Concept Map Visualizations. *International Journal on Digital Libraries* 5 (1): 18–24 (Springer).

Swan, Jacky, Harry Scarbrough, and Sue Newell. 2010. Why Don't (or Do) Organizations Learn From Projects?. *Management Learning* 41 (3): 325–344 (London: Sage Publications).

Tiropanis, Thanassis, David Millard, and Hugh C. Davis. 2012. Guest Editorial: Special Section on Semantic Technologies for Learning and Teaching Support in Higher Education. *IEEE Transactions on Learning Technologies* 5 (2): 102–103. https://doi.org/10.1109/TLT.2012.13.

Toral, S.L., María del Rocío Martínez-Torres, Federico Barrero, S. Gallardo, and M.J. Durán. 2007. An Electronic Engineering Curriculum Design Based on Concept-Mapping Techniques. *International Journal of Technology and Design Education* 17 (3): 341–356 (Springer).

Trochim, William M.K. 1989. An Introduction to Concept Mapping. *Evaluation and Program Planning* 12: 1–16.

Tseng, Shian-Shyong, Pei-Chi Sue, Jun-Ming Su, Jui-Feng Weng, and Wen-Nung Tsai. 2007. A New Approach for Constructing the Concept Map. *Computers & Education* 49 (3): 691–707 (Elsevier).

Weichhart, Georg. 2012. S-BPM Education on the Dalton Plan: An E-Learning Approach. In *S-BPM ONE—Education and Industrial Developments*, 181–193. Springer.

———. 2014a. *Der Dalton Plan im E-Learning: Transformation einer Reformpädagogik ins Web*. Trauner Verlag.

———. 2014b. Learning for Sustainable Organisational Interoperability. In *Preprints of the 19th IFAC World Congress*, August, 4280–4285. International Federation of Automation and Control.

Weichhart, Georg, and Chris Stary. 2014. Traceable Pedagogical Design Rationales for Personalized Learning Technologies: An Interoperable System-to-System Approach. *International Journal of People-Oriented Programming (IJPOP)* 3 (2): 25–55 (IGI Global).

Weske, Mathias. 2010. *Business Process Management: Concepts, Languages, Architectures*. Berlin: Springer.

Zaharieva, Maia, and Wolfgang Klas. 2004. MobiLearn: An Open Approach for Structuring Content for Mobile Learning Environments. In *Advanced Information Systems Engineering*, Lecture Notes in Computer Science, vol. 3307, ed. Haralambos Mouratidis and Colette Rolland, 114–124. Berlin and Heidelberg: Springer Berlin Heidelberg. https://doi.org/10.1007/978-3-540-30481-4_11.

Zardas, G. 2008. The Importance of Integrating Learning Theories and Pedagogical Principles in AHES (Adaptive Hypermedia Educational Systems). In *2008 Eighth IEEE International Conference on Advanced Learning Technologies*, 884–885. IEEE.

Open Access This chapter is licensed under the terms of the Creative Commons Attribution 4.0 International License (http://creativecommons.org/licenses/by/4.0/), which permits use, sharing, adaptation, distribution and reproduction in any medium or format, as long as you give appropriate credit to the original author(s) and the source, provide a link to the Creative Commons licence and indicate if changes were made.

The images or other third party material in this chapter are included in the chapter's Creative Commons licence, unless indicated otherwise in a credit line to the material. If material is not included in the chapter's Creative Commons licence and your intended use is not permitted by statutory regulation or exceeds the permitted use, you will need to obtain permission directly from the copyright holder.

9

Epilogue

As we tried to demonstrate in the previous chapters, the design of digital work systems requires stakeholder involvement in generating relevant work knowledge, starting with articulation and proceeding with sharing and aligning it in more or less structured design spaces. When looking close to transforming organizations towards digital process support, however, the ultimate goal is to develop executable processes in evolving cyber-physical environments. Does such a scenario finally mean to educate stakeholder to become skilled in programming when designing digital work places and business processes?

Although actor-centered concepts to that direction exist, such as app'ificiation (Stary 2017), for complex domains, such as additive manufacturing, more in-depth knowledge of coding is likely to be required. To the latter direction, recent work with respect to software-intensive systems of layered approach involving various levels of abstraction has been proposed (Börger 2018). It should lead from requirements engineering to coding through abstract modeling concepts available as high-level programming constructs. Such kind of specifications help to define the code in a way stakeholders intend to, and as required to execute the corresponding software system by digital systems.

Börger argues that the remaining gap cannot be closed by mere programming methods, but needs to be addressed by an appropriate modeling framework comprising a design and analysis method, and a language. In his understanding, programming languages must be supported by modeling at higher levels of abstraction than that of the programming language, as programming means programming reliable complex systems or software-intensive systems. The latter refer to systems where "the software and the machines which execute it are only a part of the overall system, where for the code executing computer(s) the other parts appear as environment—technical equipment, physical surrounding, information systems, communication devices, external actors, humans—upon which the behavior of the software components depends and which they affect" (ibid., p. 1).

Since we consider this kind of system as backbone of digital work design, we could look in how far coding such systems is supported by levels of abstraction, including requirements through high-level design to machine-executable code. As means of describing information on several layers of abstraction, Börger considers natural language, dedicated languages, and frameworks appropriate, when capturing programming-relevant knowledge. Of particular interest, he considers approaches which "to relate in a controllably reliable way real-world items and behavior (objects, events and actions) to corresponding items in a textual or graphical description, whether directly by code or by an abstract model that is transformed in a correctness preserving manner to code" (p. 3).

According to Börger, this epistemological problem has a communication, an evidence, and an experimental validation strand. Referring to the intrinsic properties of languages when resolving this problem, stakeholders need an understandable language. More important application such as

> language must allow the stakeholders to *calibrate the degree of precision* of descriptions (read: their level of abstraction) to the given problem and its application domain. Last but not least, the language must allow the software engineers to *link descriptions at different levels of abstraction*—transform models, lifting what compilers do to the given levels of abstraction—in a controllably correct and well documented way to code, using a practical refinement method that is supported by techniques for both, experimental validation and mathematical verification (whether informal, rigorous or formal and machine supported). (p. 2)

Börger promotes the term ground models referring to some 'blueprints' through which domain experts and software developers need to achieve a common understanding of a proper digital support system. This consensus serves as an essential input for validation. The intended behavior is expected to be delivered by domain experts with rigor valid in the application domain. This rigor includes describing stable domain assumption with respect to the structure and behavior of system components. That information is transformed or refined to a software system specification. It contains a sufficiently precise behavior description of the digital support system meeting the requirements as provided by the actors.

Code development requires a ground model to ensure complete and correct code (cf. Fig. 9.1). Completeness means containing all features of a system that is relevant from a behavior perspective. Correctness means conveying the meaning in a reliable way. The ground model could change in the course of code development or system evolution, leading to further development iterations. Hence, validation based on the actors' inputs is essential.

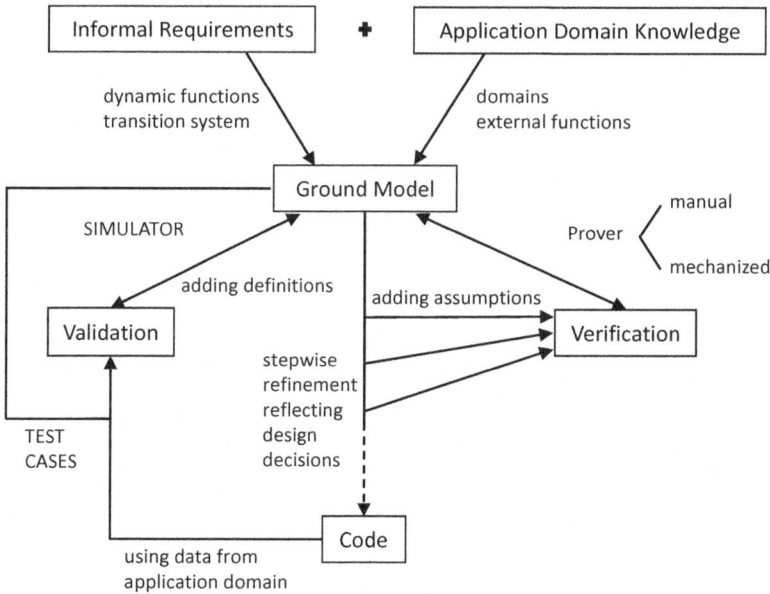

Fig. 9.1 System development involving the ground model supported by ASM (Börger and Stärk 2012)

The ground model should objectively be checkable by the respective stakeholders, in order to support articulation and alignment activities focusing on actor perspectives on work processes. Its description requires a (ground model) language which is

- generally understood
- appropriately extendable by specific application domain concepts, where needed, and
- clearly defined

It represents "a language of the kind used in rigorous scientific and engineering disciplines, made up from precise and simple but general enough basic constructs to unambiguously and directly represent arbitrary real-world facts (states of affairs) and state changing events" (Börger 2018, p. 6).

We consider the proposed organizational development framework detailed in Chap. 6 applicable to bridge the gap between articulation of stakeholder requirements and generating code. The articulation, as shown in Chap. 2, can be based on variety of formats:

- natural language which can be refined to abstract models
- graspable model entities that need to be tagged in natural language to develop a domain-relevant representation
- predefined elementary symbols representing work activities that allow complex behavior specifications based on natural language descriptions

Depending on the stakeholder capabilities and preferences, various entry points for structuring the elicitation procedure and representation of work knowledge can be selected and applied. Each of the presented formats allows further processing on a social and technical layer (Chaps. 3 and 4). For alignment and consolidation, models needs to be intelligible for all stakeholders involved in the process. They might find consensus on an abstract level or require virtual enactment to probe their model(s) of work. The latter requires executable models.

The presented concepts and instruments enable executable models to remain on an abstract, since implementation-independent level. Iterative prototyping supports is thus possible and allows for interactive validation of specific process scenarios or situations (as also advised by

(Börger 2018) for ground model inspection). Remaining on this level of description allows tracing the diagrammatic model while running its code. Behavior-centered approaches, such as Subject-oriented Business Process Management (Fleischmann et al. 2012), finally facilitate the specification of programming code, as the entire control flow and functional structure can be modeled (see Chap. 5). Still, in those cases, programming of system functions remains to be completed, either developing them from scratch or activating existing software systems. In both cases, the complexity has been reduced to a manageable system architecture.

References

Börger, Egon. 2018. Why Programming Must Be Supported by Modeling and How. In *Proceedings of ISoLA*, LNCS, vol. 11244, 1–22.
Börger, Egon, and Robert Stärk. 2012. *Abstract State Machines: A Method for High-Level System Design and Analysis*. Springer Science & Business Media.
Fleischmann, Albert, Werner Schmidt, C. Stary, Stefan Obermeier, and Egon Börger. 2012. *Subject-Oriented Business Process Management*. New York: Springer.
Stary, Christian. 2017. Contextual App'ification. In *Proceedings of ROCHI*.

Open Access This chapter is licensed under the terms of the Creative Commons Attribution 4.0 International License (http://creativecommons.org/licenses/by/4.0/), which permits use, sharing, adaptation, distribution and reproduction in any medium or format, as long as you give appropriate credit to the original author(s) and the source, provide a link to the Creative Commons licence and indicate if changes were made.

The images or other third party material in this chapter are included in the chapter's Creative Commons licence, unless indicated otherwise in a credit line to the material. If material is not included in the chapter's Creative Commons licence and your intended use is not permitted by statutory regulation or exceeds the permitted use, you will need to obtain permission directly from the copyright holder.

Ontological Glossary

This brief ontological glossary

(i) explains the meaning and use of terms in the context of this work, and
(ii) puts these terms into a mutual context by denoting their context in a diagrammatic form

Its objective is to enable an overall understanding of the addressed topics, as they stem from different fields and might have certain meaning in various technical domains. Wherever possible, the original meaning of terms has been incorporated into their explanations, in order to acknowledge their origin, while sometimes widening their scope to increase the intelligibility of the presented concepts and techniques.

For the sake of usability, we start with the textual descriptions in alphabetical order of essential terms and proceed with the ontology diagram.

- *Actors* are those entities in an organization who actually carry out business or work processes based on their knowledge.
- *Business operation* = handling all business-relevant activities.

- *Business processes* describe how actors work together and perform their work contributions in an organization to pursue a common goal. They are specifications of who is doing what with which information, material, or goods to achieve the objectives of a business.
- *Facilitators* are persons preparing and guiding the elicitation, representation, alignment, validation of work knowledge and the execution of processes. They are social caretakers with methodological accountancy.
- *Mental models* are representation to understanding how people make decisions based upon their perception of their current work situation and their prior knowledge.
- *Articulation & Alignment* = process of elicitation, representation, sharing, collective validation, enactment, and generation of knowledge.
- *Role carriers* are persons or digital system accomplishing or implementing a certain work task and showing corresponding behavior, such as accountant and information provider.
- *Situations* are snapshots of a business operation, for example, a business case, some market event.
- *Stakeholders* are persons, organizational units, or organizations relevant for a situation or business operation, for example, business partners, customers.
- *Work activities* are derived from work tasks and denote a set of specific actions, for example, send/receive messages.
- *Work knowledge* is required to perform the business and work processes, and it enables actors to make their decisions on how to continue work based on their perceptions of the environment. It captures information on *why* and *how* to operate a business, for example, cross-selling a product for increasing market share.
- *Work tasks* are derived from business operation and denote sets of work activities.

Ontological Glossary

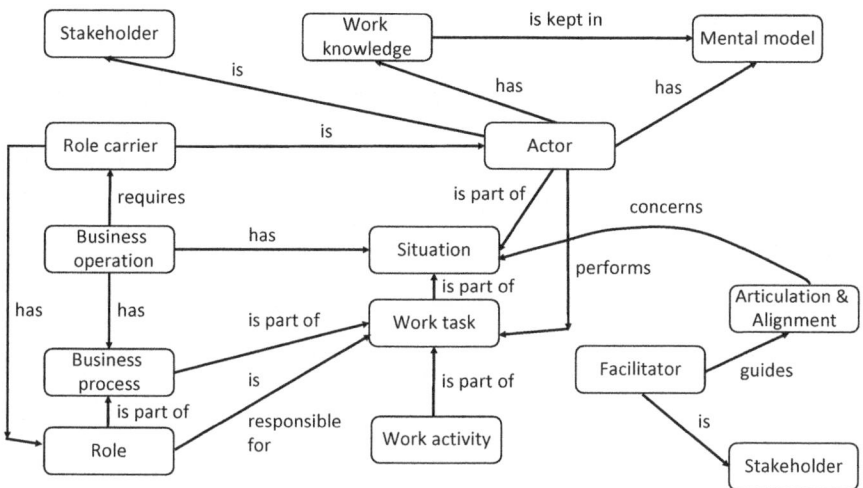

Fig. A.1 Ontology of essential terms used in this work

The ontology is presented in diagrammatic form as concept map. Thereby, nodes represent terms and direct links represent semantic relations which allow reading binary relations between two nodes (terms) as natural language expressions. For the sake of intelligibility, only single relations between terms have been used.

The diagram positions the Actor and carrier of Work Knowledge in a work-specific Role—we denote the terms of the ontology with upper case letters. Work Knowledge is represented in individual Mental Models. As these mental models influence the behavior of Actors in work-specific Situations, they are target of Articulation and Alignment. Both are guided by a Facilitator, another relevant Stakeholder for digital work design.

As a Role Carrier, an Actor is part of a Situation which characterizes a Business Operation. Thereby, Actors accomplish Work Tasks that are part of Business Processes in order to achieve the objectives of the business. As constitutive part of Business Processes, Roles denote the responsibility of Work Tasks, also being a functional part of Business Processes. Work Tasks are described by Work Activities set by Actors, including communication and interaction with other actors and digital systems.

Index

A

Actor, 1–9, 13–15, 18, 28, 30, 31, 33, 35–37, 45–50, 55, 59, 60, 68–70, 85, 87, 89–91, 96, 98, 99, 101, 102, 108, 109, 111–114, 116–118, 122, 124, 126, 133, 140, 142, 144, 148, 149, 151, 152, 160, 163, 166, 168, 169, 171, 172, 180, 181, 183, 186–188, 195–198, 200, 201, 205, 206, 214–225, 231–235, 237, 250, 251, 254–256, 259, 260, 264, 269, 275–281, 290, 291, 304, 310, 317, 326, 331, 334, 337, 393, 398–400, 402–404, 407–409, 420–422

(Process) Adaptation, 194, 217, 273, 277–278

Alignment, v, vi, 3–6, 10, 11, 13, 14, 17–19, 27, 28, 46, 51, 53, 61–71, 98, 110, 121, 122, 126, 133–173, 179–192, 233, 237, 240, 253–255, 273, 277–281, 287, 291, 292, 302, 325, 326, 348, 381–387, 389, 394, 422

Alignment practice, 6, 156–170

Alignment scheme, 145–147

Articulation, v, vi, 4, 6, 13, 18, 27, 28, 32, 34–37, 39, 46, 53, 55, 59, 62, 67, 70, 71, 83–127, 134, 144–147, 157, 158, 161–168, 171, 173, 179–192, 200, 234, 237, 269, 271, 281, 287, 290–303, 310–312, 317, 325, 326, 328, 329, 334, 339–342, 345, 349, 350, 357,

388, 390, 392, 394, 399–404, 410, 419, 422
Articulation Engineered for Organizational Learning in Interopable, Open Networks (AELION), 269–280
Articulation support, 31, 84, 271, 357–366
Articulation Work, 5, 8–12, 56–60, 227, 251–256, 302, 358, 389
Asset, 42, 56–61, 68, 69, 83, 111–113, 126

B

Behavior abstraction, 91–96, 402
Behavioral constraint, 264–269
Behavioral interface, 264, 266, 267
Behavior constraints, 3
Business operation, v, vi, 18, 70, 83, 84, 113, 124, 288, 329, 335–351, 389, 409
Business processing environment, 6, 8, 9, 18, 216, 220–225, 232, 255, 279, 281, 312
Business Process Modeling Notation (BPMN), 34, 99–101, 103, 144, 145, 157, 162, 328, 334, 345, 346, 350, 351, 360
Business rule, 258

C

Cascading perspectives, 400
Co-construction, 186, 292
Co-creation, 47, 173
Collaboration, vii, 1, 4, 19, 44, 65, 84, 97, 110, 133, 143, 144, 147, 148, 160, 164–168, 179, 184–186, 188, 189, 264, 267, 269, 306, 310, 314, 315, 339, 347, 379
Collaborative consolidation, 103, 157, 158, 162, 165–168, 342–345, 348, 349
Collaborative Model Articulation and Elicitation (CoMPArE), 145–148, 157, 336
Collaborative Model Articulation and Elicitation of Work Processes (CoMPArE/WP), 134, 142–145, 156–172, 189–192, 325, 335–351
Collective intelligence, 61–70, 112
Collective level, 118
Common understanding, 9, 10, 51, 57, 61, 68, 133, 145, 157, 159, 183, 252, 253, 421
Communication alignment, 151–170
Communication negotiation, 151
Complex Adaptive Systems (CAS), 40–43, 315
Complexity, 12, 36, 39, 42–44, 71, 99, 135, 158, 199, 205, 274, 297, 348, 351, 394, 396, 402, 404, 423
Complex system, 38–46, 50, 66–67, 69, 111, 420
Comprehand, 272, 295, 297, 299, 301, 302, 359
Comprehand Cards, 292–293, 296
Comprehand Table, 293–302
Concept maps (CMap), 15, 52, 53, 114, 159, 160, 189, 304, 305, 312, 331, 352, 355–359, 361,

362, 365, 369, 376–378, 380, 381, 383, 388, 389
Conceptual framework, 19, 257, 270
Consolidation, 14, 15, 103, 140, 144, 147, 151, 163, 164, 166, 168, 173, 184, 190–192, 214, 215, 240, 311, 338, 342–345, 349, 350, 422
Content management, 29, 91, 271, 272, 290, 372, 383, 384, 388, 389

D

Dalton Plan, 305, 306, 384, 386, 388
Diagrammatic representation, 53, 68, 403
Didactic ontology, 372
Digital technology, v, 1
Digital work design, v, vi, 18, 133, 234, 281, 317, 326, 327, 336, 352, 353, 390, 420
Digital work support (system), 1
Distributed modeling, 151, 156
Domain knowledge, 136, 326
Domain process ontology, 139–141

E

Emergence, vi, 44, 249, 349
Enactment, v, 107, 145, 158, 159, 168–170, 191, 193–215, 234–237, 280, 309, 317, 338, 344–350, 422
Evolution, 14, 224, 300, 302, 349, 421
(Automated) Execution, 58

F

Facilitation, 15, 17, 18, 71, 98, 121, 122, 126, 135, 173, 188, 195, 196, 237, 329, 348, 351
Feedback graph, 384, 386, 387

G

Ground model, 421–423

H

Healthcare, 44, 326, 327, 410
Holomap, 114, 116–118
Human-centric modeling
Human work, 1, 10, 53, 253

I

Individual articulation, 102, 147, 158, 162–165, 190, 339–342, 345, 349
Individual level, 255, 287
Intangible assets, 56–61, 68, 111, 126
Intangible transaction, 112–118, 126
Intelligibility Catcher (ICs), 305
Interactive modeling, 297, 312, 316

K

Knowledge claim, 8, 18, 223–226, 232, 233, 255, 256, 273–281, 303, 315
Knowledge Lifecycle (KLC), 5–9, 216–221, 223–227, 232, 233, 250, 251, 253–255, 273, 275,

278, 280, 288, 303, 310, 312, 315
Knowledge processing, 9, 217, 225, 288, 312
Knowledge processing environment, 6, 8, 217, 222–227, 232, 255, 273, 278
Knowledge sharing, 16, 65, 112, 158, 274, 336, 348

L

Learning chain, 287
Learning management, 272
Learning support (system), 91, 234, 255, 269, 271, 272, 279, 305, 306, 308, 326, 352–389
Lifecycle, 89, 329, 360, 376, 382

M

Memory support, 272, 274
Mental model, 5, 6, 10–13, 15, 19, 28, 37, 51–53, 55, 62, 65, 70, 71, 122, 134, 145, 161, 172, 173, 179, 180, 192, 233, 251–253, 255, 256, 269, 272, 278, 291, 292, 295, 302, 327, 335, 356
Mental model theory (MMT), 10, 227, 251, 253–255
Meta-knowledge, 274, 275, 303–307
Methodology, 15, 27, 52, 143–145, 152, 159, 163, 215, 309, 310, 317, 351
Model, v, 2, 28, 86, 134, 179–192, 325, 420

Model-centered learning, 5, 8, 10–13
Modeling, 6, 32, 86, 133, 180, 269, 290, 328, 419
Modeling cards, 170, 292
Multi-perspective modeling, 6, 13–15, 142, 143

N

Natural modeling, 16, 133, 134, 143, 145, 156–158, 168, 335–337, 341, 342, 350, 351
Negotiation, 9, 10, 14–16, 29, 51, 133, 144, 147, 151, 158, 173, 200, 252, 255, 336, 348, 349
Network, 42, 90, 110–118, 126, 287, 304, 352, 401, 402

O

Ontology, 33, 64, 135–141, 372
Ontology-based alignment, 138
Organizational design, vi, 18, 27, 122
Organizational development, vii, 113, 179, 217, 281, 287–317, 326, 335, 382, 422
Organizational learning (OL), 36, 234, 249, 253–256, 269–280, 287, 288, 327, 355, 370
Organizational learning process, 6, 277
Organizational management, 389–410
Organization of work, 45, 49, 52, 53, 68, 234, 288, 330

P

Participatory enactment, 193–215
Perspective, vi, 3, 6, 8, 9, 14, 17–19, 28, 32, 33, 35, 39, 40, 42, 44, 46–48, 50–55, 62, 63, 67–70, 83, 84, 90, 98, 110–112, 117, 118, 121, 122, 124, 126, 133–173, 183, 186, 198, 205, 215, 216, 218, 225, 235, 237, 239, 240, 251, 252, 287, 312, 313, 335, 339, 343, 352, 354–358, 380, 389, 392, 393, 395–406, 409, 421, 422
Pragmatics, 88–96, 389, 404, 409
Process adaptation, 278
Process collection, 273, 277–278
Process enactment, v, 200, 214, 215, 309
Process execution, 194, 200, 212, 218, 225, 232, 233, 251, 258, 259, 272, 275, 277, 278, 368
Process improvement, 217
Process knowledge, 107, 113, 145, 216, 219, 234, 273, 276–279, 287, 288
Process validation, 199, 215, 273, 279–280
Production Work, 56, 254
Progressive education, 361, 362, 365, 382, 384

R

(Process) Refinement, 5, 92, 98, 101, 145, 158, 168–170, 198, 273, 277–278, 309, 326, 347, 350, 365, 368, 407, 420
Reflection, vi, 13, 45–47, 49, 53, 60, 70, 109, 118, 126, 162, 171, 180, 184, 194, 206, 234, 272, 273, 277–280, 288, 290, 301, 312, 328–330, 334, 345, 349, 354, 356, 358–361
Reflective practice, 46–50, 68
Reflective practitioner, 46
Repository, 137, 140, 271, 273–276, 281, 301, 309, 310, 313, 316
Responsibility, 14, 46, 63, 88, 144, 147, 182, 184–188, 192, 197, 201, 219, 235, 259–261, 264, 333, 345, 354, 398
Role, 10, 28, 84, 139, 185, 250, 290, 328
Role carrier, 111, 118, 122, 126, 237, 240, 410
Role knowledge, 274, 276, 303–304

S

Scaffolding, 17, 98, 142, 163, 179–192, 196, 199, 201, 203, 210–214, 232, 233, 235, 240, 279, 290, 358
Semantic distance, 140, 142–145
Semantic navigation, 357, 376–381
Semantics, 16, 17, 52, 53, 60, 63, 86–89, 98–102, 107, 108, 133, 135–140, 142, 143, 146, 149, 150, 157, 160–163, 169, 188, 190, 192, 224, 225, 291–293, 296, 297, 302, 312, 328, 334–336, 339, 345, 347, 349–351, 375, 379, 389, 390
Setting the stage, 28–31, 62, 157, 159–161, 338–340, 344
Shared model, 309, 338, 342
Situation, 10, 12, 13, 16, 28, 29, 32, 33, 37, 38, 42, 43, 45–51, 55,

57, 58, 60, 61, 68, 69, 71, 85–90, 121, 122, 126, 148, 157, 159, 173, 181, 183, 197, 198, 211, 217–225, 232, 233, 239, 240, 252, 253, 255, 256, 258, 259, 269, 275–277, 279, 293, 310, 314, 325, 330, 335, 336, 349, 350, 357, 379, 383, 389–393, 395–397, 399–407, 409, 422

Situation awareness, 31–38

Situation-specific interdisciplinary team (SIT), 259, 269, 270

Social media, 28–30, 43, 91, 271, 272, 290, 372, 384, 388, 389

Socio-technical system, vi, 17–19, 31, 38, 41–43, 58, 86, 114, 143, 194, 195, 326

Stakeholder, v, vi, 9, 10, 27–32, 35, 37, 39, 40, 42, 43, 55, 62–64, 71, 83–86, 88–90, 96, 107–110, 121, 122, 126, 135, 137, 157, 173, 180, 187, 189, 192, 195, 199–201, 214, 215, 237, 239, 252, 271, 281, 288, 317, 325, 326, 328–330, 333–335, 337, 342, 344, 345, 351, 355, 356, 359, 371, 375, 389, 392, 395–404, 409, 410, 419, 420, 422

Stakeholder support, 19, 142–145, 389

Structure elaboration techniques, 51, 52, 97, 277, 278, 291, 292, 311

Subject orientation, vii, 85, 86, 89, 126, 127, 397, 409

Subject-oriented Business Process Management (S-BPM), vii, 99, 103–106, 147, 152, 155, 200, 202, 215–234, 237, 250, 251, 253, 254, 273, 280, 297, 299, 309, 310, 423

Support technology, 290

Syntactic modeling, 60, 97, 188

System-of-Systems (SoS), 28, 33, 43–45, 50, 91

T

Tabletop, 155, 294–297, 300–302, 311, 316, 317, 330, 359–361

Tabletop concept mapping, 300, 359, 360

Tangible interface, 294, 295

Tangible transaction, 112–118, 126

Techno-centric modeling, 6, 16–17, 145, 159, 335

Transactive memory, 274–276, 306, 312

V

Value network, 85, 110, 111, 113–115, 117, 118, 124–126

Value Network Analysis (VNA), 110–116

View, 1, 4, 6, 10, 11, 15, 18, 34, 50, 51, 57–59, 61, 90, 97, 110, 117, 146–148, 153, 157, 159, 160, 185, 189, 191, 205, 206, 214, 215, 218, 253, 276, 279, 287, 294, 301, 302, 307, 308, 311, 337, 341, 342, 351, 352, 356, 362, 375, 377–381, 393, 396, 400–404

Virtual enactment, 145, 158, 159, 168–170, 194, 199–203, 214,

215, 235–237, 278, 338, 344–350, 422

W

Work activity, 9, 34, 217, 232, 252, 399, 422
Work-agogy, 48, 49
Workflow execution, 202, 216, 309, 341
Workflow management system (WfMS), 220, 309
Work knowledge, vi, 8, 27, 29, 34, 35, 53, 55, 62, 69, 107, 110, 118, 121, 122, 126, 134, 137, 140, 173, 234, 237, 281, 317, 326, 328, 329, 334, 409, 419, 422
Work knowledge elicitation, 31

Work organization, vi
Work process, v, vi, 2–6, 9, 10, 12–14, 17–19, 27–29, 35, 36, 46, 49, 59, 60, 69, 70, 83, 84, 89, 97–99, 101, 103, 107, 108, 127, 133, 134, 136, 138, 145–148, 151, 152, 155, 173, 179, 180, 189, 193, 196, 198–200, 214–234, 237, 239, 249, 252, 254–262, 264, 269, 273, 275–279, 287, 290, 297, 301, 302, 309, 311, 312, 314, 315, 326, 334, 351, 399, 422
Work process instance, 259
Work situation, 2, 9, 10, 12, 28, 58, 71, 86, 87, 173, 218, 251–253, 280, 311, 329
Work task, 1–4, 28, 35, 37, 61, 71, 121, 173, 234, 330, 332

The manufacturer's authorised representative in the EU is Springer Nature Customer Service Centre GmbH, Europaplatz 3, 69115 Heidelberg, Germany. If you have any concerns regarding our products, please contact ProductSafety@springernature.com

Printed and bound by CPI Group (UK) Ltd, Croydon, CR0 4YY
23/03/2026
02076663-0014